Probability Theory and Stochastic Modelling

Volume 111

Probability Theory and Stochastic Modelling publishes cutting-edge research monographs in probability and its applications, as well as postgraduate-level textbooks that either introduce the reader to new developments in the field, or present a fresh perspective on fundamental topics.

Books in this series are expected to follow rigorous mathematical standards and all titles will be thoroughly peer-reviewed before being considered for publication.

Probability Theory and Stochastic Modelling covers all aspects of modern probability theory including:

· Gaussian processes
· Markov processes
· Random fields, point processes, and random sets
· Random matrices
· Statistical mechanics and random media
· Stochastic analysis
· High-dimensional probability

as well as applications that include (but are not restricted to):

· Branching processes and other models of population growth
· Communications and processing networks
· Computational methods in probability theory and stochastic processes, including simulation
· Genetics and other stochastic models in biology and the life sciences
· Information theory, signal processing, and image synthesis
· Mathematical economics and finance
· Statistical methods (e.g., empirical processes and MCMC)
· Statistics for stochastic processes
· Stochastic control and stochastic differential games
· Stochastic models in operations research and stochastic optimization
· Stochastic models in the physical sciences

Probability Theory and Stochastic Modelling is a merger and continuation of Springer's Stochastic Modelling and Applied Probability and Probability and Its Applications series.

Diego Armentano • Jean-Marc Azaïs •
Federico Dalmao • Yohann De Castro •
Céline Delmas • José Rafael León •
Ernesto Mordecki

Geometry of Level Sets of Random Fields, Kac–Rice Formulas, Hermite Expansions and Applications

Springer

Diego Armentano
FCEA, Instituto de Estadística
Universidad de la República
Montevideo, Uruguay

Jean-Marc Azaïs
Institut de Mathématiques de Toulouse
Université de Toulouse
Toulouse, France

Federico Dalmao
CENUR Litoral Norte, DMEL
Universidad de la República
Salto, Uruguay

Yohann De Castro
Institut Camille Jordan
École Centrale Lyon
Écully, France

Céline Delmas
UR 875 - MIAT
INRAE
Castanet-Tolosan, France

José Rafael León
FING, IMERL
Universidad de la República
Montevideo, Uruguay

Ernesto Mordecki
FCIEN, Centro de Matemática
Universidad de la República
Montevideo, Uruguay

ISSN 2199-3130 ISSN 2199-3149 (electronic)
Probability Theory and Stochastic Modelling
ISBN 978-3-032-29143-1 ISBN 978-3-032-29144-8 (eBook)
https://doi.org/10.1007/978-3-032-29144-8

This Springer imprint is published by the registered company Springer Nature Switzerland AG
The registered company address is: Gewerbestrasse 11, 6330 Cham, Switzerland

Preface

The study of level sets began with the pioneering works of Kac (1943) and Rice (1945), who initiated the investigation of the number of real roots of a random Gaussian polynomial or process. Itô (1963) subsequently provided the first general, rigorous proof of what has come to be known as the Kac–Rice formula. Kac considered the number of real roots of a random Gaussian polynomial as the degree d tends to infinity and proved that the expectation of the number of real roots, N_d, satisfies the relationship

$$\mathbb{E}(N_d) \simeq \frac{2}{\pi} \log(d).$$

One possible interpretation of the above result is the following: "A random polynomial has very few real roots". Here, a natural question arose: "What is the nature of the fluctuations around the mean?".

This was answered by Maslova (1974), who computed the equivalent of the variance and established a central limit theorem (Maslova 1975) using the η dependence approximation of Malevich (1969), a technique now known as the "Malevich method" (see also Cuzick 1976).

Initially, studies focused on the number of roots (or, more generally, on the number of crossings of a level u) of a random process $\mathbb{R} \to \mathbb{R}$, with models that were primarily Gaussian. The reference monograph on this subject is Cramér and Leadbetter (1967), which provides the first proof of the higher moments of such functionals.

The proof of the Kac–Rice formula is essentially a formula of change of variables. As the subject developed, Federer (1969) introduced general versions of this formula, namely the area and co-area formulas, which have become powerful tools for constructing both general and concise proofs of the Kac–Rice formula.

Following the paper Adler and Hasofer (1976) and the monograph Adler (1981), the community began to consider random fields starting with the study of the Euler characteristic of the excursion above some level u. A study of the perimeter or surface area of the level set was first carried out by Switzer (1976) using the isotropy assumption, and later under more general conditions in Wschebor (1982), Wschebor (1985), and Benzaquen and Cabaña (1982).

In such a case, the dimension D of the parameter space is greater than that of the image space and the level set $\mathcal{L}_u$ is generically a set of dimension $D - d$. The Kac–Rice formula computes the expectation of a measure instead of the number of roots. Additionally, some random fields are even defined on a manifold as in Azaïs and Wschebor (2004) and Auffinger et al. (2013).

The classical path for studying the behavior of the measure of nodal sets and level sets more generally consists in the successive examination of the expectation, the variance, and the central limit theorem (CLT). However, since the work carried out by Maslova, new methods have been developed. In a pioneering paper, Nualart and Peccati (2005) proposed a new method adapted to Gaussian spaces, namely the Fourth Moment Theorem. Their method is in turn related to the Hermite decomposition of random variables.

The Hermite decomposition of the measure of the level set began with Slud (1991a) and Kratz and León (2001), while the Fourth Moment Theorem was used for the first time in Estrade and León (2016) and Azaïs et al. (2016). Nowadays, it is a standard method for proving CLT for the number of roots or more generally for the measure of level sets. In addition, the Hermite method, through Arcones' inequality, allows for the computation of some bounds on moments that are not accessible through the Kac–Rice formula: the positivity of some asymptotic variances, for example. We can see, then, that these two methods clearly complement one another.

At the beginning of the twenty-first century several new models were studied, including non-Gaussian random polynomials, random series of functions, series of spherical harmonics, and shot noises, most of which are not Gaussian. However, the classical monographs on the Kac–Rice formula that have appeared successively Cramér and Leadbetter (1967), Adler (1981), Adler and Taylor (2009), Azaïs and Wschebor (2009), and Berzin et al. (2022) over the years do not cover these topics. Therefore, there is a need for a new approach that can:

- Provide a new proof of the Kac–Rice formula that is adapted to non-Gaussian cases and which, by extension, is valid under weaker conditions even in the Gaussian case
- Offer a simple presentation of the minimal notions of geometry
- Account for Hermite expansion and the Fourth Moment Theorem
- Present recent applications, including applications to Physics (see Chap. 8), High-Dimensional statistics and kernel regression (see Chap. 6), and critical points (see Chap. 5)

This book begins with four chapters, which introduce most of the methodologies employed in our study:

- Chapter 1 presents classical results on the number of roots of a stochastic process $\mathbb{R} \to \mathbb{R}$. Though this chapter mostly consists of a summary of existing studies on the Kac–Rice formula, it also presents counterexamples and explains why weak forms of the Kac–Rice formula for almost all levels (such as those presented in Zähle (1984) and Brillinger (1972)) are not sufficient. The last section of the chapter also presents recent results on the finiteness of moments.

- Chapter 2 is a central chapter that is devoted to the computation of the expectation of the $D-d$ measure of the nodal set and more generally the level set of a random field that goes from $\mathbb{R}^D$ to $\mathbb{R}^d$ with $D \geq d$. Note that when $D = d$, this level set is generically a set of isolated points. For other cases, after studying the geometry of the level set, we prove, under weak conditions, that this set is $D-d$ rectifiable. The proof begins with the case $D = d$. The result is extended to the general case using the Crofton formula. It contains very recent results, in particular a stronger version of the Bulinskaya lemma (Proposition 2.11); a non-Gaussian Kac–Rice formula (Theorem 2.17), and various examples including functions of Gaussian random fields, likelihood processes, gravitational microlensing, and shot-noise processes. We also present very recent results on the finiteness of the second moment (of the measure of the level set) and on higher moments.
- Chapter 3 is an overview of known results from existing literature. It is not designed to be read linearly: readers should instead consult the relevant passage for each given problem. The chapter presents little in terms of new findings, though some statements are a little stronger here thanks to the improvements in methodology presented in Chap. 2. Section 3.5.3 also presents a detailed calculation for random fields on the sphere which has not been undertaken elsewhere.
- Chapter 4 gives a very simple presentation of Hermite representation, the diagram formula, Arcones' inequality, and the Fourth Moment Theorem. It intentionally avoids the most complicated tools such as multiple Wiener–Itô integrals or Malliavin calculus. It ends with a central limit theorem for the measure of the level set.

The following chapters present various applications of our methodology—a crucial step, as we see it, for evaluating the importance of this book's contribution to the field. We thus look to show how easily the results of the first chapters can be applied, whilst demonstrating that the conditions of the theorems are easy to verify and that they can be applied across a number of areas in mathematics, physics and statistics.

- Chapter 5 is devoted to the study of critical points of random fields. It is a fairly consequential chapter, since it covers many aspects of the computation of moments including the second moment, which is related to the study of repulsion between critical points. Additionally, in this chapter the result of Auffinger et al. (2013) concerning large deviations of the number of critical points as the dimension of the sphere tends to infinity is generalized to the "flat case". The chapter concludes with a multivariate CLT that applies in particular to the Euler Characteristics.
- Chapter 6 shows how the Kac–Rice method can solve some problems relating to the distribution of the maximum of random fields that appear in infinite-dimensional compressed sensing.

- Chapter 7 considers the most common non-Gaussian model, the shot noise model. The existence of densities, finiteness of moments, and a CLT are all proven here.
- Chapter 8 consists of a bibliographic review of other topics such as winding numbers, random polynomials, random waves and systems of random polynomial equations.

This book aims to be as self-contained as is possible for a single volume. To fully understand it, however, readers will need to have a basic grasp of Probability Theory, including stochastic integrals. Similarly, while we have looked to include an elementary presentation of geometry-related questions, for some fundamental theorems, such as the co-area formula, readers are invited to refer to the existing literature. Some basic notions of statistics will also be necessary for Chap. 6.

All the authors of this monograph have been influenced directly or indirectly by the work of *Mario Wschebor*. This monograph can therefore be considered as a dynamic continuation of his work: since Mario passed away, further work in the field has led to a number of new perspectives, all of which pay homage to his work by perpetuating it. This book is dedicated to his memory.

Montevideo, Uruguay — Diego Armentano
Toulouse, France — Jean-Marc Azaïs
Salto, Uruguay — Federico Dalmao
Écully, France — Yohann De Castro
Castanet-Tolosan, France — Céline Delmas
Montevideo, Uruguay — José Rafael León
Montevideo, Uruguay — Ernesto Mordecki
October 2025

Competing Interests The authors have no competing interests to declare that are relevant to the content of this manuscript.

Basic Notation

An index of notation can be found at the end of this book, but for the sake of clarity, we will define a number of terms from the outset. Throughout this book, we will consider a complete probability space $(\Omega, \mathcal{U}, \mathbb{P})$. For this space, the term "random process" will denote a random function from $T \subset \mathbb{R}$ to $\mathbb{R}$, while "random field" will denote a random function from $T \subset \mathbb{R}^D$ to $\mathbb{R}^d$, $D > 1$,$D \geq d$. Extension to manifolds is also considered.

General notation	
$M^\top$; $(M_{i:})$	transpose of M; (ith line)
Id_n; $\mathrm{J}_{n,p}$; J_p	identity matrix of size d; "all one" matrix of dim. $n \times p$; $J_{p,p}$
t (and s)	the gen. value of the argument belongs, in general, to T
$u \in \mathbb{R}^d$; v	a particular value of the image; a value in the neighborhood of u.
C^k	Set of k-th differentiable functions
δ_{jk}	Kronecker symbol
`(const)`	unimportant positive constant, value can change from one occurrence to another.
$\bar{A}$; $\mathring{A}$	closure; interior of A
x^+	nonnegative part: $\sup(x, 0)$
Probability	
ϕ; ϕ_D; Φ; $\bar{\Phi}$	Standard normal density; in $\mathbb{R}^D$; cumulative distribution; $\bar{\Phi} := 1 - \Phi$.
$\mathrm{Var}(Z)$; $\mathrm{Cov}(Z, T)$	Variance-covariance matrix of a random vector; $\mathbb{E}\big[(Z - \mathbb{E}(Z))(T - \mathbb{E}(T))^\top\big]$
$c(s, t)$	Covariance function $:= \mathbb{E}(X(s)X(t))$
	$= r(s - t)$,$= \mathbf{r}(\\|s - t\\|^2) = \rho(\\|s - t\\|)$ when stationary or isotropic.
a.s.; N.D.	abbreviation for: "almost surely"; Non-Degenerate.

Multiple integrals are written this way:

$$\int_0^1 \mathrm{d}x \int_0^x \mathrm{d}y = \frac{1}{2}.$$

Basic Notation

Contents

List of Figures

Chapter 1
The Kac–Rice Formula, Dimension 1

This chapter introduces the different forms of the Kac–Rice formula in the simplest case of one-dimensional random processes. It distinguishes carefully between: formulas valid for almost every level, formulas that gives upper-bounds and the "true" Kac–Rice formula valid for all levels, and thus for any particular level, such as zero, highlighting why the former is often insufficient for practical applications. Through explicit counterexamples, the text demonstrates how the formula can fail when local extrema occur precisely at the considered level. This implies that Bulinskaya-type lemmas that exclude the existence of such local extrema are a key point in the proof. A Gaussian ad hoc proof is given while the non-Gaussian case is postponed to the next chapter.

The chapter concludes by exploring recent methods for proving the finiteness of higher moments for the number of level crossings in both Gaussian and non-Gaussian frameworks.

1.1 Introduction

As mentioned in the preface, the study of the number of crossings of a level by a random process began with the pioneering works of Kac (1943) and Rice (1945) in the case of a random process $\mathbb{R} \to \mathbb{R}$.

This chapter is devoted to the study of this topic and serves as an introduction to the more sophisticated methods of the following chapters. It builds on the classical monographs Cramér and Leadbetter (1967) and Azaïs and Wschebor (2009) on this topic by emphasizing the distinctions between different versions of the Kac–Rice formula: for almost all levels, the upper part, and for every level, the true Kac–Rice formula. More precisely, a Kac–Rice formula for almost all levels is worthless in most applications because the considered level is often zero or a fixed level given by

D. Armentano et al., *Geometry of Level Sets of Random Fields, Kac–Rice Formulas, Hermite Expansions and Applications*, Probability Theory and Stochastic Modelling 111, https://doi.org/10.1007/978-3-032-29144-8_1

the experiment, such as the threshold between black and white in a black-and-white photo. Finally, this chapter provides some explicit counterexamples.

The chapter begins by establishing the Kac–Rice formula for Gaussian processes, and subsequently looking at some non-Gaussian cases. It then goes on to address the problem of the finiteness of moments, since this question naturally arises once the Kac–Rice formula has been proven. The Gaussian and non-Gaussian tools that we present use the Kac–Rice formula or some other analytic properties.

Most of the results of this chapter are classical, with the exception of some recent results about finiteness of moments for crossings.

1.2 Formulas for Almost Every Level

Everybody knows the change of variable formula: If $f(\cdot)$ is a real C^1 function, one-to-one from $[a, b]$, $a < b$, to $[c, d]$, $c < d$, then, setting $x = f(t)$:

$$\int_c^d g(x)\,\mathrm{d}x = \int_a^b g(f(t))|f'(t)|\,\mathrm{d}t. \tag{1.1}$$

The assumption that $f(\cdot)$ is one-to-one is a strong requirement. In fact, the formula holds in more general cases, taking into account the number of roots; this is called the Banach formula (Banach 1925). Define for an interval T the number of roots of $X(t) = u$:

$$N_u(f, T) := \#\{t \in T : f(t) = u\}. \tag{1.2}$$

Of course, this number can take the value $+\infty$. We have

Proposition 1.1 (Banach Formula) *With the above definition, if $f(\cdot)$ is an absolutely continuous function defined on an open set T and if $g(\cdot)$ is a bounded non-negative Borel function, then*

$$\int_{\mathbb{R}} N_u(f, B)g(u)\,\mathrm{d}u = \int_B g(f(t))|f'(t)|\,\mathrm{d}t, \tag{1.3}$$

for every Borel subset B of T. Note that the integrals on both sides are always well defined, in particular the Morse–Sard Lemma 2.3, implies that $N_u(f, B)$ is finite for almost every u.

In fact, the result is also true under a slightly weaker hypothesis: the function $f(\cdot)$ needs only to be of bounded variation, although we will not use this result. Note of course that in the one-to-one situation of the first example

$$N_u(f, T) = \begin{cases} 1, & \text{if } u \in T = [c, d], \\ 0, & \text{otherwise,} \end{cases}$$

so that (1.1) and (1.3) agree.

This formula admits two multivariate generalizations, namely the Area formula and the Co-Area formula, that will be presented in Chap. 2. These formulas will allow us to directly obtain a Kac–Rice formula for almost every level.

Proposition 1.2 (Kac–Rice Formula for Almost Every Level) *Let $\{X(t): t \in T\}$, with T an open subset of $\mathbb{R}$, be a real-valued stochastic process with absolutely continuous paths. Assume that for each $t \in T$, the distribution of the random variable $X(t)$ has a density $p_{X(t)}(\cdot)$ and that $X'(t)$ is integrable. Then, for every Borel set $B \subset T$, we have*

$$\mathbb{E}(N_u(X, B)) = \int_B \mathbb{E}\big(\Delta(t) \mid X(t) = u\big)\, p_{X(t)}(u)\, \mathrm{d}t, \tag{1.4}$$

for almost every $u \in \mathbb{R}$, where $\Delta(t) = |X'(t)|$.

Proof Let $g(\cdot)$ be a bounded non-negative Borel test function. By the Banach formula:

$$\int_{\mathbb{R}} g(u) N_u(X, B)\, \mathrm{d}u = \int_B g(X(t))|X'(t)|\, \mathrm{d}t.$$

Take the expectation on both sides. Note that these expectations are always well-defined, since we are dealing with non-negative functions. Note that the conditional expectation: $\mathbb{E}\big(|X'(t)| \mid X(t) = u\big)$ is well defined for almost every u. Applying the Fubini Theorem, and conditioning by $X(t) = u$, we get

$$\int_{\mathbb{R}} g(u)\mathbb{E}\big(N_u(X, B)\big)\, \mathrm{d}u = \int_{\mathbb{R}} g(u)\, \mathrm{d}u \int_B \mathbb{E}\big(|X'(t)| \mid X(t) = u\big) p_{X(t)}(u)\, \mathrm{d}t.$$

By duality, since $g(\cdot)$ is arbitrary, the terms in the factor of $g(u)$ on both sides are equal for almost every u, giving us our result. □

Remark 1.1 ("Zero Is Not Almost Every u") Some practitioners consider that the above result is sufficient for all applications. This is rarely the case however: very often the considered level is $u = 0$! (as in the seminal work of Kac 1943) and a statement, valid for almost every u, is not necessarily valid for $u = 0$! It is therefore necessary to establish the formula (1.4) **for every level**. This will require a good deal of work, especially for non-Gaussian models.

Remark 1.2 Note that if the process $X(\cdot)$ is Gaussian, it is very easy to define the conditional expectation $\mathbb{E}\big(|X'(t)| \mid X(t) = u\big)$ for every u with continuity properties by using regression formulas. In the general setting, the right-hand side of (1.4) is defined, a priori, only for almost every u. Further work is therefore necessary to define the density and the conditional expectation for every u. Without this work the Kac–Rice formula for an arbitrary fixed level (as for instance with $u = 0$) makes no sense. The first versions of this result are to be found in Brillinger (1972) and Zähle (1984).

1.3 Counterexamples

Before carrying out the significant work required for establishing the Kac–Rice formula for every level, it is worth looking for some counterexamples.

Example 1.1 (The Random Gaussian Line) Let $X(t) = \xi.t$, $t \in T = [-1, 1]$, where ξ is a standard normal random variable. It is easy to check that for every $t \neq 0$

$$\mathbb{E}\big(|X'(t)| \mid X(t) = u\big) = \left|\frac{u}{t}\right|, \quad p_{X(t)}(u) = \frac{1}{|t|}\phi\Big(\frac{u}{|t|}\Big),$$

and

$$\mathbb{E}\big(N_u(X, T)\big) = \mathbb{P}\{|\xi| > |u|\}.$$

This last quantity is obtained by the right-hand side of the Kac–Rice formula for $u \neq 0$ **but not for** $u = 0$: the formula gives zero!

Example 1.2 (The Poor Shot-Noise) This example is the most basic example of a shot-noise, a case which will be studied in much greater detail later in this volume (Chap. 7).

Let $\psi(x) = \exp(-x^2/2)$, and let ξ be a standard Gaussian variable. Define

$$X(t) = \psi(t - \xi).$$

It is obvious that the number of roots N_u of $X(t) = u$ in $\mathbb{R}$ is

$$N_u = \begin{cases} 0, & \text{if } u \le 0, \text{ or } u > 1, \\ 2, & \text{if } 0 < u < 1, \\ 1, & \text{if } u = 1. \end{cases}$$

Of course, the respective expectations take the same values for the corresponding levels.

We can now compute the ingredients of the Kac–Rice formula. If $0 < u < 1$, the event $\{X(t) = u\}$ implies that $t - \xi = \pm\sqrt{-2\log(u)}$ and

$$X'(t) = \pm u\sqrt{-2\log(u)},$$

each sign with probability $1/2$. This directly implies that

$$\mathbb{E}(|X'(t)| \mid X(t) = u) = u\sqrt{-2\log(u)}.$$

On the other hand, the Banach formula (Proposition 1.1) proves the following important proposition concerning the push-forward distribution. This is a particular case of Proposition 2.6, so its proof is omitted.

Proposition 1.3 (Push-Forward Distribution) *Let Y be a random variable admitting a density f with respect to the Lebesgue measure. Let $H\colon \mathbb{R} \to \mathbb{R}$ be an absolutely continuous function. In this case, the push-forward distribution, the distribution of $Z = H(Y)$, has a law that is the sum of a measure which is supported by the critical values of H and an absolutely continuous distribution with density*

$$h(z) = \sum_{y:\in H^{-1}(z)} \frac{f(y)}{|H'(y)|}.$$

In particular, if $\mathbb{P}\{H'(Y) = 0\} = 0$, the distribution of Z is absolutely continuous, with density given by the above formula.

Let us now return to the counterexample (Example 1.2). Let $\phi(\cdot)$ be the density of ξ, the standard Gaussian density, and set $\psi_t(z) := \psi(t - z)$. Using Proposition 1.3, we see that $X(t)$ admits a density, which is given by

$$p_{X(t)}(u) = \sum_{z\in\psi_t^{-1}(u)} \frac{\phi(z)}{|\psi_t'(z)|} = \frac{\mathrm{e}^{-t^2/2}}{\sqrt{2\pi}} \frac{\exp\big(t\sqrt{-2\log(u)}\big) + \exp\big(-t\sqrt{-2\log(u)}\big)}{\sqrt{-2\log(u)}}.$$

The right-hand side of the Kac–Rice formula hence takes the value

$$\int_{\mathbb{R}} u\sqrt{-2\log(u)}.\frac{\mathrm{e}^{-t^2/2}}{\sqrt{2\pi}} \frac{\exp\big(t\sqrt{-2\log(u)}\big) + \exp\big(-t\sqrt{-2\log(u)}\big)}{\sqrt{-2\log(u)}}\,\mathrm{d}t = 2.$$

While for $u = 1$, $\mathbb{E}(|X'(t)| \mid X(t) = 0) = 0$ and the Kac–Rice formula returns zero while the correct result is 1.

Remark 1.3 Readers familiar with the Kac–Rice formula will not be surprised by this counterexample. Indeed, there is always a local maximum at the level $u = 1$ and this is a pathology that we need to eliminate using the methods of the next section.

1.4 Nonexistence of Local Extrema at a Given Level

As shown by the second counterexample, the presence of local extrema at the precise considered level may cause the Kac–Rice formula to fail. In addition, this condition is needed to establish the "Kac counting formula" below. It is the most natural way to count the number of roots of $X(t) = u$, see Kac (1943).

Lemma 1.4 (Kac's Counting Formula) *Assume that f is a function of class C^1 on $T = [a, b]$, such that $f(a) \neq u$, $f(b) \neq u$ and $\{t : t \in T,\ f(t) = u,\ f'(t) = 0\} = \emptyset$. Then*

$$N_u(f, T) = \lim_{\delta \to 0} N_{u,\delta}(f, T), \quad \textit{with } N_{u,\delta}(f, T) := \frac{1}{2\delta} \int_T \mathbb{1}_{\{|f(t)-u|<\delta\}} \left|f'(t)\right| \mathrm{d}t. \tag{1.5}$$

Proof The hypothesis implies that the roots are isolated, implying in turn that $N_u(f, T)$ is finite, say $N_u(f, T) = n$. If $n = 0$, the result is obvious since the integrand in the right-hand side of (1.5) is identically zero if δ is small enough. If $n \geq 1$, let $s_1, \ldots, s_n$ be the roots of $f(\cdot) = u$. Since $f'(s_j) \neq 0$ for every $j = 1, \ldots, n$, if $\delta > 0$ is small enough, the inverse image of the interval $(u - \delta, u + \delta)$ by the function f is the union of exactly n pairwise disjoint intervals $J_1, \ldots, J_n$, which contain respectively the points $s_1, \ldots, s_n$. The restriction of $f(\cdot)$ to each one of the intervals J_k ($k = 1, \ldots, n$) is a diffeomorphism and it is easy to check, by changing our variables, that

$$\int_{J_k} \left|f'(t)\right| \mathrm{d}t = 2\,\delta,$$

for each k. So, if $\delta > 0$ is small enough:

$$\frac{1}{2\delta} \int_T \mathbb{1}_{\{|f(t)-u|<\delta\}} \left|f'(t)\right| \mathrm{d}t = \frac{1}{2\delta} \sum_{k=1}^{n} \int_{J_k} \left|f'(t)\right| \mathrm{d}t = n,$$

and we are done. □

The formula above is rather simple but it contains a limit. The following alternative expression has been proposed by Angst and Poly (2020).

Proposition 1.5 *Let $f(\cdot)$ be a C^2 periodic function on $[0, 2\pi]$ and let $F(\cdot)$ be a C^1 function such that*

$$F(x) \to -1, \textit{ as } x \to -\infty; \quad F(x) \to 1, \textit{ as } x \to +\infty.$$

Then

$$N_0(f, [0, 2\pi]) = -\frac{1}{2} \int_0^{2\pi} F'\left(\frac{f'(x)}{f(x)}\right)\left(\frac{f'(x)}{f(x)}\right)' \mathrm{d}x.$$

Note that the conditions imposed on the function $f(\cdot)$ are stronger here.

A Formal Proof of the Kac–Rice Formula

Note that, formally, we can pass to the limit in δ in (1.5), giving

$$\mathbb{E}\big(N_u(X,T)\big) = \lim_{\delta\to 0}\frac{1}{2\delta}\int_T \mathbb{E}\big(|X'(t)|\mathbb{1}_{|X(t)-u|\le\delta}\big)\,\mathrm{d}t.$$

If the density and the conditional expectation are well defined

$$\int_T \mathbb{E}\big(|X'(t)|\mathbb{1}_{|X(t)-u|\le\delta}\big)\,\mathrm{d}t = \int_{u-\delta}^{u+\delta}\mathrm{d}v\int_T \mathbb{E}\big(|X'(t)| \mid X(t)=v\big)p_{X(t)}(v)\,\mathrm{d}t.$$

Moreover, if

$$v \mapsto \int_T \mathbb{E}\big(|X'(t)| \mid X(t)=v\big)p_{X(t)}(v)\,\mathrm{d}t \text{ is continuous,}$$

its value at u is equal to the limit of its mean value over $[u-\delta, u+\delta]$, giving

$$\mathbb{E}\big(N_u(X,T)\big) = \int_T \mathbb{E}\big(|X'(t)| \mid X(t)=u\big)p_{X(t)}(u)\,\mathrm{d}t,$$

and we are done. The remainder of this chapter and Chap. 2 will look to justify all these passages to the limit under hypotheses that are as general as possible.

The non-existence of a local maximum is very easily obtained in the Gaussian case, by a result of Ylvisaker (1968) outlined below.

Proposition 1.6 (Ylvisaker) *Let $\{Z(t) : t \in T\}$ be a real-valued Gaussian process indexed by a compact separable topological space T, having continuous paths and* $\mathrm{Var}(Z(t)) > 0$ *for all $t \in T$.*

Then, for fixed $u \in \mathbb{R}$, one has $\mathbb{P}(E_u^Z \neq \emptyset) = 0$, where E_u^Z is the set of local extrema of $Z(\cdot)$ having value equal to u.

A proof can be found in Azaïs and Wschebor (2009, Th. 1.21). Note that this result applies in any dimension (including infinite).

A second important result is Bulinskaya's lemma (Bulinskaya 1961). We will present some important extensions to the multivariate case in Chap. 2.

Proposition 1.7 (Bulinskaya) *Let $\{X(t) : t \in T\}$ be a stochastic process with paths of class C^1 defined on the interval T of the real line.*

Assume that for each $t \in T$, the random variable $X(t)$ has a density $p_{X(t)}(u)$ which is uniformly bounded as t varies in a compact subset of T. Define the set of critical points with value u of the random path $X(\cdot)$ by

$$T_u^X = \big\{t : t \in T,\ X(t) = u,\ X'(t) = 0\big\}.$$

Then,

$$\mathbb{P}(T_u^X \neq \emptyset) = 0.$$

Proof It suffices to prove that $\mathbb{P}(T_u^X \cap J \neq \emptyset) = 0$ for any compact sub-interval J of T.

Let ℓ be the length of J and $t_0 < t_1 < \cdots < t_m$ a uniform partition of J, i.e. $t_{j+1} - t_j = \ell/m$, for $j = 0, 1, \ldots, m-1$. Denote by $\omega_{X'}(\delta, J)$ the modulus of continuity of X' on the interval J, and consider the event

$$E_{\delta,\varepsilon} = \{\omega_{X'}(\delta, J) \geq \varepsilon\}.$$

Let $\varepsilon > 0$ be given; choose $\delta > 0$ so that $\mathbb{P}(E_{\delta,\varepsilon}) < \varepsilon$ and m so that $\frac{\ell}{m} < \delta$. We have:

$$\mathbb{P}(T_u^X \cap J \neq \emptyset) \leq \mathbb{P}(E_{\delta,\varepsilon}) + \sum_{j=0}^{m-1} \mathbb{P}(\left\{T_u^X \cap \left[t_j, t_{j+1}\right] \neq \emptyset\right\} \cap E_{\delta,\varepsilon}^{\complement})$$

$$< \varepsilon + \sum_{j=0}^{m-1} \mathbb{P}\Big(\left|X(t_j) - u\right| \leq \varepsilon \frac{\ell}{m}\Big) = \varepsilon + \sum_{j=0}^{m-1} \int_{|x-u| \leq \varepsilon \frac{\ell}{m}} p_{X(t_j)}(x)\, \mathrm{d}x.$$

If C is an upper bound for $p_{X(t)}(x)$, $t \in J$, $|x - u| \leq \varepsilon\, \ell$, we obtain

$$\mathbb{P}(T_u^X \cap J \neq \emptyset) \leq \varepsilon + C\varepsilon\ell.$$

Since $\varepsilon > 0$ is arbitrary, the result follows. □

1.5 Upper-Bound

An upper-bound in the Kac–Rice formula follows easily from an application of Fatou's lemma to the "Kac counter" in (1.5). However, in this section we want to be a little more general, and get rid of the condition of non-existence of extrema on the barrier. For this purpose, we need a generalization of Lemma 1.4 without this condition, one which is given in the following result thanks to Marcus (1977).

Lemma 1.8 (Extended Kac Counting Formula) *Consider $-\infty < a < b < +\infty$ and $f : T = (a, b) \to \mathbb{R}$, an absolutely continuous function with derivative $f'(t)$. Suppose that $f(t)$ is not identically equal to u on any sub-interval of (a, b). Then:*

$$N_u(f; T) \leq \liminf_{\delta \to 0} \frac{1}{2\delta} \int_T \mathbb{1}_{\{|f(t)-u|<\delta\}} \left|f'(t)\right| \mathrm{d}t.$$

Proof (Sketch) Suppose first that $N_u = n$ is finite and let $a < t_1 < \cdots < t_n < b$, be the root points, then for δ sufficiently small and because of the second hypothesis we can find $s_1, s_1', \ldots, s_n, s_n'$ such that $a < s_1 < t_1 < s_1' < \cdots < s_n < t_n < s_n' < b$, and

$$\int_{s_i}^{t_i} |f'(t)|\,\mathrm{d}t \geq \delta,$$

$$\int_{t_i}^{s_i'} |f'(t)|\,\mathrm{d}t \geq \delta, \quad i = 1, \ldots, n. \tag{1.6}$$

For example, we can start from t_1 leftward and find some $\bar{s}_1$ such that $|f(\bar{s}_1) - u| = \delta_1 > 0$, $\int_{\bar{s}_1}^{t_1} |f'(t)|\,\mathrm{d}t \geq \delta_1$. Considering the other cases, we obtain $\delta_1, \delta_1', \ldots, \delta_n, \delta_n'$. Hence, δ is the minimum of these $2n$ values. Returning to the case of $\bar{s}_1$, by the intermediate value theorem there exists a point s_1 between $\bar{s}_1$ and t_1 such that $|f(s_1) - u| = \delta$. This is sufficient for obtaining (1.6). This formula in turn implies that for such δ the right-hand side is greater than n.

Suppose now that N_u is infinite, then for each m we can choose m points $t_1, \ldots, t_m$, as above and show that for δ sufficiently small the right-hand-side is greater than m. □

Using this lemma we can state the upper-bound. Its proof is simply an application of Lemma 1.8 and Fatou's lemma.

Theorem 1.9 *Let $X(\cdot)$ be a real stochastic process defined on an open interval T. Suppose that $X(\cdot)$ has almost surely (a.s.) absolutely continuous paths; that for every t, the derivative $X'(t)$ is integrable; and that $X(t)$ is a.s. not identically equal to u on any sub-interval of (a, b). Then,*

$$\mathbb{E}\big(N_u(X, T)\big) \leq \liminf_{\delta \to 0} \frac{1}{2\delta} \int_T \mathbb{E}\big(|X'(t)| \mathbb{1}_{|X(t)-u| \leq \delta}\big)\,\mathrm{d}t. \tag{1.7}$$

Note that the right-hand side is always well-defined.

Remark 1.4 If the interval $T = [a, b]$ is closed we have to ask whether $X(t)$ admits a density, to avoid the cases $X(a) = u$ or $X(b) = u$, or, alternatively, to add 1 to the bound in (1.7). We will leave the details as an exercise for the reader.

1.6 Uni-dimensional Gaussian Case

1.6.1 Spectral Measure and Differentiability

Let us start by recalling Bochner's Theorem.

Theorem 1.10 *Let us consider a stationary Gaussian process with covariance*

$$r(s-t) := \operatorname{Cov}(X(s), X(t)).$$

Assume that this function is continuous. In this case, it is the Fourier transform of a non-negative measure F : **the spectral measure.**

A proof of Bochner's Theorem in the multidimensional case is given in Sect. 3.4.1.

This measure is symmetrical and satisfies

$$r(t) = \int_{-\infty}^{+\infty} e^{\imath\lambda t}\, dF(\lambda) = 2\int_{0}^{+\infty} \cos(\lambda t)\, dF(\lambda).$$

The moments of this measure are called the spectral moments and are traditionally denoted by $\lambda_0, \lambda_2, \ldots$.. Note that, due to symmetry, odd moments vanish.

A Gaussian process $X(\cdot)$ is said to be differentiable in quadratic mean if for all $t \in \mathbb{R}$

$$\frac{X(t+h) - X(t)}{h} \text{ converges in quadratic mean, as } h \to 0.$$

The limit, that is denoted by $X'(\cdot)$, is a Gaussian process. In the present section we consider Gaussian processes that are differentiable in quadratic mean.

For stationary Gaussian processes, differentiability in quadratic mean is equivalent to the finiteness of the second spectral moment λ_2. Differentiability in quadratic mean is a little weaker than a.s. differentiability of the sample paths, see Azaïs and Wschebor (2009, Ch. 1).

1.6.2 Main Result

Before stating our main result, let us present an extension argument that will be useful in many situations.

Remark 1.5 (Extension Argument) Both sides of the Kac–Rice formula are measures on the Borel subsets B of T. On the left-hand side

$$\mathbb{E}\big(N_u(X, B)\big),$$

is the expectation of the counting measure. On the right-hand side

$$\int_B \mathbb{E}\big(|X'(t)| \mid X(t) = u\big) p_{X(t)}(u)\, dt,$$

is a measure with density with respect to the Lebesgue measure. If they coincide on a generating class of sets, for instance, intervals, open sets, etc., they necessarily coincide on the whole Borel σ-algebra. As a result, to obtain the result for every Borel set, we need only prove the Kac–Rice formula for a generating class of the Borel σ-algebra.

In the present chapter, our main result is limited to Gaussian processes. Non-Gaussian random processes will be considered in the next chapter.

Theorem 1.11 (Gaussian Kac–Rice Formula) *Let $\{X(t)\colon t \in T\}$, with T a compact interval of the real line, be a Gaussian process. Let $m(\cdot)$ and $c(\cdot,\cdot)$ be respectively its expectation and covariance functions. Suppose that $m(\cdot)$ is C^1, that $c(\cdot,\cdot)$ is C^2, and that* $\mathrm{Var}(X(t)) > 0$ *for all $t \in T$. Then, for every Borel subset B of T and for every level $u \in \mathbb{R}$,*

$$\mathbb{E}\big(N_u(X,B)\big) = \int_B \mathbb{E}\big(|X'(t)| \mid X(t) = u\big) p_{X(t)}(u)\,\mathrm{d}t.$$

Theorem 1.11 has a generalization to higher moments: the order k Kac–Rice formula, see Theorem 1.12.

Remark 1.6 The condition "$m(\cdot)$ is C^1 and $c(\cdot,\cdot)$ is C^2" is a sufficient condition for having differentiability in quadratic mean (Azaïs and Wschebor 2009, Prop. 1.13), and in practice it is the only easy way to prove it.

Remark 1.7 For stationary processes, the result is very simple. Firstly, the expectation is constant and thus continuous. Secondly, the covariance function is C^2 if and only if the second spectral moment λ_2 is finite, see Azaïs and Wschebor (2009, Ex. 1.4). Suppose the process is standardized

$$\mathbb{E}(X(t)) = 0, \quad \mathrm{Var}(X(t)) = 1,$$

as we will assume throughout. When $\lambda_2 < +\infty$, Theorem 1.11 reads

$$\mathbb{E}\big(N_u\big) = |B|\sqrt{\lambda_2}\,\frac{\mathrm{e}^{-u^2/2}}{\pi}, \tag{1.8}$$

where $|B|$ is the Lebesgue measure of B. We will leave the details as an exercise for the reader. In addition, if $\lambda_2 = +\infty$, it is proved in Proposition 1.15 that $\mathbb{E}\big(N_u\big) = +\infty$, so that the Kac–Rice formula still holds true.

In conclusion (1.8) **is always true for standardized Gaussian stationary processes.**

Proof of Theorem 1.11 Here we can use an approximation argument. We can first suppose that B is a compact interval and without loss of generality, we can assume that $B = [0,1]$ and use a polygonal dyadic approximation. Let $X_n(t)$ be the polygonal line that passes through the points $\big(i2^{-n}, X(i2^{-n})\big)$, $i = 0,\ldots,2^n$.

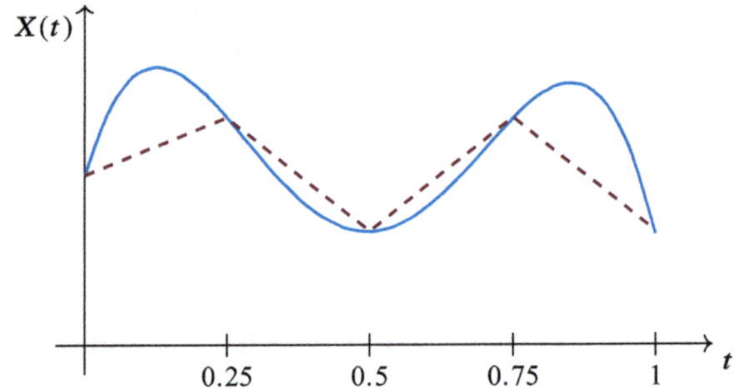

Fig. 1.1 Polygonal approximation

In other words $X_n(t)$ performs the linear interpolation of the process $X(\cdot)$ after discretization with mesh 2^{-n} (Fig. 1.1).

Because of the hypothesis, with probability 1, $X(i2^{-n}) \neq u$ and it is easy to see that the Kac counting formula (Lemma 1.4) still applies to $X_n(\cdot)$. Let $N_{u,\delta,n}$ be the "Kac counter" associated to δ, the process X_n, and the set B. Clearly, we have

$$N_{u,\delta,n} \leq 2^n.$$

This implies that we can pass to the limit, using the dominated convergence theorem, to establish that

$$\begin{aligned}\mathbb{E}(N_{u,n}) &= \lim_{\delta\to 0} \frac{1}{2\delta} \int_B \mathbb{E}\big(\mathbb{1}_{\{|X_n(t)-u|<\delta\}}\, |X_n'(t)|\big)\, \mathrm{d}t \\ &= \lim_{\delta\to 0} \int_B \mathrm{d}t \frac{1}{2\delta} \int_{u-\delta}^{u+\delta} \mathbb{E}\big(|X_n'(t)| \mid X_n(t) = x\big) p_{X_n(t)}(x)\, \mathrm{d}x. \qquad (1.9)\end{aligned}$$

Note that $X_n(t)$ converges uniformly to $X(t)$ and this implies, by compactness, that $\mathrm{Var}(X_n(t))$ is lower bounded by a positive constant for n sufficiently large. This implies in turn the continuity and boundedness of $p_{X_n(t)}(x)$ with respect to t and x. The expectation and the variance of $X_n(t)$ and $X_n'(t)$ converge uniformly to those of $X(t)$ and $X'(t)$, implying that

$$\int_B \mathbb{E}\big(|X_n'(t)| \mid X_n(t) = x\big) p_{X_n(t)}(x)\, \mathrm{d}t,$$

is continuous as a function of x. As a consequence, the limit in δ in (1.9) exists, yielding

$$\mathbb{E}(N_{u,n}) = \int_B \mathbb{E}\big(|X_n'(t)| \mid X_n(t) = u\big) p_{X_n(t)}(u)\, \mathrm{d}t.$$

It remains to let $n \to \infty$. According to Ylvisaker's Theorem (Proposition 1.6), with probability one there exists no local extrema at the level u. This implies that

$$N_{u,n} \uparrow N_u, \ a.s.,$$

so we can thus apply the monotone convergence theorem. Concerning the right-hand side, the uniform convergence of the expectation and variance of $X_n(t)$ implies the convergence of this term to

$$\int_B \mathbb{E}\big(|X'(t)| \mid X(t) = u\big) p_{X(t)}(u)\,\mathrm{d}t.$$

We can now use the approximation argument (Remark 1.5). The compact intervals $B \subset T$ are a generating class for the Borel σ-algebra, giving the result. □

Remark 1.8 (Complement to Theorem 1.9) Note that the extension argument does not apply to the bound given by (1.7), which is not true for every Borel set. **This is because the bound is not a measure**. However, in the particular case where

$$u \mapsto \int_T \mathbb{E}\big(|X'(t)| \mid X(t) = u\big) p_{X(t)}(u)\,\mathrm{d}t$$

is continuous, the above formula reads

$$\mathbb{E}\big(N_u(X, T)\big) \leq \int_T \mathbb{E}\big(|X'(t)| \mid X(t) = u\big) p_{X(t)}(u)\,\mathrm{d}t. \tag{1.10}$$

Here, since both sides are measures, this inequality is also true for every Borel set in T.

1.6.3 The Non-Gaussian Case

The first result in this direction appears in the paper by Marcus (1977), which uses the polygonal method introduced by Itô (1963). The conditions in this paper are complicated, since some regularity conditions of the polygonal approximation $X_n(\cdot)$ are needed. The most difficult to check is the continuity of the map

$$v \mapsto \int_I \mathbb{E}\big(|X_n'(t)| \mid X_n(t) = v\big) p_{X_n(t)}(v)\,\mathrm{d}t.$$

We owe the second result to Wschebor (1985), see also Azaïs and Wschebor (2009, Th. 3.4.). It also involves complicated conditions. For the simplest case $k = 1$ (Kac–Rice formula of order 1) the conditions are, for x in a neighborhood of u:

- the paths are of class C^1,
- $X(t), X'(t)$ has a joint density $p_{X(t),X'(t)}$,
- the function

$$I(x) := \int_B \mathrm{d}t \int_{\mathbb{R}} |x'| p_{X(t),X'(t)}(x, x')\, \mathrm{d}x' \ \text{ admits a continuous version,}$$

- the function

$$(t, x) \mapsto \int_{\mathbb{R}} |x'| p_{X(t),X'(t)}(x, x')\, \mathrm{d}x' \ \text{ is continuous,}$$

- the integral

$$\int_{\mathbb{R}^2} |x_1' - x_2'| p_{X(t),X'(t_1)),X'(t)}(x, x_1', x_2')\, \mathrm{d}x_1'\, \mathrm{d}x_2'$$

  tends to zero as $(t_1 - t_2) \to 0$, uniformly in x, in a neighborhood of u and for t in a compact subset of B.

Even in the simple case of a χ^2 process, these conditions are difficult to verify. We hope that the multi-dimensional methods of Chap. 2 will be easier to apply.

1.7 Higher Moments

The proof for higher moments given below clearly shows that the Kac–Rice formula for almost all levels as well as the upper bound apply to higher moments. To avoid a lengthy exposition, we will limit ourselves to the "True Kac–Rice formula" i.e. the formula for every level. Similar results are available for the Kac–Rice formula for almost every level and for the majorization.

Theorem 1.12 (Gaussian Kac–Rice Formula of Order k**)** *Let $X(\cdot)$ be a Gaussian process defined in an open set T. Assume that $X(\cdot)$ satisfies the hypotheses of Theorem 1.11. Assume that for every set of pairwise distinct points $t_1, \dots, t_k$ in T, $k > 1$, the joint distribution of $X(t_1), \dots, X(t_k)$, is N.D. Then for every Borel set $B \subset T$ and for every level $u \in \mathbb{R}$:*

$$\mathbb{E}\Big(N_u(B)\big(N_u(B) - 1\big) \dots \big(N_u(B) - k + 1\big)\Big)$$
$$= \int_{B^k} \mathbb{E}\big(|X'(t_1) \dots X'(t_k)| \mid X(t_1) = u, \dots, X(t_k) = u\big)$$
$$p_{X(t_1),\dots,X(t_k)}(u, \dots, u)\, \mathrm{d}t_1 \dots \mathrm{d}t_k. \qquad (1.11)$$

Let $u_1, \dots, u_k$ be pairwise different levels. Then,

$$\mathbb{E}\big(N_{u_1}(B)\dots N_{u_k}(B)\big) = \int_{B^k} \mathbb{E}\big(|X'(t_1)\dots X'(t_k)|\,\big|X(t_1) = u_1, \dots, X(t_k) = u_k\big)$$

$$p_{X(t_1),\dots,X(t_k)}(u_1, \dots, u_k)\,\mathrm{d}t_1 \dots \mathrm{d}t_k. \qquad (1.12)$$

A statement analogous to (1.12) is also true for pairwise non-intersecting sets $B_1, \dots, B_k$ contained in T, see (1.13) below.

Remark 1.9 A striking fact here is the appearance of $\mathbb{E}\big(N_u(B)(N_u(B) - 1)\dots(N_u(B) - k + 1)\big)$: the factorial moment. It only appears when we consider identical levels and identical intervals. The Kac–Rice formula gives the intensity of the point process of the roots with respect to the Lebesgue measure. Thus it does not count the irregular part located on the diagonal, as explained below.

Proof In the first case, formula (1.11) for one u and one B, we define the Cartesian product

$$C_u^k := C_u \times \dots \times C_u, \qquad \text{with } C_u := \{t \in B : X(t) = u\}.$$

The key point here is that this point process in $\mathbb{R}^k$ has a singular part on the diagonal:

$$D_k(B) = \Big\{(t_1, \dots, t_k) \in B^k : \exists j \neq j' : t_j = t_{j'}\Big\}$$

and that the quantity $N_u^{[k]}$ is just the cardinality of C_u^k that lies outside this diagonal: the value of the absolutely continuous part of the point process. To prove the result, by additivity, it suffices to consider pairwise distinct closed sub-intervals $B_1, \dots, B_k$ of T and to prove that

$$\mathbb{E}\big(N_u(B_1)\dots N_u(B_k)\big)$$

$$= \int_{B_1\times\dots\times B_k} \mathbb{E}\big(|X'(t_1)\dots X'(t_k)|\,\big|\,X(t_1) = u, \dots, X(t_k) = u\big)$$

$$p_{X(t_1),\dots,X(t_k)}(u, \dots, u)\,\mathrm{d}t_1 \dots \mathrm{d}t_k. \qquad (1.13)$$

Define the random field

$$Z(t_1, \dots, t_k) := \big(X(t_1), \dots, X(t_k)\big) \quad T^k \mapsto \mathbb{R}^k.$$

Then

$$N_u(B_1)\cdots N_u(B_k) = \#\{((t_1, \dots, t_k) \in B_1\times\dots\times B_k : Z((t_1, \dots, t_k) = (u, \dots, u).$$

Then it suffices to use the argument of the proof of Theorem 1.11. □

1.8 Free Lunch

There exist two situations where we can overcome all the technicalities of the non-Gaussian cases. The first one uses the following bound on the "Kac Counter",

$$N_{u,\delta} := \frac{1}{2\delta}\int_T \mathbb{1}_{\{|f(t)-u|<\delta\}} \left|f'(t)\right| \mathrm{d}t \le N_0(X') + 1,$$

which is a direct consequence of the intermediate value theorem.

Proof Assume that $N_0(X')$ is finite (otherwise there is nothing to prove), $N_0(X') = n$, say. We can then partition the interval T into $n+1$ intervals in which $X(\cdot)$ is monotone. The contribution of each of these intervals to the Kac counter is less than or equal to 1. □

The result is the following:

Theorem 1.13 *Suppose that a.s.*

- $X'(\cdot)$ *is absolutely continuous,*
- $X'(t)$ *is a.s. not identically equal to* 0 *on any open sub-interval of* B,
- $\forall t \in T$, $X''(t)$ *is integrable,*
- $X(t)$ *admits a density* $p_{X(t)}(v)$ *which is uniformly bounded (in* t *and* v*),*
- *the integral*

$$\int_T \mathbb{E}\big(X''(t) \mid X'(t) = x'\big) p_{X'(t)}(x')\,\mathrm{d}t$$

is locally bounded for x' *in a neighborhood of zero.*

Then, for all Borel sets $B \subset T$,

$$\mathbb{E}(N_u(X, B)) = \lim_{\delta\to 0}\frac{1}{2\delta}\int_B \mathbb{E}\big(|X'(t)|\mathbb{1}_{|X(t)-u|\le\delta}\big)\,\mathrm{d}t. \tag{1.14}$$

If in addition $X'(t)$ *has a finite expectation for every* $t \in T$, *define for* $|v-u| < \varepsilon$

$$R(v) := \int_T \mathbb{E}\big(|X'(t)| \mid X(t) = v\big) p_{X(t)}(v)\,\mathrm{d}t.$$

Then (1.14) *can be rewritten as*

$$\mathbb{E}(N_u(X, B)) = \lim_{\delta\to 0}\frac{1}{2\delta}\int_{u-\delta}^{u+\delta} R(v)\,\mathrm{d}v. \tag{1.15}$$

This method has been used for smoothed diffusions by Azaïs (1989) and, in a certain manner, for shot-noise processes by Biermé and Desolneux (2012).

Proof Applying Theorem 1.9 to $X'(\cdot)$ we can establish that

$$\mathbb{E}\big(N_0(X', T)\big) < +\infty.$$

Then we can use Bulinskaya's Lemma (Proposition 1.7) to establish that a.s. there exist no local extrema at a given level u so we can apply the Kac counting formula (Proposition 1.4) and the dominated convergence theorem to proceed to the limit and get the result. □

The second situation uses uniform integrability.

Theorem 1.14 *Suppose that the sample paths of* $X(\cdot)$ *are* C^1*, that the density of* $X(t)$ *at point* v *is uniformly bounded for* $t \in T$ *and for* $|v - u| < \varepsilon$*, and that* $\sup_{|v-u|<\varepsilon} \mathbb{E}(N_v(T)^2) < \infty$*. Then* (1.14) *or* (1.15) *hold true.*

Remark 1.10 The methods of Sect. 1.9 may be useful to verify the last condition in this theorem.

Proof (Sketch) The proof uses uniform integrability to justify the passage to the limit as $\delta \to 0$ in the Kac counting formula (Proposition 1.4). The rest of the proof is identical to the case above. □

This method is very easy to use for random polynomials because the number of roots is bounded, see e.g. Kac (1943) and Nicolaescu (2023a). It has also been used by Armentano et al. (2023) for shot-noise processes.

1.9 Finiteness of Moments of the Number of Roots

The two issues—the Kac–Rice formula and finiteness of moments of the number of roots—are very much connected. For instance:

- The Kac–Rice formula of order one is, in general, valid under conditions that imply (without further hypotheses) that the expectation is finite. Furthermore, in some cases, necessary and sufficient conditions for the finiteness of the expectation of the measure of level sets can be obtained from the Kac–Rice formula (see Chap. 2).
- The Kac–Rice formula of order two can be used to obtain the finiteness of the second moment of the measure of level sets (see Sect. 1.9.1).
- In the other direction, as explained in Theorem 1.14, the finiteness of the second moment is a tool to establish the validity of the Kac–Rice formula.

1.9.1 Classical Gaussian Results

Let us begin with two simple and classical results for Gaussian stationary processes. The first one is a new presentation of results from Geman (1972) and Kratz and León (2006a).

Proposition 1.15 *Let $X(\cdot)$ be a Gaussian stationary process with a continuous covariance function and positive variance. Let λ_2 be its second spectral moment. Without loss of generality, we assume that the process is standardized ($\mathbb{E}(X(t)) = 0$; $\text{Var}(X(t)) = 1$). Then, for any interval T, we have*

$$\mathbb{E}\big(N_u(X,T)\big) = |T|\sqrt{\lambda_2}\frac{\mathrm{e}^{-u^2/2}}{\pi},$$

where both sides of the above equality are simultaneously finite or infinite.

Proof In view of Theorem 1.11, it remains only to prove that if λ_2 is infinite, then $\mathbb{E}\big(N_u(X,T)\big) = +\infty$. As this is a stationary process, we can assume that $T = [0, t]$. Recall that the variance is implicitly assumed to be 1. Using Ylvisaker's Theorem (Proposition 1.6), there is a.s. no maximum on the barrier u, and as a consequence, as n tends to infinity, we obtain

$$\sum_{k=1}^{2^n} \mathbf{1}_{\{X(\frac{(k-1)t}{2^n}-u)(X(\frac{kt}{2^n})-u)<0\}} \uparrow N_u(X,T).$$

By the monotone convergence theorem, using stationarity

$$\mathbb{E}[N_u(X,T)] = \lim_{n\to\infty} 2^n \mathbb{E}[\mathbf{1}_{\{(X(0)-u)(X(\frac{t}{2^n})-u)<0\}}].$$

But

$$\mathbb{E}[\mathbf{1}_{\{(X(0)-u)(X(\frac{t}{2^n})-u)<0\}}] = \mathbb{P}\{X(0) > u, X(t_n) < u\} + \mathbb{P}\{X(0) < u, X(t_n) > u\},$$

where t_n stands for $t/2^n$. We will analyze the first term above, the consideration of the second one is similar. If we carry out the regression

$$X(t_n) = r(t_n)X(0) + \sqrt{1 - r^2(t_n)}Z,$$

where Z is a standard normal random variable independent of $X(0)$. This implies that

$$\{X(0) > u, X(t_n) < u\} = \left\{u < X(0) \leq \frac{u - \sqrt{1 - r^2(t_n)}Z}{r(t_n)}\right\}.$$

The probability in the first term above, multiplied by the factor 2^n, can be written, using the finite-increments formula, as

$$2^n \int_{-\infty}^{0} \varphi(z)\,\mathrm{d}z \int_{u}^{\frac{u-\sqrt{1-r^2(t_n)}z}{r(t_n)}} \varphi(x)\,\mathrm{d}x$$

$$= \frac{2^n t_n^2}{r(t_n)} \int_{-\infty}^{0} |z|\varphi(z) \frac{\sqrt{1-r^2(t_n)}}{t_n^2} \varphi(\theta_n(z,u))\,\mathrm{d}z + o(1), \qquad (1.16)$$

where $\theta_n(z,u)$ belongs to $(u, \frac{u-\sqrt{1-r^2(t_n)}z}{r(t_n)})$ and thus tends to u, and it is easy to check that the extra term tends to zero

On the other hand, if we use Fatou's Lemma:

$$\liminf_{n\to\infty} \frac{1-r(t_n)}{t_n^2} \geq \int_0^{\infty} \liminf_{n\to\infty} \frac{(1-\cos(\lambda t_n))}{t_n^2}\,\mathrm{d}\mu(\lambda) = +\infty.$$

This implies that the expression in (1.16) tends to infinity, concluding the proof. □

The second classical result concerns the moment of order two. Let us begin by proving the validity of the Kac–Rice formula of order 2.

Proposition 1.16 *Let $X(\cdot)$ be a stationary Gaussian process. Assume that it is a.s. differentiable in quadratic mean, and non-constant, i.e. $0 < \lambda_2 < +\infty$. Then, for T small enough, the second order Kac–Rice formula given by* (6.39) *holds.*

Proof To be able to apply Theorem 1.12, we need only to prove that the distribution of $(X(s), X(t))$ is N.D. for $s \neq t$. The finiteness of the second spectral moment λ_2 implies, assuming as usual that the process has been standardized, that

$$r(t) = \mathbb{E}(X(s)X(s+t)) = 1 - \frac{\lambda_2}{2}t^2 + o(t^2).$$

This implies that, when $t > 0$ is sufficiently small, the covariance $|r(t)| < 1$, which is sufficient to get the result. □

The following result gives sufficient conditions for non-degeneracy of the finite-dimensional distribution of the process and its derivatives, proving the validity of the Kac–Rice formula of any order

Proposition 1.17 *Let $\{X(t), t \in \mathbb{R}^d\}$ be a real centered stationary Gaussian process with paths of class C^k, $k = 0, 1, \ldots$. Suppose that the support of the spectral measure has an accumulation point. Then, for any choice of distinct parameter values $t_1, \ldots, t_n$, the joint distribution of the random variables*

$$X(t_1), \ldots, X(t_n), X'(t_1), \ldots, X'(t_n), \ldots, X^{(k)}(t_1), \ldots, X^{(k)}(t_n) \qquad (1.17)$$

is N.D.

Proof Let Σ be the variance-covariance matrix of the vector (1.17). Consider the quadratic form

$$F(z) := z^\top \Sigma z = \int_{\mathbb{R}} \left| \sum_{j=1}^{n} e^{\imath t_j \lambda} + \cdots + \sum_{j=1}^{n} (\imath.\lambda)^k e^{\imath t_j \lambda} \right|^2 dF(\lambda).$$

We want to prove that the matrix Σ is N.D. To that end, when we consider z such that $F(z) = 0$, we want to prove that $z = 0$. The integrand in the expression above is positive. It must vanish on every point of the support of the spectral measure. But the function

$$\lambda \mapsto \sum_{j=1}^{n} e^{\imath t_j \lambda} + \cdots + \sum_{j=1}^{n} (\imath.\lambda)^k e^{\imath t_j \lambda}$$

is analytic and must vanish on a locally infinite set of points. So it is zero. □

The Kac–Rice formula of order two, when it is valid, reads:

$$\mathbb{E}\big(N_u(N_u - 1)\big) = \int_{T^2} \mathbb{E}_C\big(|X'(s)X'(t)|\big) p_{X(s),X(t)}(u, u)\, ds\, dt, \tag{1.18}$$

where $\mathbb{E}_C$ denotes the conditional expectation to $C := \{X(s) = X(t) = u\}$. Indeed, in (1.18) the integrand is degenerate on the diagonal $s = t$ and the formula does not directly give the finiteness of the second moment. The following is a precise study of the behavior near the diagonal.

Let us introduce the notation

$$\sigma^2(\tau) := \mathrm{Var}_C(X'(0)) := \mathrm{Var}(X'(0)|X(0), X(\tau)).$$

Proposition 1.18 (The Geman Condition) *The following statements are equivalent.*

(a) $\mathbb{E}(N_u([0, T])^2)$ *is finite for some* u *and* T.
(b) $\mathbb{E}(N_u([0, T])^2)$ *is finite for all* u *and all finite* T.
(c) The integral $\int \frac{\sigma^2(\tau)}{\tau}\, d\tau$ *converges at zero.*

Remark 1.11 Integrating by parts in (c) we get the classical Geman's condition by using the following lemma, whose proof, as well as the proofs of the next lemmas, are given at the end of the section.

Lemma 1.19 *There is an equivalence between the convergence at zero of the following two integrals*

$$\int \frac{\lambda_2 + r''(\tau)}{\tau}\, d\tau, \quad \textit{and} \quad \int \frac{\sigma^2(\tau)}{\tau}\, d\tau.$$

Before proving the proposition, we will need some extra notation and two lemmas.

Lemma 1.20 *For τ sufficiently small we can set the following definitions, with the following relations:*

(a) $\mu_{1,\tau,u} := \mathbb{E}(X'(\tau) \mid X(0) = X(\tau) = u) = \frac{r'(\tau)u}{1+r(\tau)}$.
(b) $\mu_{2,\tau,u} := \mathbb{E}(X'(0) \mid X(0) = X(\tau) = u) = -\mu_{1,\tau,u}$.
(c) $\sigma^2(\tau) = \mathrm{Var}(X'(0) \mid X(0), X(\tau)) = \lambda_2 - \frac{(r'(\tau))^2}{1-r^2(\tau)}$.
(d) $\det(\mathrm{Var}(X(0); X(\tau))) = 1 - r^2(\tau) \asymp \tau^2$, *where* $x \asymp y$ *means* $x \le (\mathtt{const})y$ *and* $y \le (\mathtt{const})x$.
(e) *If the fourth spectral moment* λ_4 *satisfies* $\lambda_2^2 < \lambda_4 < +\infty$, *then* $\frac{|\mu_{1,\tau,u}|}{\sigma(\tau)} \le (\mathtt{const})u$.

Lemma 1.21 *Assume that* $|m_1|, |m_2| \le K$ *for some constant* K *and that* $(Y_1; Y_2) \overset{d}{=} N\big((m_1; m_2), \left(\begin{smallmatrix} 1 & \rho \\ \rho & 1 \end{smallmatrix}\right)\big)$. *Then* $\mathbb{E}|Y_1Y_2| \asymp 1$, *where the two constants implied in the symbol* $\asymp$ *depend on* K.

Proof of Proposition 1.18 First, we have to consider the particular case $\lambda_4 = \lambda_2^2$. This corresponds to the sine-cosine process: $X(t) = \xi_1 \sin(wt) + \xi_2 \cos(wt)$, where ξ_1, ξ_2 are independent standard normals. In this case, a direct calculation shows that (a)–(c) hold.

We now consider the other cases assuming that $\lambda_2^2 < \lambda_4$. We start from (c), assuming that

$$\int_0^T \frac{\sigma^2(\tau)}{\tau}\, d\tau < +\infty \qquad \text{with } T \text{ sufficiently small.}$$

The expectation of the number of crossings is finite because the second spectral moment is also finite (see Proposition 1.15). We therefore need only work with the second factorial moment. The Kac–Rice formula, which is valid thanks to Proposition 1.16, can be written as

$$\begin{aligned} \mathbb{E}\big(N_u([0,T])(N_u([0,T])-1)\big) &= \frac{1}{\pi}\int_0^T (T-\tau)\mathbb{E}_C\big(|X'(0)||X'(\tau)|\big)\frac{e^{-\frac{u^2}{1+r}}}{\sqrt{1-r^2}}\, d\tau. \\ &\le (\mathtt{const}) \int_0^T \mathbb{E}_C\Big(\Big|\frac{X'(0)}{\sigma(\tau)}\Big|\Big|\frac{X'(\tau)}{\sigma(\tau)}\Big|\Big)\frac{\sigma^2(\tau)}{\tau}\, d\tau, \end{aligned} \tag{1.19}$$

using Lemma 1.20 (d). By Lemma 1.20 (e), $\frac{X'(0)}{\sigma(\tau)}$ and $\frac{X'(\tau)}{\sigma(\tau)}$ have a bounded conditional mean, then applying now Lemma 1.21:

$$\mathbb{E}(N_u([0,T])(N_u([0,T]) - 1) \le (\mathtt{const}) \int_0^T \frac{\sigma^2(\tau)}{\tau}\, d\tau. \tag{1.20}$$

This gives the finiteness of the second moment for T sufficiently small. For arbitrary T, we can apply the Minkowski inequality. This concludes (b)

In the other direction, we can start from (a) with $u = 0$ and T sufficiently small (which is weaker than (b)) in order to prove (c). Again we can consider the second factorial moment and apply the Kac–Rice formula to establish that

$$\mathbb{E}(N_u([0,T])(N_u([0,T])-1) \geq (\texttt{const}) \int_0^{T/2} \mathbb{E}_C\left(\left|\frac{X'(0)}{\sigma(\tau)}\right|\left|\frac{X'(\tau)}{\sigma(\tau)}\right|\right)\frac{\sigma^2(\tau)}{\tau}\,d\tau.$$

To conclude the proof, we can simply apply Lemma 1.21 in the other direction. □

Remark 1.12 We can also obtain (1.20) with an explicit constant by use of the Cauchy–Schwarz inequality.

Remark 1.13 A by-product of the proof above is that for Condition (c), the expression $\mathbb{E}(N_u([0,T])^2)$ is continuous as a function of u.

Proof of Lemma 1.19 Let us consider the integral:

$$\int_0^\delta \frac{\sigma^2(\tau)}{\tau}\,d\tau.$$

For τ small enough

$$\frac{\sigma^2(\tau)}{\tau} \sim \left(\frac{1}{\lambda_2}\right)^{\frac{3}{2}} \frac{\lambda_2(1-r^2(\tau)) - (r'(\tau))^2}{\tau^3}.$$

Then, integrating by parts

$$\int_0^\delta \frac{\lambda_2(1-r^2(\tau)) - (r'(\tau))^2}{\tau^3}\,d\tau$$
$$= \frac{\lambda_2(1-r^2(\delta)) - (r'(\delta))^2}{2\delta^2} + \int_0^\delta \frac{r'(\tau)}{\tau}\left(\frac{-\lambda_2 r(\tau) - r''(\tau)}{\tau}\right)d\tau.$$

Hence, we need to consider the second term, which is equal to

$$-\lambda_2 \int_0^\delta \frac{r'(\tau)}{\tau}\left(\frac{r(\tau)-1}{\tau}\right)d\tau - \int_0^\delta \frac{r'(\tau)}{\tau}\left(\frac{r''(\tau)+\lambda_2}{\tau}\right)d\tau.$$

As the first term is evidently convergent, the above sum is convergent if and only if

$$\int_0^\delta \frac{r''(\tau)+\lambda_2}{\tau}\,d\tau < \infty.$$

□

Proof of Lemma 1.20 For the sake of brevity, we will write r, r', r'' instead of $r(\tau), r'(\tau), r''(\tau)$. Items (a)–(d) are proved in Chap. 3.

To prove (e) we first need to study the behavior of $r'/\sigma(\tau)$ near to zero. We need to consider two cases.

The first one is when the fourth spectral moment λ_4 is finite: we have $r(t) = 1 - \lambda_2 t^2/2 + \lambda_4 t^4/(4!) + o(t^4)$. By using a Taylor expansion of fourth order on the numerator and the denominator of the fraction $(\frac{r'u}{(1+r)\sigma(\tau)})^2$, we obtain

$$\Big(\frac{r'u}{(1+r)\sigma(\tau)}\Big)^2 \to \frac{\lambda_2^2 u^2}{\lambda_4 - \lambda_2^2} \leq (\texttt{const})\, u^2, \qquad \text{giving (e).}$$

Consider now the second case: $\lambda_4 = +\infty$.

Given that $r''(\tau) - r''(0) = \int_0^\infty (1 - \cos(\tau\lambda))\lambda^2 dF(\lambda)$, we have by Fatou's lemma

$$\begin{aligned}\liminf_{\tau\to 0} \frac{r''(\tau) - r''(0)}{\tau^2} &\geq \int_0^{+\infty} \liminf_{\tau\to 0} \frac{1-\cos(\tau\lambda)}{\tau^2\lambda^2} \lambda^4\, \mathrm{d}F(\lambda)\\ &= 2\int_0^\infty \lambda^4\, \mathrm{d}F(\lambda) = +\infty. \end{aligned} \tag{1.21}$$

Since $1 + r$ tends to 2, it suffices to study the behavior of

$$\frac{r'^2}{\sigma^2(\tau)} \simeq \frac{\lambda_2^3}{\frac{\lambda_2(1-r^2)-r'^2}{\tau^4}}.$$

Note that $\lambda_2(1 - r^2(\tau)) - r'^2(\tau) = 2\lambda_2(1 - r(\tau)) - r'^2(\tau) + O(\tau^4)$. Furthermore by using the l'Hospital rule

$$\lim_{\tau\to 0} \frac{2\lambda_2(1 - r(\tau)) - r'^2(\tau)}{\tau^4} = \lim_{\tau\to 0} \left(\frac{-r'(\tau)}{2\tau}\right)\left(\frac{r''(\tau) - r''(0))}{\tau^2}\right) = +\infty,$$

because of (1.21) and since we know that $\frac{-r'(\tau)}{2\tau} \to \frac{\lambda_2}{2}$. Thus $\frac{r'(\tau)}{\sigma(\tau)} \to 0$. These two results imply that (e) holds. □

Proof of Lemma 1.21 We can write

$$Y_2 - m_2 = \rho(Y_1 - m_1) + \sqrt{1-\rho^2} Z_1,$$

where Z_1 is a standard Gaussian independent of Y_1. Thus

$$\begin{aligned} Y_1 Y_2 &= (m_1 m_2 + \rho) + (m_2 + \rho m_1)(Y_1 - m_1)\\ &+ m_1\sqrt{1-\rho^2} Z_1 + \rho((Y_1 - m_1)^2 - 1) + \sqrt{1-\rho^2}(Y_1 - m_1) Z_1. \end{aligned}$$

This formula shows that $\mathbb{E}|Y_1Y_2|$ is a continuous function of (m_1, m_2, ρ) and, by compactness, it is upper bounded.

In the other direction, setting $Y_i = m_i + \bar{Y}_i$, $i = 1, 2$, we have

$$Y_1Y_2 = [m_1m_2] + [m_1\bar{Y}_2 + m_2\bar{Y}_1] + [\bar{Y}_1\bar{Y}_2],$$

where the three brackets are in different Itô–Wiener chaos and thus the joint variance is the sum of the variance of each term. This implies that

$$\mathrm{Var}(Y_1Y_2) \geq \mathrm{Var}(\bar{Y}_1\bar{Y}_2) = \mathbb{E}(\bar{Y}_1^2\bar{Y}_2^2) = \mathbb{E}\Big(\big(H_2(\bar{Y}_1)+1\big)\big(H_2(\bar{Y}_2)+1\big)\Big) = 1+2\rho^2,$$

where the function $H_2(y) = y^2 - 1$ is the second Hermite polynomial, and also where we have used Mehler's formula (see Sect. 4.4): $\mathbb{E}(H_i(\bar{Y}_1)H_j(\bar{Y}_2) = \delta_{ij}\rho^i i!$. Note that $H_0(y) = 1$.

As a consequence

$$\mathrm{Var}(Y_1Y_2) \geq 1.$$

and Y_1Y_2 is never a.s. constant. Thus $\mathbb{E}|Y_1Y_2| > 0$ and, by compactness, is lower bounded. □

1.9.2 General Methods

The Kac–Rice formula is not in fact a relevant tool for studying the finiteness of the moments of orders $k > 2$. The conditions are difficult to establish and the results difficult to handle because of singularities on the numerous diagonals of the set T^d. The main tool to bound moments comes from elsewhere: a very regular function that has many zeros must be very flat. If this function is a sample path of a stochastic process, such that $X(t)$ admits a density, this event happens with a very small probability. What follows is a systematic exploitation of this observation. It is extracted from Armentano et al. (2023).

Theorem 1.22 *Consider a stochastic process $X(\cdot)$, a compact real interval T and N_u, the number of roots of $X(t) = u$ in T. Assume that $X(\cdot)$ satisfies the following conditions.*

Assumption (A.1.1) *The sample paths of $X(\cdot)$ are $C^k(T)$ for some $k \geq 2$.*

Assumption (A.1.2) *For some $m = 1, 2, \ldots$, there exists a constant D_m such that*

$$\mathbb{E}\left(|X^{(k)}|_\infty^m\right) \leq D_m.$$

Assumption (A.1.3) *There exist* $0 \le h \le k$ *and a constant* $C > 0$*, such that the joint density of the random vector* $X(t), X'(t), \dots, X^{(h)}(t)$ *is bounded by* C *uniformly in* $t \in T$*, and the argument in a neighborhood of* $(u, 0, \dots, 0)$*.*

Then, for $p = 1, 2, \dots$ *such that*

$$p < \left(k - \frac{h}{2} - \frac{1}{1+h}\right)\left(\frac{1}{m} + \frac{1}{1+h}\right)^{-1}, \tag{1.22}$$

the p*-th moment of the number of crossings* N_u *is finite.*

Assume, in addition:

Assumption (A.1.4) *The density in* ***(A.1.3)*** *is bounded by* C *uniformly in* $t \in T$ *and the argument* $(u_1, \dots, u_{h+1}) \in \mathbb{R}^{h+1}$*.*
Then

$$\mathbb{E}(N_u^p) \le (k-1)^p + D_m \cdot E_{\alpha,k,p} + C|T|^{(h+1)(k-h/2)} \cdot D_{\alpha,k,h,p},$$

where α *is any real number such that* $\frac{p}{m} < \alpha < k - \frac{h}{2} - \frac{1+p}{1+h}$*,*

$$E_{\alpha,k,p} = pk^p \sum_{a=0}^{\infty} \frac{(a+1)^p}{a^{m\alpha}}, \tag{1.23}$$

and

$$D_{\alpha,k,h,p} = \frac{2^{(h+1)(1+h/2-k)}}{k! \cdot (k-h)!} \sum_{a=1}^{\infty} (a+1)^{p-1} a^{1-(h+1)(k-h/2-\alpha)}.$$

Some remarks are in order at this stage. First let us note that $h = 0$ and $m = 1$ are allowed, and the hypotheses are therefore not as restrictive as they may seem at first sight. Naturally, the more information is available, the more precise results can be obtained.

Remark 1.14 (Large m**)** The limit case, when m can be arbitrarily large (as is the case for Gaussian processes), corresponds to α close to zero. The limit condition then becomes

$$p < \left(k - \frac{h}{2}\right)(h+1) - 1. \tag{1.24}$$

A particular case of this situation arises when for some $\beta > 0$, it holds that

$$\mathbb{E}\left(e^{\beta|X^{(k)}|_\infty}\right) \le D_\beta,$$

for some constant D_β, i.e. we have exponential moments. In this case, the constant $E_{\alpha,k,p}$ in (1.23) can be replaced by

$$\tilde{E}_{\alpha,k,p} = p(k-1)^{p-1} \sum_{a=1}^{\infty} \frac{(a+1)^{p-1}}{e^{\beta a^\alpha}}.$$

Remark 1.15 (Large k**)** When the process X has $C^\infty(I)$ paths, if **(A.1.2)** holds true for $m = 1$ with arbitrary k, and **(A.1.3)** holds true for some $h = 0, 1, \dots,$ then, all the moments of N_u are finite.

Remark 1.16 ($h = 0$) The case $h = 0$ and $m = 1$ corresponds to (3.23) on p. 82 in Azaïs and Wschebor (2009), obtained originally in Lemma 1.2 in Nualart and Wschebor (1991). When $h = 0$ and $m > 1$ Theorem 1.22 assures that the p-th moment is finite for

$$p < \frac{m}{m+1}(k-1).$$

Remark 1.17 A generalization of Theorem 1.22 to random fields is given in Sect. 2.10.2.

Before proving the theorem we establish three preliminary lemmas. Recall that $f^{(0)} = f$.

Lemma 1.23 *Consider a function* $f\colon T \mapsto \mathbb{R}$ *of class* $C^k(T)$ *for an interval* T *with middle point* $m(T)$. *If* $f - u$ *has* k *roots in* T *and* $0 \le h \le k$, *we have*

$$|f(m(T)) - u| \le \frac{|f^{(k)}|_\infty}{k!} \left(\frac{|T|}{2}\right)^k,$$

$$|f'(m(T))| \le \frac{|f^{(k)}|_\infty}{(k-1)!} \left(\frac{|T|}{2}\right)^{k-1}, \dots,$$

$$|f^{(h)}(m(T))| \le \frac{|f^{(k)}|_\infty}{(k-h)!} \left(\frac{|T|}{2}\right)^{k-h}.$$

Proof The proof is based on the Lagrange remainder form for polynomial interpolation. That is, let $g : T \mapsto \mathbb{R}$ be C^k, $t_1, \dots, t_k \in T$ and P be the only polynomial of degree $k-1$ such that $g(t_j) = P(t_j)$, $j = 1, \dots, k$. Then, for $t \in T$ we have

$$g(t) - P(t) = \frac{1}{k!} \prod_{j=1}^{k} (t - t_j) \cdot g^{(k)}(\xi),$$

for some ξ such that $\min\{t_1, \dots, t_k, t\} < \xi < \max\{t_1, \dots, t_k, t\}$, see Lemma 5.2, p. 135 in Azaïs and Wschebor (2009). Now, for $g = f - u$, $t = m(T)$ and using as

$t_1, \ldots, t_k$ the roots of $f - u$, we get $P = u$ and so the first inequality follows. The proofs of the other inequalities follow in a similar way. □

Lemma 1.24 *Let Z be a random variable taking non-negative integer values. Then,*

$$\mathbb{E}(Z^p) \leq (k-1)^p + p \sum_{\ell=k}^{\infty} \ell^{p-1} \mathbb{P}(Z \geq \ell).$$

Proof We have

$$\begin{aligned}
\mathbb{E}\left(Z^p\right) &= \sum_{\ell=1}^{\infty} \ell^p \mathbb{P}(Z = \ell) \\
&= \sum_{\ell=1}^{k-1} \ell^p \mathbb{P}(Z = \ell) + \sum_{\ell=k}^{\infty} \ell^p \mathbb{P}(Z = \ell) \\
&= \sum_{\ell=1}^{k-1} \ell^p \mathbb{P}(Z = \ell) + \sum_{\ell=k}^{\infty} \ell^p [\mathbb{P}(Z \geq \ell) - \mathbb{P}(Z \geq \ell + 1)] \\
&= \sum_{\ell=1}^{k-1} \ell^p \mathbb{P}(Z = \ell) + (k-1)^p \mathbb{P}(Z \geq k) + \sum_{\ell=k}^{\infty} [\ell^p - (\ell - 1)^p] \mathbb{P}(Z \geq \ell) \\
&\leq (k-1)^p + p \sum_{\ell=k}^{\infty} \ell^{p-1} \mathbb{P}(Z \geq \ell),
\end{aligned}$$

concluding the proof. □

Lemma 1.25 *Assume that conditions* **(A.1.1)** *to* **(A.1.3)** *of Theorem 1.22 hold true. Let $\ell \geq k$ and define a and r such that $\ell = a(k-1) + r$ with $1 \leq r \leq k-1$. Then, for $B > 0$*

$$\mathbb{P}(N_u \geq \ell, |X^{(k)}|_\infty \leq B) \leq aC(2B)^{h+1} \prod_{i=0}^{h} \left(\frac{|T|}{2a}\right)^{k-i} \frac{1}{(k-i)!},$$

where C is the bound in **(A.1.3)**.

When **(A.1.4)** *does not hold, the inequality is true only for ℓ that are sufficiently large.*

Proof We divide T into a equal sub-intervals T_j, $j = 1, \ldots, a$, s.t. at least one of them has k zeros. After this, we use a union bound:

$$
\begin{aligned}
\mathbb{P}(N_u \geq \ell, |X^{(k)}|_\infty \leq B) &= \mathbb{P}(N_u \geq a(k-1)+r, |X^{(k)}|_\infty \leq B) \\
&\leq \mathbb{P}\left(\bigcup_{j=1}^{a}\{N_u(T_j) \geq k\}, |X^{(k)}|_\infty \leq B\right) \\
\leq \sum_{j=1}^{a}\mathbb{P}\Bigg(|X(m(T_j))-u| \leq \frac{B}{k!}\left(\frac{|T|}{2a}\right)^k, &\bigcap_{i=1}^{h}|X^{(i)}(m(T_j))| \leq \frac{B}{(k-i)!}\left(\frac{|T|}{2a}\right)^{k-i}\Bigg) \\
&\leq aC\prod_{i=0}^{h}\frac{2B}{(k-i)!}\left(\frac{|T|}{2a}\right)^{k-i},
\end{aligned}
$$

obtaining the result. Of course, when $h = 0$, the intersection in the equation above does not appear. □

Proof of Theorem 1.22 Assume **(A.1.4)** for the moment, then from Lemma 1.24, we have

$$
\mathbb{E}(N_u^p) \leq (k-1)^p + \sum_{\ell=k}^{\infty} p\ell^{p-1}\mathbb{P}(N_u \geq \ell). \tag{1.25}
$$

In order to bound the summand in the r.h.s. of (1.25), for given ℓ consider a and r as in Lemma 1.25, and $\alpha > 0$ to be defined later. We will use the following bound:

$$
\mathbb{P}(N_u \geq \ell) \leq \mathbb{P}(N_u \geq \ell, |X^{(k)}|_\infty \leq a^\alpha) + \mathbb{P}(|X^{(k)}|_\infty > a^\alpha).
$$

Then using Lemma 1.25 for the first summand and Chebyshev's inequality for the second, we obtain

$$
\begin{aligned}
\mathbb{E}(N_u^p) &\leq (k-1)^p + p(k-1)^{p-1}\sum_{a=1}^{\infty}(a+1)^{p-1}\mathbb{P}(N_u \geq \ell) \\
&\leq (k-1)^p \\
&+ p(k-1)^{p-1}\Bigg(C2^{h+1}\left(\frac{|T|}{2}\right)^{(h+1)(k-h/2)}\prod_{i=0}^{h}\frac{1}{(k-i)!} \\
&\cdot\sum_{a=1}^{\infty}(a+1)^{p-1}a^{(h+1)(\alpha+h/2-k)+1} + D_m\sum_{a=1}^{\infty}(a+1)^{p-1}a^{-m\alpha}\Bigg).
\end{aligned}
$$

The conditions for the simultaneous convergence of the two series above are

$$
\frac{p}{m} < \alpha < k - \frac{h}{2} - \frac{1+p}{1+h}.
$$

Under condition (1.22), there always exists an adequate α, concluding the proof of Theorem 1.22.

When **(A.1.4)** does not hold, the calculation above is true only for ℓ (or a) that are sufficiently large, and this does not change the condition of convergence of the series. □

1.9.3 Examples

Below, we present four examples. The first is a simple one that shows both the power and the limitations of Theorem 1.22. The other three provide new results on the finiteness of the moments of the number of crossings: the first one is the classical case of Gaussian processes; the second one considers Chi-square processes derived from Gaussian random variables; and the third one considers regularized diffusion as considered in Nualart and Wschebor (1991).

1.9.3.1 Conditional Sine-Cosine Process

Let us begin with a toy example. Consider a random variable ω, such that $\mathbb{E}\,(\omega^M) < \infty$ and $\mathbb{E}\,(\omega^{M+1}) = \infty$. This variable is the random frequency of a process $X(\cdot) = \{X(t)\colon 0 \leq t \leq 2\pi\}$ that we construct as

$$X(t) = \xi_1 \sin \omega t + \xi_2 \cos \omega t,$$

where ξ_1 and ξ_2 are two standard normal independent random variables, also independent of ω. Conditionally on the value of ω, the process $X(\cdot)$ is a Gaussian sine-cosine process. It is clear that the density of $X(t)$ is standard normal for each t. Nevertheless, it should be observed that $X(t)$ is, in general, not a Gaussian process. The number N_0 of roots of $X(t)$ on $[0, 2\pi]$ coincides with the number of roots of the process

$$\tilde{X}(t) = \frac{\xi_1}{\sqrt{\xi_1^2 + \xi_2^2}} \sin \omega t + \frac{\xi_2}{\sqrt{\xi_1^2 + \xi_2^2}} \cos \omega t = \cos(\omega t - \theta),$$

where $\theta = \arctan(\xi_1/\xi_2)$ is defined a.s. Therefore, N_0 satisfies $|N_0 - 2\omega| \leq 2$. This means that, for the considered process, the maximal finite moment of $N_0[0, 2\pi]$ is M. To apply Theorem 1.22 we compute

$$|X^{(k)}|_\infty \leq \omega^k \sqrt{\xi_1^2 + \xi_2^2}.$$

Note that $\sqrt{\xi_1^2 + \xi_2^2}$ follows a Rayleigh distribution. Then, based on the independence of the three random variables above, we have

$$\mathbb{E}\left(|X^{(k)}|_\infty^m\right) \leq H_m \mathbb{E}(\omega^{mk}),$$

where H_m is the finite moment of the Rayleigh distribution. So we can apply Theorem 1.22 with $h = 0$, $k = M$ and $m = 1$. This demonstrates that the moments of order $p \leq (M - 1)/2$ are finite.

This result should be compared with the true result, which is M.

1.9.3.2 Stationary Gaussian Process

Consider a stationary Gaussian process with spectral measure supported in a set with an accumulation point. By Proposition 1.17 we know that for all h $X(t), X'(t), \ldots, X^{(h)}(t)$ admit a joint density. By the Borell–Sudakov–Tsirelson Theorem (see, for example Azaïs and Wschebor 2009, Th. 2.8), for all m:

$$\mathbb{E}\left(|X^{(k)}|_\infty^m\right) < +\infty.$$

It is then possible to take $h = k$ and m arbitrarily large, and Formula (1.22) gives a finite moment of order

$$p < k/2(1 + k) - 1.$$

This result improves the previous one obtained in Nualart and Wschebor (1991), that is recovered in the case $h = 0$ and $m = 1$.To visualize it, see Table 1.1.

1.9.3.3 Chi-Square Process

Consider a Chi-square process $Y(\cdot) = \{Y(t) : t \in \mathbb{R}\}$ given by

$$Y(t) = |X(t)|^2 = \sum_{i=1}^{n} X_i(t)^2,$$

Table 1.1 If $X(\cdot)$ is C^k, following Theorem 1.22 we obtain that $\mathbb{E}(N_u^p)$ is finite for values of k and p above

k	2	3	4
p	1	≤ 4	≤ 8

where $X(\cdot) = \{X(t)\colon t \in \mathbb{R}\}$ with $X(t) = (X_1(t), \dots, X_n(t))$ is an $\mathbb{R}^n$ valued random process whose coordinates are n independent copies of a stationary Gaussian process with $C^k(\mathbb{R})$ paths and variance 1. First observe that the case $n = 1$ admits an ad-hoc treatment. If $n = 1$ for $u = 0$ we have $N_0^Y[0, T] = N_0^X[0, T]$, and for $u > 0$ we have $N_u^Y[0, T] = N_{\sqrt{u}}^X[0, T] + N_{-\sqrt{u}}^X[0, T]$. For $n \geq 2$ the level $u = 0$ is not interesting, as for this level the process is differentiable and non-negative and has a bounded density, so Bulinskaya's Lemma (Proposition 1.7) gives that a.s. the number of crossings is zero. Accordingly, we can consider $n \geq 2$ and $u > 0$. Observe that, excluding the uninteresting case where $\lambda_2 = \mathrm{Var}X'(t) = 0$, the joint density of the random variables $X(t)$, $X'(t)$ is bounded, as they form a pair of independent Gaussian random variables.

It is not difficult to check the hypothesis of Theorem 1.22 in two different situations: $n = 2$ and $n \geq 3$. We can immediately see that $Y(\cdot)$ has C^k paths and that $|Y|_\infty$ has moments of every order, when considering a finite interval $I \subset \mathbb{R}$. Observe now that $Y(0)$ has a $\chi^2(n)$ density that is bounded for $n \geq 2$. We can then apply Theorem 1.22 with the given k, arbitrary m, and $h = 0$. Based on (1.22) we obtain the finiteness of the moments of the crossings, for $u > 0$, of order

$$p = k - 2,$$

which is relevant only in the case $k \geq 3$. A more refined analysis can now be carried out. Regarding the derivative we have

$$Y'(t) = 2\sum_{i=1}^{n} X_i(t)X_i'(t).$$

We see that conditional to $X(t)$, the random variable $Y'(t)$ has a Gaussian distribution with variance $4\lambda_2 Y(t)$. Its conditional density $p_{Y'|X=x}$ is bounded by

$$\texttt{(const)}\big(Y(t)\big)^{-1/2}.$$

Recall that $\texttt{(const)}$ denotes a meaningless constant whose value may change from line to line.

Let now B_1 and B_2 be two Borel sets of $\mathbb{R}$ with measure $|B_1|$ and $|B_2|$. Writing Y and Y' for $Y(t)$ and $Y'(t)$, we have

$$\begin{aligned}
\mathbb{P}(Y' \in B_1, Y \in B_2) &= \int_{\mathbb{R}^n} \mathrm{d}x \int_{B_1} \mathrm{d}y p_{Y'|X=x}(y) \mathbb{1}_{\{Y \in B_2\}} p_X(x) \\
\leq \texttt{(const)} \int_{\mathbb{R}^n} |B_1| Y^{-1/2} \mathbb{1}_{\{Y \in B_2\}} p_X(x)\,\mathrm{d}x &= \texttt{(const)}|B_1| \int_{B_2} y^{-1/2} p_Y(y)\,\mathrm{d}y \\
&= \texttt{(const)}|B_1| \int_{B_2} y^{n/2-3/2} \mathrm{e}^{-y/2}\,\mathrm{d}y, \qquad (1.26)
\end{aligned}$$

where $p_X(\cdot)$ and $p_Y(\cdot)$ are the density functions of X and Y respectively. Note that we have used the explicit expression of the density of a $\chi^2(n)$ distribution. When $n \geq 3$, the integral in (1.26) is bounded, giving

$$\mathbb{P}\left(Y' \in B_1, Y \in B_2\right) \leq (\texttt{const})|B_1||B_2|.$$

This proves that the process satisfies the hypotheses of Theorem 1.22 with $h = 1$. The conclusion is that, for $n \geq 3$, the number of crossings of Y has a finite moment of order

$$p = 2k - 3.$$

1.9.3.4 Regularized Processes

In this example, we consider the number of crossings with a level u of a regularized diffusion. We depart from a diffusion $X(\cdot) = \{X(t) : t \geq 0\}$ defined as the solution of the stochastic differential equation

$$X(t) = x_0 + \int_0^t b(s, X(s))\,\mathrm{d}s + \int_0^t \sigma(s, X(s))\,\mathrm{d}W(s), \quad t \geq 0.$$

Here $W = \{W(s) : s \geq 0\}$ is a standard Brownian motion, $x_0 \in \mathbb{R}$, and $b, \sigma : [0, \infty) \times \mathbb{R} \mapsto \mathbb{R}$ are Lipschitz w.r.t. the second variable, that is, for $T > 0$ there exists $K_T > 0$ such that for $x, y \in \mathbb{R}$ and $0 \leq s \leq T$ it holds that

$$|b(s, x) - b(s, y)| + |\sigma(s, x) - \sigma(s, y)| \leq K_T|x - y|.$$

Assume also the linear growth condition

$$|b(s, x)| + |\sigma(s, x)| \leq K_T(1 + |x|).$$

Then, the above equation has a unique strong solution that satisfies

$$\sup_{0 \leq t \leq T} \mathbb{E}(X(t)^2) \leq H_T < \infty$$

for a constant H_T, see Borodin (2017, Th. 7.1). The regularized diffusion is defined by

$$X_\Psi(t) = (\Psi * X)(t),$$

where $\Psi : \mathbb{R} \to [0, \infty)$ is a C^∞-function with support contained in $[-1, 1]$ that integrates to 1. Thus, X_Ψ is obtained from $X(\cdot)$ by path-wise convolution with Ψ. Let us consider two cases:

Case I. We assume that $b = 0$ and $\sigma : \mathbb{R}^+ \times \mathbb{R} \mapsto \mathbb{R}$ is strictly positive and C^3.

Case II. The volatility σ is strictly positive and C^3 as in Case I, and there exist constants B_T, c_T and C_T such that

$$|b(s,x)| \leq B_T, \quad 0 < c_T \leq \sigma(s,x) \leq C_T$$

for all $0 \leq s \leq T$ and $x \in \mathbb{R}$.

We will start with Case I. With respect to the hypotheses of Theorem 1.22 we see that X_Ψ is C^∞ since it inherits the regularity of Ψ. Furthermore, for large enough a (depending on Ψ) the random variable $X_\Psi(t)$ has a uniformly bounded density on an interval $[a, T]$ for all finite T, see Lemma 3.1 in Nualart and Wschebor (1991). Besides this, a direct computation gives

$$|X_\Psi^{(h)}|_\infty \leq c \cdot |X|_\infty,$$

with $c = \int |\Psi^{(h)}(u)|\,\mathrm{d}u$ and the infinite norm is taken on the interval $[a, T]$. Hence, it is sufficient to bound the moments of $|X|_\infty$. Now, the Burkholder–Davis–Gundy inequality (Protter 2005, Th. 48) gives

$$\mathbb{E}(|X|_\infty^2) \leq C_1 \mathbb{E}([X]_T),$$

where $[X]_T$ is the quadratic variation of X on $[0, T]$. Now, by Protter (2005, Th. II-29), we have

$$\begin{aligned}
[X]_T &= \left[\int_0^T \sigma(s, X(s))\,\mathrm{d}W(s)\right]_T = \int_0^T \sigma(s, X(s))^2\,\mathrm{d}s \\
&\leq K_T^2 \int_0^T (1 + |X(s)|)^2\,\mathrm{d}s \leq 2K_T^2 \int_0^T (1 + |X(s)|^2)\,\mathrm{d}s.
\end{aligned}$$

Taking expectations, we have

$$\mathbb{E}([X]_T) \leq 2K_T^2 \int_0^T (1 + \mathbb{E}(X(s))^2)\,\mathrm{d}s \leq 2K_T^2 T(1 + H_T) < \infty.$$

Hence, by Theorem 1.22 with $h = 0$, $m = 2$ and arbitrary k, we obtain that

$$\mathbb{E}(N_u^p) < \infty,$$

for all p, as obtained in Nualart and Wschebor (1991).

For Case II, we apply Girsanov's theorem (Borodin 2017). We then have two SDEs

$$\begin{aligned}
\mathrm{d}X_0(t) &= \sigma(t, X(t))\,\mathrm{d}W(t) \\
\mathrm{d}X(t) &= b(t, X(t))\,\mathrm{d}t + \sigma(t, X(t))\,\mathrm{d}W(t).
\end{aligned}$$

Consider the density process

$$\rho(t) = \exp\left(\int_0^t \frac{b(s, X(s))}{\sigma(s, X(s))} \mathrm{d}W(s) - \frac{1}{2} \int_0^t \left(\frac{b(s, X(s))}{\sigma(s, X(s))} \right)^2 \mathrm{d}s \right).$$

Then, Girsanov's Theorem states that for an arbitrary function $F : C([0, T], \mathbb{R}) \mapsto \mathbb{R}$ of the trajectories of the process, we have $\mathbb{E}\,(F(X)) = \mathbb{E}\,(\rho(T)F(X_0))$. If F is the p-power of the number of crossings on $[0, T]$, we have

$$\begin{aligned} \mathbb{E}\left(N_u([0, T], X)^p\right) &= \mathbb{E}\left(\rho(T) N_u([0, T], X_0)^p\right) \\ &\le \mathbb{E}\left(\rho(T)^2\right)^{1/2} \mathbb{E}\left(N_u([0, T], X_0)^{2p}\right)^{1/2}, \end{aligned}$$

where we indicate the process X or X_0 in the notation and we apply Hölder's inequality. To bound the first expectation in the r.h.s. consider

$$\begin{aligned} \rho(t)^2 = \exp&\left(2 \int_0^t \frac{b(s, X(s))}{\sigma(s, X(s))} \mathrm{d}W(s) - 2 \int_0^t \left(\frac{b(s, X(s))}{\sigma(s, X(s))} \right)^2 \mathrm{d}s \right) \\ &\times \exp\left(\int_0^t \left(\frac{b(s, X(s))}{\sigma(s, X(s))} \right)^2 \mathrm{d}s \right). \end{aligned}$$

Then,

$$\begin{aligned} \rho(t)^2 \le \exp&\left(2 \int_0^t \frac{b(s, X(s))}{\sigma(s, X(s))} \mathrm{d}W(s) - 2 \int_0^t \left(\frac{b(s, X(s))}{\sigma(s, X(s))} \right)^2 \mathrm{d}s \right) \\ &\times \exp\left(\left(\frac{B_T}{c_T} \right)^2 T \right), \end{aligned}$$

and

$$\mathbb{E}\left(\rho(T)^2\right) \le \exp\left(\left(\frac{B_T}{c_T} \right)^2 T \right) < \infty.$$

As $\mathbb{E}\left(N_u([0, T], X_0)^{2p}\right)$ is finite by Case I, the finiteness of $\mathbb{E}\,(N_u([0, T], X)^p)$ for all p follows.

Examples with shot noise processes are given in Chap. 7.

Chapter 2
Multidimensional Methods

Serving as the theoretical core of the book, this chapter extends the Kac–Rice methodology to multidimensional random fields mapping from $\mathbb{R}^D$ to $\mathbb{R}^d$ (where $D \geq d$). It introduces essential differential and integral geometry concepts, including the Area, Co-Area, and Crofton formulas. A major contribution of this chapter is the establishment, when $D > d$, of a weaker version of the Bulinskaya condition. This permits a proof of the Kac–Rice formula for all levels for Gaussian and non-Gaussian random fields. The proof differs depending on whether the dimensions are equal or not. In the first case we compute the expectation of the number of points of the level set, in the second case we compute that of the Hausdorff measure of the level set. We demonstrate the power of these multidimensional tools through diverse examples, including functions of Gaussian fields, the geometric analysis of likelihood processes, and models of gravitational stochastic microlensing.

The chapter ends with recent results on finiteness of the second moment and higher moments of the volume of the level set.

2.1 Introduction

This chapter presents multidimensional versions of the Kac–Rice formula. One of its original features is that it considers simultaneously the case where the dimension D of the parameter space and the dimension d of the image space are equal, as well as the case in which $D > d$.

In 1957, in order to help model the surface of the sea, Longuet-Higgins (1957) established a Kac–Rice formula for the expectation of the length of the level set of a random field $X(\cdot) : \mathbb{R}^2 \mapsto \mathbb{R}$. His proof uses an approach which is intuitive and which nonetheless yields several very interesting ideas. Another interesting undertaking is to be found in a paper by Brillinger (1972), who studies the zeros of a system of random equations and in which the celebrated Area formula was used

D. Armentano et al., *Geometry of Level Sets of Random Fields, Kac–Rice Formulas, Hermite Expansions and Applications*, Probability Theory and Stochastic Modelling 111, https://doi.org/10.1007/978-3-032-29144-8_2

for the first time. Using this formula, Brillinger was able to establish a Kac–Rice formula for almost all levels (see Proposition 2.8 below). In his paper, he remarks that, as long as both sides of the equation are continuous, the equality for all levels is satisfied. However, his conditions for establishing this continuity are very strong, as we see in his Corollary 3.1.

We owe the first convincing application of the Kac–Rice formula for obtaining the expectation of the number of zeros of random fields when $D = d > 1$ to Adler and Hasofer (1976) (in fact they consider critical points that are zeros of the derivatives), with two different general proofs having been published in Adler and Taylor (2009) and Azaïs and Wschebor (2005a).

Another case arises when the dimensions disagree: $D > d$. In such a case, the level set is generically a set of dimension $D - d$. The definition and the computation of the expectation of its measure constitutes a fairly difficult problem. The first attempts to address it seem to date back to Wschebor (1982) and Cabaña (1986). The subject was subsequently developed by Kratz and León (1997) and a first general proof was given in Azaïs and Wschebor (2009) with a detailed version in Berzin et al. (2022). All the cited works, and all the results of this chapter, except for marginal results, are limited to the computation of the moments of the measure of the level set. The computation of other quantities is a problem that remains largely unsolved. However, the computation of moments provides a lot of geometrical information.

In the last few decades, new applications of these methods have been developed. High-dimensional statistics and machine learning have allowed us to begin to address the problem of how to study the so-called landscape of random fields: the sought-after parameter is the argument minimum of a so-called empirical risk function and the computational complexity of such an optimization program is depicted by the number of critical points with a given index whose value function is close to the minimum. In some cases, this problem is related to the Ising model (Auffinger et al. 2013), see Chaps. 5 and 6.

An outline of the chapter is as follows: We begin with a presentation of geometrical tools for differential geometry and integral geometry. The proofs of the celebrated Area formula, Co-Area formula, and the Crofton formula can be found in classical monographs such as Morgan (2016). The first main result concerns the Bulinskaya-type lemma on the regularity of the level set. The new proof, using local time, Proposition 2.11, considerably extends existing results even in the Gaussian case. More specifically, if the paths are C^1 and if the density $p_{X(t)}$ is bounded uniformly in t then a.s. the irregular part (the set of points where the Jacobian does not have full rank) of the level set has a zero σ_{D-d} measure. When $D = d$ this means that the irregular part of the level set is a.s. empty. When $D > d$, since the regular part of the level set is a manifold of dimension $D - d$, it implies that the level set is a.s. a $D - d$ rectifiable set.

The chapter also contains a new proof of the true Kac–Rice formula using the method of Azaïs and Wschebor (2009) extended to non-Gaussian Random fields, see Theorem 2.17. By the true Kac–Rice formula, we mean a Kac–Rice formula that is valid **for every level**. This implies that there is no ambiguity regarding a particular level set, say 0: this is in contrast to some approaches that prove a Kac–Rice formula only for almost all level sets. For example, for a random field $\mathbb{R}^2 \to \mathbb{R}$ studies of anisotropy such as Berzin (2021) use, as a statistical observation, the length of the level curve at a certain level given by experiment. For such a purpose a Kac–Rice formula for almost every level is not sufficient.

Here, the proof is first conducted for the case $D = d$, before being extended to the case $D > d$ through the Crofton formula. Many examples are given, including functions of Gaussian processes, zeros of the likelihood function, modified Euler Characteristics, etc. The chapter also presents several extensions: random fields defined on manifolds, Theorem 2.20; expectation of an integral with weights on the level set, Theorem 2.17; and higher moments, Theorem 2.21.

2.2 Differential Geometry in a Nutshell

2.2.1 Riemannian Sub-manifolds

A C^k $(k \geq 2)$ compact manifold M of dimension D will be defined as an embedded compact manifold, that is, a manifold included in the Euclidean space $\mathbb{R}^N$, where $N \geq D$ is the dimension of the ambient space. By a result of Nash (1956) this entails no loss of generality provided that $k > 2$.

For details on our presentation we suggest the reader to consult the books Guillemin and Pollack (2010) or Spivak (1965).

Recall that $M \subset \mathbb{R}^N$ is a C^k manifold of dimension D if for every point $t \in M$ there exists an open set $U \subset \mathbb{R}^N$, with $t \in U$, such that $U \cap M$ is C^k-diffeomorphic with an open set of $\mathbb{R}^D$. More precisely, there exists an open set $W \subset \mathbb{R}^D$, and a C^k diffeomorphism[1] $\phi : U \cap M \mapsto W$. This map is called a local *chart* of M at t. Its inverse, which is a map defined on an open set of $\mathbb{R}^D$ onto the manifold, is called a *parametrization.*

Note that, while Nash's theorem ensures the generality of our presentation, the manifolds we will consider are naturally embedded: they are the sphere, flat torus, Stiefel manifolds, etc. In such a case, the framework of embedded manifolds is the relevant framework for effective computations. Note that other approaches exist, based on abstract manifolds (Adler and Taylor 2009).

The *tangent space* at $t \in M$, denoted by $\mathbb{T}_t M$, is defined as the set of derivatives at t of smooth curves (at least C^1) on M passing through t. It is a linear subspace of dimension D included in $\mathbb{R}^N$.

On the tangent space, we can use the restriction of the classical Euclidean inner product.

This gives M the structure of a Riemannian manifold: a manifold equipped with a family of scalar products on the tangent spaces with some regularity. Again, when

[1] A C^k diffeomorphism is a one-to-one and onto C^k map, with C^k inverse.

$k > 2$, the Nash C^k embedding theorem says that every Riemannian manifold can be represented in this way.[2]

2.2.2 Smooth Maps on Manifolds

A real-valued map $F : M \mapsto \mathbb{R}$ is a C^k map if its composition with local parametrization is a C^k map. (Note that this definition is independent of the parametrization.) Similarly, one may define a C^k map $F : M \mapsto \mathbb{R}$ as a local restriction of a C^k map $\bar{F}$ defined on an open neighborhood of M.

Given a C^1 map $F : M \mapsto \mathbb{R}$ we define its gradient at $t \in M$ (denoted by $\nabla F(t)$), with respect to the restriction of the classical Euclidean inner product $\langle \cdot, \cdot \rangle$, as the unique vector in $\mathbb{T}_t M$ such that

$$\langle \nabla F(t), v \rangle = \frac{\mathrm{d}}{\mathrm{d}s}\bigg|_{s=0} F(\gamma(s)), \qquad v \in \mathbb{T}_t M, \tag{2.1}$$

where γ is any smooth curve such that $\gamma(0) = t$, and $\gamma'(0) = v$. That is, the derivative at 0 of the smooth curve $F \circ \gamma$. It is the projection on the tangent space $\mathbb{T}_t M$ of the Euclidean gradient $\bar{F}'(t)$ of any extension $\bar{F}(\cdot)$ of $F(\cdot)$.

$\nabla F(t)$ is a vector belonging to $\mathbb{T}_t M$. It can be viewed as a vector in $\mathbb{R}^N$. It can also be written in a convenient basis of $\mathbb{T}_t M$. One way to do that is to use a non-elementary argument using the Levi-Civita connection. It can provide, locally in the neighborhood of t, a family of bases with "nice" properties.[3] In that case we can assume that $\nabla F(t) \in \mathbb{R}^D$.

Suppose now that $F(\cdot)$ takes its values in $\mathbb{R}^d$, then the gradient can be computed coordinate-wise to form a Jacobian matrix of dimensions $d \times N$ or $d \times D$ according to the point of view. It will still be denoted by $\nabla F(t)$.

2.2.3 Different Formulations of the Hessian

Let $F : M \mapsto \mathbb{R}$ be a C^2 map. An important step is the computation of the number of its critical points. The application of the Kac–Rice formula demands the computation of the Riemannian Hessian $\nabla^2 F(t)$ (the Hessian along the manifold).

[2] For every Riemannian manifold of class $k > 2$ there exists an isometric embedding, of class C^k, into Euclidean space. Thus, it can be represented as a sub-manifold of Euclidean space equipped with the induced Euclidean metric.

[3] Strictly speaking the Levi-Civita connection is a way of differentiating with no torsion with respect to the metric (here the induced Euclidean metric).

In this section we will study the Riemannian Hessian from different perspectives, and give useful alternative formulations for practical applications. As we will see in this monograph, we will take advantage of each of the different formulations depending on the context.

2.2.3.1 Hessian via Quadratic Form

The Hessian of a C^2 function can be viewed as a symmetric matrix, a quadratic form or a bilinear form. Note that in the Euclidean case, its entries are just the second derivatives

$$F''(t) := \left(\frac{\partial^2 F(t)}{\partial t_i \partial t_j}\right).$$

On a manifold $M \subset \mathbb{R}^N$, the Riemannian Hessian $\nabla^2 F$ is defined as a quadratic form by

$$\nabla^2 F(t_0)[v] = \langle \frac{\mathrm{d}}{\mathrm{d}s}\bigg|_{s=0} \nabla F(\gamma(s)), v \rangle, \qquad v \in \mathbb{T}_{t_0} M, \tag{2.2}$$

where, again, γ is any smooth curve such that $\gamma(0) = t$, and $\gamma'(0) = v$. Note that if, in addition, $\gamma''(0) \perp \mathbb{T}_{t_0} M$, for example if $\gamma(\cdot)$ is a geodesic, we have

$$\nabla^2 F(t_0)[v] = \frac{\mathrm{d}^2}{\mathrm{d}s^2}\bigg|_{s=0} F(\gamma(s)).$$

2.2.3.2 Hessian via Retractions

We define the retraction, a mapping from the tangent space onto the manifold whose Jacobian at the origin is the identity.

Definition 2.1 A *retraction* at $t_0 \in M$, is a local parametrization $\psi : U \subset \mathbb{T}_{t_0} M \mapsto M$, where U is a relative open set in $\mathbb{T}_{t_0} M$ containing 0, such that

(i) $\psi(0) = t_0$;
(ii) $D\psi(0) = \mathrm{Id}_D$, where we naturally identify the tangent space of $\mathbb{T}_{t_0} M$ at 0 with itself.

Two retractions are very important. The first is the inverse of the projection on the tangent space, the second is the exponential map based on geodesics.

The interest of retractions is explained in the following lemma (Absil et al. 2008, Proposition 5.1).

Lemma 2.1 *Let $M \subset \mathbb{R}^N$ be smooth manifold, with the inherited Riemannian structure from $\mathbb{R}^N$. Let $F : M \mapsto \mathbb{R}$ be a C^2 function. Let $\psi : U \subset \mathbb{T}_{t_0}M \mapsto M$ be a retraction at $t_0 \in M$. Then,*

$$\nabla^2 F(t_0) = \nabla^2(F \circ \psi)(0).$$

A first consequence of this lemma is that the Riemannian Hessian can be computed using ordinary differentiation on $\mathbb{R}^D$.

If we use the Levi-Civita connection then (2.2) reads

$$\nabla^2 F := \nabla\nabla F.$$

In most cases, this definition does not in fact help much for effective computation. Besides the retraction method, we can use tables of known results.

2.2.3.3 Hessian via Weingarten Map or Second Fundamental Form

Let us assume that $\overline{F}$ is a local extension in a neighborhood $U \subset \mathbb{R}^N$ around $t_0 \in M$. For each $t \in U$, let us denote by

$$\Pi_t : \mathbb{R}^N \mapsto \mathbb{T}_t M$$

the orthogonal projection onto the tangent space $\mathbb{T}_t M$. If we fix the canonical basis of $\mathbb{R}^N$, then this projection induces a smooth map

$$\Pi : M \mapsto \mathbb{S}_N, \quad t \mapsto \Pi_t,$$

from the manifold M onto the space of symmetric $N \times N$ matrices.

It is an exercise to check that the Riemannian gradient field $\nabla F(t) \in \mathbb{T}_t M$ coincides with $\Pi_t \overline{F}'(t)$, so by (2.2) and the chain rule one has

$$\nabla^2 F(t_0)[v] = \left\langle D\Pi(t_0)\overline{F}'(t_0), v\right\rangle + \langle \overline{F}''(t_0)v, v\rangle, \tag{2.3}$$

where $D\Pi(t_0)$ is the directional derivative of the map Π at t_0 in the direction of $v \in \mathbb{T}_{t_0}M$, that is, from (2.1)

$$D\Pi(t_0)v = \left.\frac{\mathrm{d}}{\mathrm{d}s}\right|_{s=0} \Pi_{\gamma(s)},$$

where $\gamma(s)$ is any curve on M such that $\gamma(0) = t_0$ and $\gamma'(0) = v$.

Note that the second term in (2.3) is simply the tangent projection of the Euclidean Hessian. The Riemannian Hessian is obtained by adding a corrective term. Tables of such correcting terms can be obtained, for example, in Absil et al. (2013)

We can use also the second fundamental form of the embedded manifold. See, for example, do Carmo (2016).

Lemma 2.1 is useful when we are dealing with an explicit formula for a geodesic. This is the case when we work with the sphere or Stiefel's manifolds.

Example 2.1 (The Case of the Sphere) Let $M = rS^{N-1}$ be the sphere with radius r. In this case, we have, in matrix notation

$$\nabla^2 F(t) = \Pi_t \bar{F}''(t)\Pi_t - (1/r)\,\mathrm{Id}_{N-1}\bar{F}'_N(t), \tag{2.4}$$

choosing outward orientation for the normal derivative $\bar{F}'_N(t)$. This corrects a typo in formula (6.24) in Azaïs and Wschebor (2009). There are several ways to obtain this result. Note that because of the invariance of the problem, thanks to isometry we can use the "*Santa Claus method*": locating ourselves at the north pole. By scaling we can assume that $r = 1$.

- Using the retraction given by the inverse of the projection of the tangent space, we parametrize the sphere by

$$t_1 e_1 + \cdots + t_{N-1}e_{N-1} + \sqrt{1 - t_1^2, \ldots, -t_{N-1}^2}.$$

  This is the method used in Auffinger et al. (2013) and implicitly in Azaïs and Wschebor (2004). After some algebra, we obtain (2.4)
- The simplest approach is perhaps to use geodesics. Let $\mu \in S^{N-2}$ be a direction in the tangent space. The geodesic with derivative μ at the north pole (a great circle) is

$$\phi \mapsto \sin(\phi)\mu + \cos(\phi)e_N.$$

  So we consider

$$G(\phi) := F(\sin(\phi)\mu + \cos(\phi)e_N) =: F().$$

  We have

$$\begin{aligned} G''(\phi) &= -\sin(\phi)F'_\mu() + \cos^2(\phi)F''_\mu() - 2\cos(\phi)\sin(\phi)F''_{\mu,N}() \\ &\quad - \cos(\phi F'_N() - \sin^2(\phi)F''_N(), \\ G''(0) &= F''_\mu - F'_N, \end{aligned}$$

  which directly implies (2.4).
- We can use known results on the Weingarten map or on the second fundamental form.

The case of the Stiefel manifold is more involved and we refer to Edelman et al. (1998).

2.3 Basic Notions of Integral Geometry and the Kac–Rice Formula for Almost Every Level

2.3.1 *Basic Results of Integral Geometry*

2.3.1.1 Area and Co-area Formulas

This section presents, without proof, classical results of integral geometry that are needed in the next sections. For details and proofs, see Morgan (2016), which is an excellent introduction to the book by Federer (1969); see also Berzin et al. (2022).

Proposition 2.2 (Area Formula) *Let f be a C^1 function defined on an open subset U of $\mathbb{R}^d$ taking values in $\mathbb{R}^d$. Let g: $\mathbb{R}^d \mapsto \mathbb{R}$ be continuous and bounded. Then,*

$$\int_{\mathbb{R}^d} g(u)N_u(f, B)\mathrm{d}u = \int_B |\det(f'(t))| g(f(t))\mathrm{d}t, \tag{2.5}$$

for any Borel subset B of U, whenever the integral on the right-hand side is well defined.

Remark 2.1 Let us begin with a few remarks.

1. In the cited references, the statement is given, in its full generality, for Lipschitz functions, which is a weaker assumption than C^1, see Federer (1969, Th. 3.2.5).
2. Suppose that M and V are two C^1 Riemannian manifolds with common dimensions d and $f : M \mapsto V$ is a C^1 function. We can define the Riemannian gradient $\nabla f(t)$ at t (using differentiation along curves), and the Riemannian volumes σ_M and σ_V on M and V respectively. Then (2.5) takes the form:

$$\int_V g(u)N_u(f, B)\mathrm{d}\sigma_V(u) = \int_B |\det(\nabla f(t))| g(f(t))\mathrm{d}\sigma_M(t), \tag{2.6}$$

   for any Borel subset B of M, whenever the integral on the right-hand side is well defined.
3. If $g(\cdot)$ is non-negative, which will be sufficient for our applications, the integral on the right-hand side of (2.5) or (2.6) is always well-defined.
4. By Sard's lemma below, the set of critical values of f has zero Lebesgue measure. As a consequence, for compact B, using the local inversion theorem (as we will do in detail in Sect. 2.6), we know that $N_u(f, B)$ is finite for almost every u.

5. By another approximation argument, (2.5) can be rewritten as

$$\int_{\mathbb{R}^d} \sum_{t \in f^{-1}(u)} h(t, u) \mathrm{d}u = \int_{\mathbb{R}^d} |\det(f'(t))| h((t, f(t)) \mathrm{d}t, \tag{2.7}$$

for every bounded Borel-measurable function h such that the right-hand side of (2.7) is well-defined.

Lemma 2.3 (Morse–Sard) *Let f be a C^ℓ function defined on an open subset $W \subset \mathbb{R}^D$ taking values on $\mathbb{R}^d$, $D \geq d$. Then, if $\ell \geq D - d + 1$, the set of critical values has Lebesgue measure zero in $\mathbb{R}^d$.*

An example of application is the following. Suppose that $f : \mathbb{R}^d \mapsto \mathbb{R}^d$ is a C^1 function, then the set of critical values has Lebesgue measure zero.

Proposition 2.4 (Co-Area Formula) *Let f be a C^1 function defined on an open subset W of $\mathbb{R}^D$ taking values in $\mathbb{R}^d$, $D \geq d$. Let $g : \mathbb{R}^d \mapsto \mathbb{R}$ be continuous and bounded. Then,*

$$\int_{\mathbb{R}^d} g(u) \sigma_{D-d}\big(f^{-1}(u) \cap B\big) \mathrm{d}u = \int_B \Delta(t) g(f(t)) \mathrm{d}t, \tag{2.8}$$

for any Borel subset B of W, whenever the integral on the right-hand side is well defined. We have set $\Delta(t) := \big[\det\big(f'(t)(f'(t))^\top\big)\big]^{1/2}$.

Remark 2.2

1. As for the Area formula, the Co-Area Formula, can be extended to manifolds. We will skip the details here.
2. Clearly, this extends the Area formula, since, if $d = D$, $\sigma_0(f^{-1}(u) \cap B)$ is the number points of the level set on B, i.e., $N_u(B)$.
3. Remark 2.1 after Proposition 2.2 applies here in the same way. This means that if we replace the function $g(u)$ in (2.8) by any measurable function $h(t, u)$, we obtain the weighted Co-Area formula:

$$\int_{\mathbb{R}^d} \mathrm{d}u \int_{R^D} h(t, u) \sigma_{D-d}(f^{-1}(u), \mathrm{d}t) = \int_{\mathbb{R}^D} \Delta(t) h((t, f(t)) \mathrm{d}t, \tag{2.9}$$

whenever the right-hand side is well defined.

2.3.1.2 Integral Geometric Measure

Suppose that we are in the Euclidean space $\mathbb{R}^D$. Given $m < D$, we consider the Grassmannian manifold $\mathbb{G}_{D,m}$, that is, the collection of m-dimensional linear subspaces of $\mathbb{R}^D$.

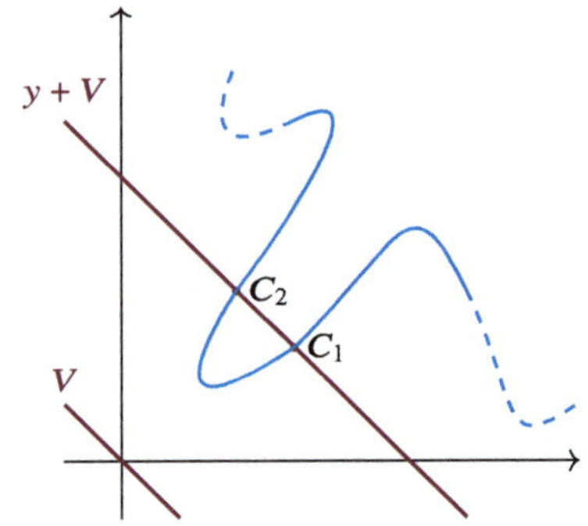

Fig. 2.1 The integral geometric measure of the blue curve ($D = 2, m = 1$) is the integral of the number of red intersections

Definition 2.2 Let $\mathrm{d}\mathbb{G}_{D,m}$ be the Haar measure: the uniquely defined orthogonally invariant probability measure on this space. Let B be a Borel set in $\mathbb{R}^D$. We define the *m-integral geometric measure* of B by

$$\mathcal{I}_{D,m}(B) := c_{D,m} \int_{V \in \mathbb{G}_{D,D-m}} \mathrm{d}\mathbb{G}_{D,D-m}(V) \int_{y \in V^\perp} \mathrm{d}\lambda_m(y) \, \# \{B \cap \{y + V\}\} . \tag{2.10}$$

We naturally identify the element of the Grassmannian $V \in \mathbb{G}_{D,D-m}$ with the associated subspace of co-dimension m on $\mathbb{R}^D$. The constant $c_{D,m}$ is defined by

$$c_{D,m} := \pi^{1/2} \frac{\Gamma\left(\frac{D+1}{2}\right)}{\Gamma\left(\frac{m+1}{2}\right)\Gamma\left(\frac{D-m+1}{2}\right)} . \tag{2.11}$$

See Fig. 2.1.

The integrand in (2.10) is measurable (see for example Morgan 2016, p. 13), and since it is non-negative, the integral is always well defined, finite or infinite. This measure is also called the *Favard measure*.

2.3.1.3 Crofton's Formula

A Borel set can always be decomposed into the disjoint union of a rectifiable set and a purely non-rectifiable set. This last class of sets always has a null integral geometric measure, (Morgan 2016, 3.17 Structure Theorem), and this implies that the Hausdorff measure of a set is always greater than or equal to the integral geometric measure. They coincide for rectifiable sets: this is the object of Crofton's formula.

Definition 2.3 A set $B \in \mathbb{R}^D$ is said to be m-rectifiable, $m \leq D$, if it is the countable union of C^1 manifolds of dimension m and a set of σ_m measure zero.

Crofton's formula is a fundamental formula of integral geometry.

Theorem 2.5 (Crofton's Formula) *If $B \subset \mathbb{R}^D$ is m-rectifiable, then Crofton's formula (Morgan 2016, p. 31) yields*

$$\sigma_m(B) = \mathcal{I}_{D,m}(B). \tag{2.12}$$

Proposition 2.6 (Push-Forward Distribution) *Let $Y \in \mathbb{R}^D$ be a random variable admitting a density $p(y)$ w.r.t. the Lebesgue measure and $f : \mathbb{R}^D \mapsto \mathbb{R}^d$, $d \leq D$, be a C^1 function. Set $U = f(Y)$ and define R_U and I_U as the sets of regular and irregular values, respectively, of $f(\cdot)$:*

$$I_U := \{u \in \mathbb{R}^d : \exists y \in \mathbb{R}^D; u = f(y); \Delta(y) = 0\},$$

$$R_U := \{u \in \mathbb{R}^d : \exists y \in \mathbb{R}^D; u = f(y); \Delta(y) \neq 0\}.$$

Note that I_U and R_U may have a non-empty intersection. Next, let us turn to the Push-forward distribution: the distribution of $U = f(Y)$ is the sum of a measure which is supported by the set I_U and an absolutely continuous distribution on the set R_U of regular values of f with density

$$h(u) = \int_{y \in f^{-1}(u)} \frac{p(y)}{\Delta(y)} d\sigma_{D-d}(y).$$

In particular, if $\mathbb{P}\{\Delta(Y) = 0\} = 0$, the distribution of U is absolutely continuous with density given by the formula above.

Note that if $f(\cdot)$ is C^{D-d+1}, the Morse–Sard lemma (Lemma 2.3) implies that I_U has zero Lebesgue measure.

Proof Define

$$I_Y := \{y \in \mathbb{R}^D : \Delta(y) = 0\}$$

$$R_Y := I_Y^c.$$

The distribution of Y (and consequently that of U) can be split into two distributions:

- one is supported by I_Y and induces for U a distribution on the set I_U of irregular values of $f(\cdot)$
- the other one is supported by R_Y and induces for U a distribution which is supported on R_U, the set of regular values of $f(\cdot)$.

We will now study the distribution of U restricted to the set R_U of regular values. Let $g(\cdot)$ be a bounded Borel function defined on $\mathbb{R}^d$ and with support on R_U, then

$$\mathbb{E}(g(U) = \mathbb{E}\big(g(f(Y))\big) = \int_{R_Y} g(f(y))\Delta(y)\frac{p(y)}{\Delta(y)}\mathrm{d}y.$$

Applying (2.9) with $h(y, f(y)) = \frac{p(y)}{\Delta(y)}\mathbb{1}_{f(y)\in R_U}$ yields

$$\mathbb{E}(g(U)) = \int_{\mathbb{R}^d} g(u)\mathrm{d}u \int_{y\in f^{-1}(u)} \frac{p(y)}{\Delta(y)}\mathrm{d}\sigma_{D-d}(y),$$

giving the result. □

2.3.1.4 Continuity of Integrals

In addition, we need the following lemma, which is a consequence of the implicit function theorem.

Lemma 2.7 (Continuity of Integrals over a Level Set) *Let K be a compact set of $\mathbb{R}^D$, $H : K \mapsto \mathbb{R}^d$, $D \geq d$, a C^1 function and $f : \mathbb{R}^D \mapsto \mathbb{R}^d$ a continuous (respectively differentiable) function. Let $y \in \mathbb{R}^d$ such that*

$$\big\{x \in \mathcal{L} : \Delta(x) = 0\big\} = \emptyset,$$

where $\mathcal{L} := \{t \in K : H(t) = y\}$, $\Delta(x) := \sqrt{\det(H'(x)H'(x)^{\top})}$. We can define the relative interior of the level $\mathcal{L}$, $\mathring{\mathcal{L}}$ by $\mathring{\mathcal{L}} := \{t \in \mathring{K} : H(t) = y\}$. Then, $\mathring{\mathcal{L}}$ is a C^1 manifold and

$$y \mapsto \int_{H^{-1}(y)} f(x)\mathrm{d}\sigma_{D-d}(x) \textit{ is continuous (resp. differentiable) at } y.$$

Proof The proof is split into several steps.

Step 0 Let us first recall the general framework of the implicit function theorem: F is a C^1 function of (x, z) and (x_0, z_0) is a solution of $F(x, z) = 0$ such that the partial Jacobian of F w.r.t. z at (x_0, z_0) is not zero. There is therefore an open neighborhood V_1 of x_0, an open neighborhood $V_{1,2}$ of (x_0, z_0) and a differentiable function $\phi(\cdot)$ such that

$$\{(x, z) \in V_{1,2} \text{ and } F(x, z) = 0\} \text{ is equivalent to } \{x \in V_1 \text{ and } z = \phi(x)\}.$$

Step 1 The level set is a manifold.
Let us consider the level set $\mathcal{L} := \mathcal{L}_y := H^{-1}(y)$.

At each point t^* of $\mathcal{L}$, the Jacobian matrix $H'(t^*)$ is of full rank d and there exists at least one subset $\mathcal{S}$ of size d, of $\{1, \dots, D\}$, such that the corresponding extracted matrix is invertible. Indexed by $\mathcal{S}$ we can build a finite covering of $\mathcal{L}$. In the following, we consider only the subset of $\mathcal{L}$ where $\mathcal{S}$ consists of the last d coordinates. The other cases are equivalent. Let us set $\overleftarrow{t} := (t_1, \dots, t_{D-d})$ and $\overrightarrow{t} := (t_{D-d+1}, \dots, t_D)$.
Using the implicit function theorem with $x = \overleftarrow{t}$, $z = \overrightarrow{t}$, $F(x, z) = H(t) - y$, we see that $\overleftarrow{t}$ defines a parameterization (the inverse of a chart) given by $(\overleftarrow{t}, \phi(\overleftarrow{t}))$, as a consequence $\mathcal{L}$ is a manifold.

Step 2 A covering of the graph $\{t, H(t)\}$ in the neighborhood of the level set. We will still limit ourselves to the case where the set $\mathcal{S}$ consists of the last d coordinates. In the same way as in Step 1, we can consider a particular point $t^* \in \mathcal{L}$ and apply the implicit function theorem, now for $\bar{y}$ in some neighborhood of y with

$$x = (\overleftarrow{t}, \bar{y}); \; z = \overrightarrow{t}; \; F(x, z) = H(t) - \bar{y}.$$

The open neighborhood $V_{1,2}$ can be chosen as a product $W_1 \times W_2$ for the two variables $\overleftarrow{t}$ and $\bar{y}$. The set $W_{t^*} := \{(\overleftarrow{t}, \phi(\overleftarrow{t}, y)); \overleftarrow{t} \in W_1\}$ is open in the relative topology of the compact manifold $\mathcal{L}$, containing t^*. When t^* varies, the W_{t^*} constitute a covering of which we can extract a finite covering say $W_{t_1^*}, \dots, W_{t_n^*}$ using the compactness of $\mathcal{L}$. To each t_i^* is associated an interval of the type W_2 for y. This interval will be denoted by $W_{t_i^*,2}$

Let W_y be the finite intersection of the intervals $W_{t_i^*,2}$. Let us consider the neighborhood of $\mathcal{L}$ defined by

$$\{t, H(t) : H(t) \in W_y\}.$$

The results above imply that this neighborhood is a C^1 manifold. The differentiability of

$$\bar{y} \mapsto \sigma_{D-d}(\mathcal{L}_{\bar{y}})$$

follows directly. This gives the result. □

A longer version of this proof can be found in Berzin et al. (2022, Theorem 4.1.2) or in Cabaña (1986).

2.3.2 A Kac–Rice Formula for Almost Every Level

From the Co-Area formula we directly obtain a Kac–Rice formula for almost every level. Its proof is identical to that of Proposition 1.2. We define:

Assumption (A.2.1)

- $X(\cdot)$ has C^1 paths, we denote by $X'(t)$ its derivative of dimension $d \times D$;
- for each $t \in T : \Delta(t) := \sqrt{\det\left(X'(t)(X'(t))^\top\right)}$ is integrable;
- for each $t \in T : X(t)$ admits a density $p_{X(t)}(\cdot)$ in $\mathbb{R}^d$.

We have

Proposition 2.8 (Kac–Rice Formula for Almost Every Level) *Let $X : T \mapsto \mathbb{R}^d$ be a random field, T an open (possibly unbounded) subset of $\mathbb{R}^D$. Under Assumption **(A.2.1)**, for every Borel set $B \subset T$, define the level set restricted to B by*

$$\mathcal{L}_u(B) := \{t \in B : X(t) = u\}.$$

We have

$$\mathbb{E}(\sigma_{D-d}\mathcal{L}_u(B)) = \int_B \mathbb{E}\big(\Delta(t) \mid X(t) = u\big)\, p_{X(t)}(u)\, \mathrm{d}t, \tag{2.13}$$

***for almost every** $u \in \mathbb{R}^d$.*

Note that, when $D = d$, the Hausdorff measure σ_0 counts the number of roots of $X(t) = u$. Of course the C^1 assumption can be weakened by assuming that $X(\cdot)$ has Lipschitz paths, but explicit examples are missing.

2.4 A Bulinskaya-Type Lemma

Most of the results of this and the following two sections are inspired by Armentano et al. (2025).

2.4.1 Differentiability

The random fields we will consider are assumed to have paths of class C^1. For convenience they will be assumed to be defined on an open set, say T. Sometimes compactness is needed and we will consider a compact subset of T that, in most cases, by an abuse of notation, will still be denoted T.

Remark 2.3 For stationary Gaussian random fields, the fact that the paths are C^1 implies differentiability in quadratic mean. In fact, this assumption is a little stronger than differentiability in quadratic mean, see Azaïs and Wschebor (2009).

Roughly speaking the Bulinskaya lemma holds, under certain regularity conditions, if there are more random equations than variables. There are no solutions a.s. A precise statement can be found in Azaïs and Wschebor (2009, Proposition 6.11)

Proposition 2.9 *Let $Y(\cdot) = \{Y(t) : t \in W\}$ be a random field with values in $\mathbb{R}^{m+k}$, m and k being positive integers. Let W an open subset of some linear space $\mathbb{R}^D$. Let $u \in \mathbb{R}^{m+k}$ and B a subset of W.*

We assume that $Y(\cdot)$ and W satisfy the following conditions:

- *the paths $Y(\cdot)$ are of class C^1,*
- *for each $t \in W$, the random vector $Y(t)$ has a density and there exists a constant C such that*

$$p_{Y(t)}(x) \leq C$$

for $t \in I$ and x in some neighborhood of u,
- *the Hausdorff dimension of B is less than or equal to m.*

It is hence almost sure that there is no point $t \in I$ such that $Y(t) = u$.

A proof of this proposition can be found in Berzin et al. (2022, Proposition 4.3). As we will see, this lemma can be used to infer the geometry of the level set.

2.4.2 The Geometry of the Level Set

It is sometimes written that, if the paths of $X(\cdot)$ are C^1, the level set is always a manifold: Ibragimov and Zaporozhets (2013). This can be true for almost every level using the Morse–Sard lemma, if the random field is sufficiently differentiable. Yet it is not true in full generality, as shown by the following counterexamples:

Let $D = 2$ and $d = 1$, for example, and consider

$$X(t_1, t_2) = \xi(t_1^2 - t_2^2), \qquad (t_1, t_2) \in \mathbb{R}^2,$$

where ξ is standard normal. Take $u = 0$, the nodal set is a.s.

$$\mathcal{L}_0 = \{|t_1| = |t_2|\} \subset \mathbb{R}^2,$$

which is obviously not a manifold.

The counterexample above is very particular, but inspired by it we can propose the following:

$$X(t_1, t_2) = \xi_1(t_1 - \xi_2)^2 + \xi_3(t_2 - \xi_4)^2, \qquad (t_1, t_2) \in \mathbb{R}^2,$$

where the ξ_i's are, for example, independent standard normal. The following facts are easy to check:

- $X(\cdot) \in C^\infty$
- For every $t \in \mathbb{R}^2$ the distribution of $X(t)$ and of $X'(t)$ are non-degenerate.
- With probability $1/2$ the nodal set is the point (ξ_2, ξ_4).
- Otherwise it is again the union of two intersecting straight lines passing through (ξ_2, ξ_4). It is not a manifold.

This example shows that, in the non-Gaussian case, the regularity of the level set can be subtle. In contrast, the Gaussian case is well understood; see in particular Lerario and Stecconi (2019) for a geometric approach based on transversality, relating Bulinskaya's lemma and the Kac–Rice formula.

Our first result concerns a Bulinskaya-type lemma on the regularity of the level set $\mathcal{L}_u$ for C^1 random fields. It proves that, under minimal hypotheses on the density $p_{X(t)}$ of the random vector $X(t)$, the irregular part of the level set satisfies

$$\sigma_{D-d}\{t \in T : X(t) = u, \ \Delta(t) = 0\} = 0, \text{ a.s.} \tag{2.14}$$

This is called "the weak Bulinskaya condition". By the implicit function theorem, the regular part of the level set is a C^1 manifold so the whole level set $\mathcal{L}_u$ is a.s. rectifiable.

When $D > d$, classical proofs of the Kac–Rice formula demand the "strong Bulinskaya condition"

$$\mathbb{P}\{\exists t \in U, X(t) = u, \Delta(t) = 0\} = 0, \tag{2.15}$$

which can be attained by the following version of Proposition 2.9.

Proposition 2.10 *Let $X(\cdot) = \{X(t), t \in W\}$ be a random field, defined on an open subset W of $\mathbb{R}^D$ and with values in $\mathbb{R}^d$. Let $u \in \mathbb{R}^d$.*

We assume:

- *the paths of $X(\cdot)$ are of class C^2,*
- *for each $t \in W$, the pair $(X(t), X'(t))$ has a joint density $p_{X(t),X'(t)}(x, x')$ in $\mathbb{R}^d \times \mathbb{R}^{D.d}$, which is bounded for (t, x') varying in a compact subset of $W \times \mathbb{R}^{D.d}$ and x in some neighborhood of u.*

Then the strong Bulinskaya condition (2.15) *holds true.*

Proof Apply Proposition 2.9 to the random field

$$Y(t, \lambda) = \big(X(t) : (X'(t))^\top \lambda\big)$$

defined for $(t, \lambda) \in W \times S^{D-1}$ with values in $\mathbb{R}^d \times \mathbb{R}^d$. □

Our main contribution is to prove that the "weak Bulinskaya condition" (2.14) is sufficient. In addition we prove (2.14) under minimal conditions. Note that when $D = d$, the two conditions, weak and strong Bulinskaya, are equivalent.

Proposition 2.11 (Bulinskaya-Type Lemma) *Let T be an open subset of $\mathbb{R}^D$ and let $X : T \mapsto \mathbb{R}^d$ be a random field with paths of class C^1. Assume that the density $p_{X(t)}(\cdot)$ of $X(t)$ satisfies*

$$\int_T p_{X(t)}(v)\mathrm{d}t \leq C \text{ for all } v \text{ in some neighborhood } V_u \text{ of } u. \tag{2.16}$$

Then the weak Bulinskaya condition (2.14) *holds true.*

Before proving Proposition 2.11, let us start by recalling a few definitions. Given $x \in \mathbb{R}^\ell$ and $\rho > 0$, we denote by $B_\ell(x, \rho) \subset \mathbb{R}^\ell$ the ball centered at x of radius ρ. For a given $K \subset \mathbb{R}^D$, recall that $\sigma_\ell(K)$ is the ℓ-dimensional Hausdorff measure given by

$$\sigma_\ell(K) = \lim_{\varepsilon \to 0} \sigma_\ell^\varepsilon(K),$$

where σ_ℓ^ε is the ℓ-dimensional Hausdorff pre-measure on $\mathbb{R}^D$ given by

$$\sigma_\ell^\varepsilon(K) := W_\ell \inf \left\{ \sum_i {\rho_i}^\ell, \text{ for a covering } \{B_D(x_i, \rho_i)\} \text{ of } K, \ \rho_i < \varepsilon \right\}, \tag{2.17}$$

where $W_\ell := \mathrm{Leb}_\ell(B_\ell(0, 1))$ is the Lebesgue measure of the unit ball in $\mathbb{R}^\ell$. Let us recall that the ℓ-dimensional Hausdorff measure on $\mathbb{R}^\ell$ coincides with the Lebesgue measure Leb_ℓ, on Borel sets of $\mathbb{R}^\ell$.

We now need a technical Lemma.

Lemma 2.12 *Let K be a compact set of $\mathbb{R}^D$ and assume $d < D$. Suppose $\sigma_{D-d}(K)$ is finite. We therefore have constants C_1 and C_2 such that, for ε sufficiently small:*

- *there exists an "ε-packing" (a collection of disjoint balls with center belonging to K and with radius ε), with cardinality $l(\varepsilon)$ with*

$$l(\varepsilon) \geq C_1 \varepsilon^{d-D} \sigma_{D-d}(K)$$

- *The parallel set $K^{+\varepsilon} = \bigcup_{x \in K} B_D(x, \varepsilon)$ satisfies*

$$\sigma_D(K^{+\varepsilon}) \geq W_D C_2 \varepsilon^d \sigma_{D-d}(K),$$

where W_D is the Lebesgue measure of the unit ball in $\mathbb{R}^D$.

Proof Recall that the premeasure $\sigma^{\varepsilon}_{D-d}$, given in (2.17), satisfies $\sigma^{\varepsilon}_{D-d}(K) \uparrow \sigma_{D-d}(K)$ as $\varepsilon \downarrow 0$. Suppose $\sigma_{D-d}(K) = a > 0$ (otherwise there is nothing to prove). Then for ε sufficiently small $\sigma^{3\varepsilon}_{D-d}(K) > a/2$. We perform a covering of K by induction using balls centered at points in K and of radius exactly 3ε. To do so we start from one point of K and at each step we define a new ball such that the center (i) belongs to K and (ii) does not belong to the union of the preceding balls. By compactness the needed number of such balls, say $\ell(\varepsilon)$, is finite and by definition of the Hausdorff pre-measure

$$\ell(\varepsilon)(3\varepsilon)^{D-d} W_{D-d} > a/2.$$

Let $t_1, \ldots, t_{\ell(\varepsilon)}$ be the centers of the balls, the balls $B_D(t_i, \varepsilon)$ constitute an ε packing, giving the first assertion. The union of the balls, which has a volume of

$$W_D C_{D,d} a \varepsilon^d,$$

is included in the parallel set $K^{+\varepsilon}$, giving the second assertion. □

Now we can turn to the proof of Proposition 2.11.

Proof of Proposition 2.11 Assume, with no loss of generality, that $u = 0$.

Define the density of the occupation measure: LT, the *local time*, as

$$\mathrm{LT} := \liminf_{\delta \to 0} \mathrm{LT}(\delta) := \liminf_{\delta \to 0} \frac{1}{W_d(\delta)} \sigma_D\left(\{t \in T : \|X(t)\| \le \delta\}\right),$$

where $W_d(\delta)$ is the volume of the ball $B_d(0, \delta)$ with radius δ on $\mathbb{R}^d$. Note that since we used lim inf, the definition of the local time is given in a generalized sense. By Fubini's theorem, the hypothesis on the density and Fatou's lemma:

$$\mathbb{E}\big(\mathrm{LT}\big) \le (\mathtt{const})\sigma_D(T). \tag{2.18}$$

This implies that LT is an almost surely finite non-negative random variable.

Let $M := \max_{t \in T} \lambda_{max}(X'(t))$, where λ_{max} is the greatest singular value. Let

$$N(\varepsilon) := \sup_{t \in T;\ \|v\| < \varepsilon} \frac{\|X(t+v) - X(t) - X'(t)v\|}{\|v\|}.$$

Then, by compactness, $N(\varepsilon)$ and M are almost surely finite non-negative random variables and $N(\varepsilon) \to 0$, as $\varepsilon \to 0$.

Let $\tilde{\mathcal{L}}_0$ the irregular part of the level set $\mathcal{L}_0$, i.e.

$$\tilde{\mathcal{L}}_0 := \{t \in T : X(t) = 0,\ \Delta(t) = 0\}. \tag{2.19}$$

Let $t \in \tilde{\mathcal{L}}_0$, let rk$(\cdot)$ denote the rank, then $X(t) = 0$, $\mathrm{rk}(X'(t)) = k$ for some $k \in \{0, 1, \dots, d-1\}$.

Let $v_1 \in V_1 := \ker(X'(t))$, where V_1 has dimension $D-k$. Let $v_2 \in V_2 := V_1^{\perp}$, where V_2 has dimension k.

Then we have

$$\begin{aligned}\|X(t+v_1+v_2)\| &\leq \|X(t+v_1+v_2) - X(t+v_1)\| + \|X(t+v_1)\| \\ &\leq M\|v_2\| + \|v_1\| \cdot N(\|v_1\|).\end{aligned}$$

Suppose ε is such that $N(\varepsilon) < 1$, and choose v_1 and v_2 such that

$$\|v_1\| \leq \frac{1}{\sqrt{2}}\varepsilon, \quad \|v_2\| \leq \frac{1}{\sqrt{2}}\varepsilon \cdot N(\varepsilon). \tag{2.20}$$

Then, we have:

$$\left\| X\left(t + \begin{pmatrix} v_1 \\ v_2 \end{pmatrix}\right)\right\| \leq \frac{1}{\sqrt{2}}(M+1)\varepsilon \cdot N(\varepsilon).$$

The condition given in (2.20) defines a polydisc with volume

$$(\mathtt{const})\varepsilon^{D-k}\big(\varepsilon \cdot N(\varepsilon)\big)^k, \tag{2.21}$$

where the constant depends on D, d and k. The polydisc is included in $\tilde{\mathcal{L}}_0^{+\varepsilon}$ and thus in T for ε sufficiently small.

Let us define the event $Z_{D,d} = \{\omega : \sigma_{D-d}(\tilde{\mathcal{L}}_0) > 0\}$, and assume by contradiction that

$$\mathbb{P}(Z_{D,d}) > 0.$$

In the case $D = d$, given $\omega \in Z_{d,d}$, let $\delta = \delta_{\omega,\varepsilon} = \big((M+1)\varepsilon \cdot N(\varepsilon)\big)$. Then the approximated local time LT(δ) is greater than

$$\frac{1}{(\varepsilon \cdot N(\varepsilon))^d}(\mathtt{const})\varepsilon^{d-k}\big(\varepsilon \cdot N(\varepsilon)\big)^k = (\mathtt{const})N(\varepsilon)^{k-d},$$

which tends to infinity as $\varepsilon \to 0$, contradicting the fact that LT is an almost surely finite non-negative random variable (cf. (2.18) for $D = d$).

Now let us consider the case $D > d$. When $\omega \in Z_{D,d}$, by Lemma 2.12 and (2.21), the approximated local time LT(δ) for $\delta = (M+1)\varepsilon \cdot N(\varepsilon)$ satisfies

$$\mathrm{LT}(\delta) = \frac{1}{\lambda_d(B_d(0,\delta))}\sigma_D\{t \in T : \|X(t)\| \leq (M+1)\varepsilon \cdot N(\varepsilon)\}$$

$$\geq \frac{(\texttt{const})}{\lambda_d(B_d(0,\delta))}\ell(\varepsilon)\varepsilon^{D-k}\big(\varepsilon \cdot N(\varepsilon)\big)^k$$
$$\geq (\texttt{const})((M+1)\varepsilon N(\varepsilon))^{-d}\varepsilon^{D-k}\big(\varepsilon \cdot N(\varepsilon)\big)^k \times \varepsilon^{d-D}$$
$$= (\texttt{const})(M+1)N(\varepsilon)^{k-d},$$

which again tends to infinity as $\varepsilon \to 0$, giving the contradiction. □

2.5 Kac-Type Formulas and an Upper-Bound

This section is limited to the case $D = d$. We consider a random field $X(\cdot)$ $T \subset \mathbb{R}^d \mapsto \mathbb{R}^d$. We assume that a.s. the paths are C^1 and we consider the number N_u of roots in T:

$$N_u := \#\{t \in T : X(t) = u\}$$

and the number of regular roots:

$$\bar{N}_u := \#\{t \in T : X(t) = u;\ \Delta(t) \neq 0\}. \tag{2.22}$$

The local inversion theorem implies that $\{t \in T : X(t) = u;\ \Delta(t) \neq 0\}$ is a set of isolated points, and is thus locally finite.

Lemma 2.13 *With the notation above, let $X(\cdot)$ be a random field as above. Let T be an* ***open*** *set. Then,*

$$\bar{N}_u(f,T) \leq \liminf_{\delta \to 0} N_{u\delta}(f,T) := \liminf_{\delta \to 0} \frac{1}{W(\delta)} \int_T \mathbb{1}_{\{|f(t)-u|<\delta\}}\,\Delta(t)\mathrm{d}t,$$

where $\bar{N}_u$ is the number of regular roots defined in (2.22). *Recall that $W(\delta)$ is the volume of the ball with radius δ in $\mathbb{R}^d$.*

Proof (Sketch) Let R_u be the set of regular roots, i.e.

$$R_u := \{t \in T : f(t) = u;\ \Delta(t) > 0\}.$$

Consider a root $t \in R_u$. According to the local inversion theorem there exists a neighborhood V_t and a critical radius δ_t such that $f(\cdot)$ is a diffeomorphism from V_t to the ball $\mathcal{B}(t,\delta_t)$. By the change of variable formula

$$\frac{1}{W(\delta_t)} \int_{V_t} \mathbb{1}_{\{\|f(t)-u\|<\delta_t\}}\,\Delta(t)\mathrm{d}t = 1. \tag{2.23}$$

Note that the roots in R_u are isolated but, since T is an open set, this fact does not imply that $\bar{N}_u$ is finite. When $\bar{N}_u$ is finite, (2.23) clearly gives the result. When $\bar{N}_u = +\infty$, it suffices to consider m roots and to observe that (2.23) implies that for δ sufficiently small, $N_{u\delta}(f, T) \geq m$. Since m is arbitrary, this gives the result. □

A direct consequence of this lemma, Fatou's lemma, and the extension argument is

Proposition 2.14 *Let $\{X(t) : t \in S\}$, S an open set in $\mathbb{R}^d$, be a random field with values in $\mathbb{R}^d$ and C^1 paths. Then,*

- *for every open subset T of S:*

$$\mathbb{E}\big(\bar{N}_u(X, T) \leq \liminf_{\delta \to 0} \frac{1}{W(\delta)} \int_T \mathbb{E}\big(\Delta(t)\mathbb{1}_{\|X(t)-u\|\leq\delta}\big)\mathrm{d}t.$$

- *If, in addition, for all $t \in S$, $X'(t)$ is integrable and $X(t)$ admits a density, $p_{X(t)}$ such that*

$$\int_T \mathbb{E}(\Delta(t) \mid X(t) = u) p_{X(t)}(u)\mathrm{d}t,$$

is continuous for every open set T.

Then, for every Borel set B

$$\mathbb{E}\big(\bar{N}_u(X, B) \leq \int_B \mathbb{E}(\Delta(t) \mid X(t) = u) p_{X(t)}(u)\mathrm{d}t. \tag{2.24}$$

- *If, in addition, the density $p_{X(t)}(v)$ is uniformly bounded for v in a neighborhood of u:*

$$\mathbb{E}\big(N_u(X, B) \leq \int_B \mathbb{E}(\Delta(t) \mid X(t) = u) p_{X(t)}(u)\mathrm{d}t. \tag{2.25}$$

Finally, for the next section, we need a statement of a classical Kac lemma.

Lemma 2.15 *Let $F(\cdot)$ a deterministic C^1 function, and T a compact set, with the notation of Lemma 2.13, suppose that*

- $N_u(f, \partial T) = 0$
- $\{t \in \mathring{T} : f(t) = u; \det(f'(t)) = 0\} = \emptyset.$

Then, for $\delta > 0$ sufficiently small

$$N_u(f, T) = N_{u\delta}(f, T) = \frac{1}{W(\delta)} \int_T \mathbb{1}_{\{|f(t)-u|<\delta\}} \Delta(t)\mathrm{d}t.$$

Recall that $W(\delta)$ is the volume of the ball with radius δ in $\mathbb{R}^d$.

Proof As per our hypotheses, all the roots of $f(t) = u$ are interior and isolated. They are finitely many, say n. Since the case $n = 0$ is trivial, we will consider the case $n > 0$. As in the proof of Lemma 2.13 we can find n interior neighborhoods $V_1, \dots, V_n$ of these points such that, for δ sufficiently small, their complement does not contribute to $N_{u\delta}(f, T)$ and f is one-to-one. The result is then a direct consequence of the change-of-variable formula. □

2.6 The Kac–Rice Formula for All Levels

We can now present the main result of this chapter: the Kac–Rice formula for all levels. Let $X(\cdot)$ be a random field from $S \subset \mathbb{R}^D$ to $\mathbb{R}^d$, $D \geq d$. Note that the case $D = 1$ is relevant since we will establish here some results that are not contained in those of Chap. 1.

Before beginning we need the following proposition concerning the continuity of the number of roots for the C^1 topology. It is inspired by the proof in dimension 1 by Angst and Poly (2022).

Let us consider the set C of C^1 functions defined on a compact hyper-rectangle $T \subset \mathbb{R}^d$ and taking values in $\mathbb{R}^d$. This space is equipped with the following canonical norm

$$\|f\| = \sup_{t \in T}(\|f(t)\| + \|f'(t)\|),$$

where we have chosen, once and for all, a particular norm in $\mathbb{R}^d$ and in the space of $d \times d$ matrices.

Proposition 2.16 *Let T be a compact set of $\mathbb{R}^d$. Suppose $X_n(\cdot), n = 1, 2, \dots$ is a sequence of random fields defined on T and belonging to the space C defined above. Suppose that the sequence $X_n(\cdot)$ converges a.s. to $X(\cdot)$ with the above-mentioned topology. Suppose that in addition, a.s.*

Assumption (A.2.2)

(i) $N_u(X, \partial T) = 0\,,$
(ii) $\{t \in T : X(t) = u;\ \Delta(t) = 0\} = \emptyset.$

Then

$$N_u(X_n, T) \to N_u(X, T) \text{ a.s. as } n \to \infty.$$

Note that, of course, we obtain the same result if we replace a.s. convergence by weak convergence.

Proof In this proof, we use the operator norm for $\|X'(t)\|$. For each realization, let m be the number of solutions of $X(t) = u$ in T. Since T is compact, (ii) implies that m is finite. The case $m = 0$ is direct. We consider the case $m > 0$.

Let $x_1, \ldots, x_m$ be the solutions of $X(t) = u$ for $t \in T$. By hypothesis, these solutions are interior points of T.

Let $B(x, \delta)$ denote the open ball with center x and radius δ in $\mathbb{R}^d$. By the inverse function theorem, for sufficiently small δ and for all $i = 1, \ldots, m$, $X(\cdot)$ is a diffeomorphism from $B(x_i, \delta)$ to some open set of $\mathbb{R}^d$. Set

$$W := \cup_{i=1}^{n} B(x_i, \delta).$$

We may assume that there exists $\alpha > 0$ such that $\|(X'(t))^{-1}\| \leq \alpha$, for all $t \in W$.

By uniform convergence in the C^1 topology, we may find $N \in \mathbb{N}$ such that

$$\|X_n - X\| < \min(\delta, (4\alpha)^{-1}), \quad \text{for all} \quad n > N \tag{2.26}$$

and $X_n(\cdot)$ has no solution of $X_n(t) = u$ on $T \setminus W$ for $n > N$.

Let us define the auxiliary C^1 functions $\widetilde{X}_{n,i} : B(x_i, \delta) \mapsto \mathbb{R}^d$, given by

$$\widetilde{X}_{n,i}(t) = t - (X'(x_i))^{-1}(X_n(t) - u), \qquad (i = 1, \ldots, m).$$

Note that X_n has a solution of $X_n(t) = u$ on $B(x_i, \delta)$ if and only if the map $\widetilde{X}_{n,i}$ has a fixed point.

By strengthening the conditions on δ, we can even demand the following

$$\forall t \in B(x_i, \delta) : \|X'(t) - X'(x_i)\| < (4\alpha)^{-1}, \qquad (i = 1, \ldots, m). \tag{2.27}$$

From (2.26) and (2.27), for all $t \in \widetilde{B}(x_i, \delta)$,

$$\begin{aligned}
\|(\widetilde{X}_{n,i})'(t)\| &= \|I_d - (X'(x_i))^{-1} X_n'(t)\| \\
&\leq \|(X'(x_i))^{-1}\| \cdot \|X'(x_i) - X_n'(t)\| \\
&\leq \alpha \cdot \big(\|X'(x_i) - X'(t)\| + \|X'(t) - X_n'(t)\|\big) \\
&< \alpha \left(\frac{1}{4\alpha} + \frac{1}{4\alpha} \right) = \frac{1}{2},
\end{aligned}$$

for all $n > N$. We conclude that $\widetilde{X}_{n,i}$ is a contraction $B(x_i, \delta) \mapsto B(x_i, \delta)$. The fixed point theorem implies that $\widetilde{X}_{n,i}$ has exactly one fixed point.

So, for large enough n, the number of solutions is m, concluding our proposition. □

The main result of this chapter is as follows:

Theorem 2.17 (Kac–Rice Formula for the Expectation) *Let $X : T \mapsto \mathbb{R}^d$ be a random field, T an open subset of $\mathbb{R}^D$, $(D \geq d)$. Assume that:*

Assumption (A.2.3)

(i) *The sample paths of $X(\cdot)$ are a.s. C^1.*
(ii) *For each $t \in T$, $X(t)$ admits a continuous density $p_{X(t)}(v)$ which is bounded uniformly in $v \in R^d$ and t in any compact subset of T.*
(iii) *For **every** $v \in \mathbb{R}^d$, for every $t \in T$, the distribution of $\{X(s), s \in T\}$ conditional on $X(t) = v$ is well defined as a probability and is continuous, as a function of v, with respect to the C^1-topology.*

*Then, for every Borel set B contained in T, and for **every level** u, one has*

$$\mathbb{E}\left(\sigma_{D-d}\left(\mathcal{L}_u(B)\right)\right) = \int_B \mathbb{E}\left(\Delta(t) \mid X(t) = u\right) p_{X(t)}(u)\mathrm{d}t. \tag{2.28}$$

Remark 2.4

1. In full generality the condition on the density (ii) and on the conditional distribution (iii) can be localized: it is sufficient that they hold true for all v in some neighborhood of the considered level u.
2. Another way of presenting (2.28) is to say that $B \mapsto E\left(\sigma_{D-d}\left(\mathcal{L}_u(B)\right)\right)$ defines a Borel measure whose density is given by $\mathbb{E}\left(\Delta(t) \mid X(t) = u\right) p_{X(t)}(u)$. In the particular case of Gaussian processes, it is easy to prove that this measure is finite on compact sets: it is a Radon measure.

2.6.1 Proof of Theorem 2.17 for $D = d$

Readers now familiar with the difficulties entailed by the proofs of the Kac–Rice formula know that three problems may arise.

- The roots $X(t) = u$ can be associated with a Jacobian matrix $X'(t)$ which is nearly or exactly singular, namely, $\Delta(t) = 0$;
- Some roots can be located on the boundary of the set under consideration;
- The quantities considered may be too large, causing non-integrability.

All these problems can be overcome by the Bulinskaya lemma and a **bounding and tapering argument** followed by a monotone convergence argument. This (without tapering) is a routine argument in stochastic calculus where, for example, one often localizes stopping times. The tapering is necessary here to keep the continuity, which is a key part of the argument. We introduce our notation:

Let $\mathcal{F}$ be the continuous, monotone, piecewise linear function: $\mathbb{R}^+ \mapsto [0, 1]$ such that: (i) It vanishes on $[0, 1/2]$; (ii) It takes the value 1 on $[1, +\infty)$. Consider

- $F_n(x) := \mathcal{F}(nx)$, a tapered version of the indicator function $\mathbb{1}_{x>1/n}$;
- $G_n(x) := 1 - \mathcal{F}(x/n)$, a tapered version of the indicator function $\mathbb{1}_{x<n}$ (Fig. 2.2).

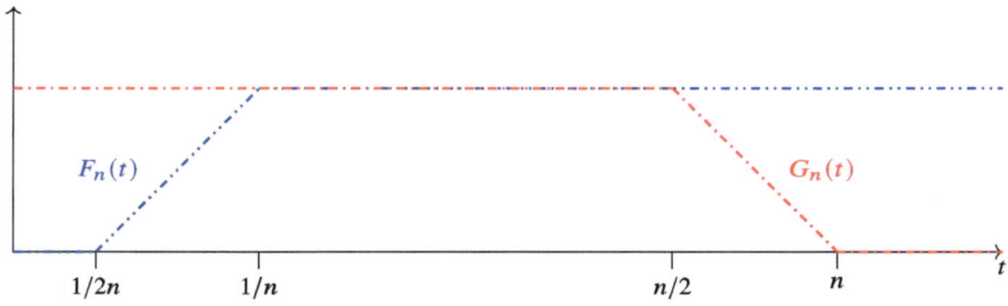

Fig. 2.2 Tapered functions

For the moment consider a compact hyper-rectangle R included in T. Let v be a point in $\mathbb{R}^d$. By application of Proposition 2.9, a.s. there is no root of $X(t) = v$ on the boundary of R, which has dimension $d-1$. The bounding and tapering argument is introduced by considering

$$C_v^n(R) := \sum_{s \in R: X(s)=v} F_n(\Delta(s)) G_n(\Delta(s)), \tag{2.29}$$

$$Q_v^n(R) := C_v^n(R) G_n(C_v^n(R)), \tag{2.30}$$

where dist(s) is the distance of the point s to the boundary of R. In (2.29), when the summation index set is empty, we put $C_v^n(R) = 0$. Let $g : \mathbb{R}^d \mapsto \mathbb{R}^+$ be continuous with compact support. We apply the Area formula, Proposition 2.2 for the function

$$h(t, v) = F_n(\Delta(t)) G_n(\Delta(t)) G_n(C_v^n(R)) g(v)$$

to get that

$$\int_{\mathbb{R}^d} g(v) Q_v^n(R) \mathrm{d}v = \int_R \Delta(t) F_n(\Delta(t)) G_n(\Delta(t))\ G_n(C_{X(t)}^n(R))\ g(X(t)) \mathrm{d}t.$$

Taking expectations on both sides provides

$$\int_{\mathbb{R}^d} g(v)\ \mathbb{E}(Q_v^n(R)) \mathrm{d}v$$
$$= \int_{\mathbb{R}^d} g(v) \mathrm{d}v \int_R \mathbb{E}\big[\Delta(t) F_n(\Delta(t)) G_n(\Delta(t)) G_n(C_v^n(R)) / X(t) = v\big] p_{X(t)}(v) \mathrm{d}t.$$

Note that the conditional expectation is well defined, since the variable is bounded. Since this equality holds for any continuous g with bounded support, it follows that

$$\mathbb{E}(Q_v^n(R)) = \int_R \mathbb{E}\big[\Delta(t)F_n(\Delta(t))G_n(\Delta(t))G_n(C_v^n(R))/X(t) = v\big]p_{X(t)}(v)\mathrm{d}t, \tag{2.31}$$

for almost every v.

Let us prove that the left-hand side of (2.31) is a continuous function of v. Outside the compact set,

$$\{t \in R : \Delta(t) \geq 1/2n; \operatorname{dist}(s) \geq 1/2n\},$$

the contribution to the sum (2.29) defining $C_v^n(R)$ is zero, for any $v \in \mathbb{R}^d$. Let us consider now a particular level u, using the local inversion theorem, the number of points $t \in R$ such that $X(t) = u;\ \Delta(t) \geq 1/2n$, $\operatorname{dist}(s) \geq 1/2n$ say k, is finite and all these points are interior.

If k is non-zero, $X(t)$ is locally invertible in k neighborhoods $V_1, \ldots, V_k \subset R$ around these k points. For v in some (random) neighborhood of u, there is exactly one root of $X(s) = v$ in each V_1, ..., V_k and the contribution to $C_v^n(R)$ of these points can be made arbitrarily close to the one corresponding to $v = u$. Outside the union of V_1, ..., V_k, $X(t) - u$ is bounded away from zero in R, so that the contribution to $C_v^n(R)$ vanishes if v is sufficiently close to u.

Therefore the function $v \mapsto Q_v^n(R)$ is a.s. continuous at $v = u$. On the other hand, it is obvious from its definition that $Q_v^n(R) \leq n$ and an application of the Lebesgue dominated convergence theorem implies the continuity of $\mathbb{E}(Q_v^n(R))$ as a function of v.

Let us look at the right-hand side of (2.31). The key point is that, because of the bounding and tapering argument, irregular points and points on the boundary do not contribute to C_v^n. As a consequence Proposition 2.16 applies to C_v^n without Assumption (**A.2.2**). Hypotheses (ii) and (iii) and the boundedness of the quantities considered due to the presence of G_n imply that

$$\mathbb{E}\big[\Delta(t)F_n(\Delta(t))G_n(\Delta(t))G_n(C_v^n(R))/X(t) = v\big]$$

is a bounded continuous function of v. Since $p_{X(t)}(v)$ is continuous as a function of v and uniformly bounded in t, the right-hand side of (2.31) is continuous and equality in (2.31) holds **for every v** $\in \mathbb{R}^d$.

We can now pass to the limit as $n \to +\infty$, applying the monotone convergence theorem. This gives the equivalent of (2.28) for the number of regular roots in $B = \mathring{R}$. Applying Proposition 2.11, for $D = d$, we get (2.28) for such a B. Since the class of open relatively compact rectangles generates the Borel σ-algebra, and both sides of (2.28) are measures on Borel subsets B of T, then by classical extension argument we get (2.28) for every Borel set B.

2.6.2 Proof of Theorem 2.17 for $D > d$

First note that, under our hypotheses and because of Proposition 2.11, a.s. the level set $\mathcal{L}_u$ is rectifiable. The proof is a direct application of the Crofton formula, Formula (2.10) and Theorem 2.17 for $D = d$. By Fubini's theorem

$$\mathbb{E}\left(\sigma_{D-d}(\mathcal{L}_u(B))\right) = c_{D,D-d} \int_{V \in \mathbb{G}_{D,d}} \mathbb{G}_{D,d}\mathrm{d}(V) \int_{y \in V^\perp} \mathrm{d}\lambda_{D-d}(y)\, \mathbb{E}\#\left\{B \cap \ell_{V,y}\right\},$$

where

$$\mathbb{E}\#\left\{B \cap \ell_{V,y}\right\} = \mathbb{E}\Big(N_u\big(X, B \cap (V + y)\big)\Big).$$

Let V be as above and let π_V be the projector (of dimension $d \times D$) on V. We can parametrize the affine space $(V + y)$ as

$$(V + y) = \{(\pi_V)^\top s + y,\, y \in \mathbb{R}^d\}.$$

Applying the Kac–Rice formula, for $D = d$, and again Fubini's theorem:

$$\begin{aligned}
&\mathbb{E}\left(\sigma_{D-d}(\mathcal{L}_u(B))\right) \\
&= c_{D,D-d} \int_{V \in \mathbb{G}_{D,d}} \mathrm{d}\mathbb{G}_{D,d}(V) \int_B \mathbb{E}\big(|\det(X'(t))(\pi_V)^\top| / X(t) = u\big) p_{X(t)}(u)\mathrm{d}t.
\end{aligned}$$

After a new inversion of integral, we get the result as soon as the following identity is proved:

$$c_{D,D-d} \int_{V \in G(D,d)} |\det(X'(t)(\pi_V)^\top)| \mathrm{d}\mathbb{G}_{D,d}(V) = \Delta(t),$$

which follows directly from the next Proposition 2.19.

We begin with a lemma about the computation of Gaussian determinants.

Lemma 2.18 *Let M be a $D \times d$ standard Gaussian matrix (its entries are i.i.d. N(0,1) r.v.), $D \geq d$. Then*

$$\mathbb{E}(\Delta(M)) = \mathbb{E}\sqrt{\left|\det MM^\top\right|} = 2^{d/2} \frac{\Gamma\left(\frac{D+1}{2}\right)}{\Gamma\left(\frac{D-d+1}{2}\right)} =: L_{D,d}.$$

Proof Let us write $M^\top = [R_1, \ldots, R_d]$. We use an interpretation of $\Delta(M)$ as the d-volume of the parallelotope generated by $R_1, \ldots, R_d$:

$$|\Delta(M)| = \|R_1\| \prod_{j=2}^{d} \mathrm{dist}(R_j, S_{j-1}).$$

With $S_{j-1} = \mathrm{Span}(R_1, \ldots, R_{j-1})$. Because of the invariance properties $\mathrm{dist}(R_j, S_{j-1})$, is distributed as a $\chi(D - j + 1)$ with expectation $\sqrt{2}\frac{\Gamma\left(\frac{D-j+2}{2}\right)}{\Gamma\left(\frac{D-j+1}{2}\right)}$ and all these variables are independent. This yields

$$\mathbb{E}\big(\Delta(M)\big) = 2^{d/2}\frac{\Gamma\left(\frac{D+1}{2}\right)}{\Gamma\left(\frac{D-d+1}{2}\right)} = L_{D,d}. \tag{2.32}$$

□

Proposition 2.19 *Let $M \in \mathbb{R}^{d\times D}$. Then*

$$\sqrt{\left|\det MM^\top\right|} = c_{D,D-d} \cdot \int_{V\in\mathbb{G}_{D,d}} |\det(M(\pi_V)^\top)| d\mathbb{G}_{D,d}(V). \tag{2.33}$$

Proof Using singular value decomposition, measure and the determinant under isometries, we can assume, without loss of generality, that

$$M = \begin{pmatrix} \lambda_1 & & & \\ & \ddots & & 0_{d\times(D-d)} \\ & & \lambda_d & \end{pmatrix},$$

where the λ_i are the singular values of M. Using the linearity of the determinant, we can even assume that $M = \big(\mathrm{Id}_d\ 0_{d\times(D-d)}\big)$ so that (2.33) is proved up to a multiplicative constant. To determine it, we use the particular case of a standard Gaussian matrix M.

Concerning the left-hand side, Lemma 2.18 implies directly that

$$\mathbb{E}(\Delta(M)) = L_{D,d}.$$

As for the second, the rows of $M(\pi_V)^\top$ are the columns of $\pi_V M^\top$ and they consist of d independent standard Gaussian vectors in V. Applying again Lemma 2.18 in V we get

$$\int_{V\in\mathbb{G}_{D,d}} |\det(M(\pi_V)^\top)| d\mathbb{G}_{D,d}(V) = L_{d,d} = \frac{\Gamma\left(\frac{d+1}{2}\right)}{\sqrt{\pi}},$$

giving

$$c_{D,D-d} = \pi^{1/2}\frac{\Gamma\left(\frac{D+1}{2}\right)}{\Gamma\left(\frac{D-d+1}{2}\right)\Gamma\left(\frac{d+1}{2}\right)},$$

which agrees with (2.11). □

2.6.3 Generalization: Manifolds, Higher Moments, Weights

In this section we limit the exposition to the Kac–Rice formula for all levels. Similar statements and proofs exist for the Kac–Rice formula for almost all levels and for the majorization part, but we omit them for brevity.

Theorem 2.17 has a lot of extensions: we can consider (a) the "non-flat case" where the parameter set T is now a D-dimensional manifold as a sphere; (b) the case of the computation of higher moments; (c) adding weights to the surface measure of the level set. Of course all the combinations of (a), (b), and (c) are possible. We give a non-exhaustive list of possible statements.

2.6.3.1 Manifolds

Let M be a C^1 manifold of dimension D embedded in $\mathbb{R}^{D'}$, $D' > D$. Let $X(\cdot)$ be a random field defined on M and taking values in $\mathbb{R}^d$ (or in another d-dimensional Riemannian manifold). For B a Borel set in M, for $u \in \mathbb{R}^d$ we define the level set

$$\mathcal{L}_u := \{t \in B : X(t) = u\}.$$

Roughly speaking this set is again of dimension $D - d$ so it is natural to consider the Hausdorff measure:

$$\sigma_{D-d}(\mathcal{L}_u).$$

Through a partition of unity argument, the "weak" Bulinskaya property can be localized inside a chart and Proposition 2.11 can be extended to our case. As a consequence the level set is a.s. still rectifiable and we can use the scheme of the proof of Theorem 2.17. The main modifications are the following:

- Using the Area formula for manifolds
- Replacing the compact hyper-rectangle R by a compact set which is the adherence of its interior

to get

Theorem 2.20 *The result (the validity of the Kac–Rice formula) of Theorem 2.17 can be extended to a random field defined on a manifold of dimension D and with values in a manifold of dimension d. Define the generalized Jacobian by*

$$\Delta(t) := \sqrt{\det(\nabla X(t) \nabla X(t)^{\top})},$$

where $\nabla X(t)$ is the Riemannian gradient defined by differentiation along curves. Under Assumption ***(A.2.3)***

$$\mathbb{E}\left(\sigma_{D-d}\left(\mathcal{L}_u(B)\right)\right) = \int_B \mathbb{E}\left(\Delta(t) \mid X(t) = u\right) p_{X(t)}(u)\mathrm{d}\sigma(t), \tag{2.34}$$

where σ is the surface measure on S.

Note that in practice the computation of $\det(\nabla X(t))$ must be performed using an orthogonal basis of the tangent space at t.

2.6.3.2 Higher Moments

Theorem 2.21 (Kac–Rice Formula for Higher Moments) *Let $k \geq 2$ be an integer*

Let $X : T \mapsto \mathbb{R}^d$ be a random field, T an open subset of $\mathbb{R}^D$ and $u \in \mathbb{R}^d$ a fixed point. Assume that:

Assumption (A.2.4)

(i) The sample paths of $X(\cdot)$ are a.s. C^1,

*(ii) For **every** $v_1, \ldots, v_k \in \mathbb{R}^d$, for every pairwise different $t_1, \ldots, t_k \in T$, the distribution of $\{X(s), s \in T\}$ conditional on $X(t_1) = v_1, \ldots, X(t_k) = v_k$ is well defined as a probability and is continuous, as a function of $v_1, \ldots, v_k$ with respect to the weak topology on the space $C^1(T)$, see Proposition 2.16.*

(iii) For every pairwise different $t_1, \ldots, t_k \in T$, $X(t_1), \ldots, X(t_k)$ admits a continuous density: $p_{X(t_1),\ldots,X(t_k)}(v_1, \ldots, v_k)$ which is bounded uniformly for $v_1, \ldots, v_k \in (\mathbb{R}^d)^k$ and $t_1, \ldots, t_k$ in any compact subset of $T^k \setminus \mathcal{D}$, where $\mathcal{D}$ is the diagonal in T^k:

$$D = \{(t_1, \ldots, t_k) \in T^k : \exists i \neq j, t_i = t_j\}.$$

Then

- *In the case $D = d$, for every Borel subset of T:*

$$\mathbb{E}\left[\left(N_u(X, B)\right)\left(N_u(X, B) - 1\right) \ldots \left(N_u(X, B) - k + 1\right)\right]$$
$$= \int_{B^k} \mathbb{E}_C\left(\Delta(t_1) \ldots \Delta(t_k)\right)\; p_{(X(t_1),\ldots,X(t_k))}(u, \ldots, u)\mathrm{d}t_1 \ldots \mathrm{d}t_k, \tag{2.35}$$

where $\mathbb{E}_C$ is the expectation concerning the condition

$$C := \{X(t_1), \ldots, X(t_k) = u, \ldots, u\}.$$

- *In the case $D > d$ or when the considered levels $u_1, \ldots u_k$ are pairwise different or when the considered Borel subsets $B_1, \ldots B_k$ are pairwise distinct the Kac–Rice formula of order k reads as*

$$\mathbb{E}\left(\sigma_{D-d}\left(\mathcal{L}_{u_1}(B_1)\right)\ldots\sigma_{D-d}\left(\mathcal{L}_{u_k}(B_k)\right)\right)$$
$$= \int_{\prod B_i} \mathbb{E}_C\left(\Delta(t_1)\ldots\Delta(t_k)\right)\ p_{(X(t_1),\ldots,X(t_k))}(u_1,\ldots,u_k)\mathrm{d}t_1\ldots\mathrm{d}t_k, \tag{2.36}$$

where $\mathbb{E}_C$ *denotes the expectation conditional on* $X(t_1),\ldots,X(t_k) = u_1,\ldots,u_k$.

Proof The result is obtained by applying Theorem 2.17 to the random field

$$Z(t_1,\ldots,t_k) := \left(X(t_1),\ldots,X(t_k)\right), \quad T^k \mapsto \mathbb{R}^{dk}.$$

By additivity, we can limit our attention to the case of pairwise disjoints sets $B_1,\ldots,B_k$. In such a case:

$$\sigma_{D-d}\left(\mathcal{L}_{u_1}(B_1)\right)\ldots,\sigma_{D-d}\left(\mathcal{L}_{u_k}(B_k)\right)$$
$$= \sigma_{k(D-d)}\{(t_1,\ldots,t_k)\in B_1\times\cdots\times B_k : Z(t_1,\ldots,t_k) = (u_1,\ldots,u_k).$$

Applying Theorem 2.17 to the random field $Z(\cdot)$ gives the result. □

To check the hypotheses of Theorem 2.21, we can use a generalization of Proposition 1.17. Its proof is analogous.

Proposition 2.22 *Let* $\{X(t), t \in \mathbb{R}^d\}$ *be a real centered stationary Gaussian process with paths of class* C^k, $k = 0, 1, \ldots$. *Suppose that* $X(\cdot)$ *admits a spectral density* $f(\lambda)$ *such that for any* $\alpha > 0$, *there exist some constant* C_α *such that*

$$0 < f(\lambda) \le C_\alpha \|\lambda\|^\alpha \textit{ for every } \lambda \neq 0.$$

Then, for any choice of distinct parameter values $t_1, \ldots., t_n$, *the joint distribution of the random variables*

$$X(t_1), \ldots., X(t_n), X'(t_1), \ldots, X'(t_n), \ldots., X^{(k)}(t_1), \ldots, X^{(k)}(t_n)$$

does not degenerate.

2.6.3.3 Weights

In the case $D = d$, suppose that we want to count the roots with a non-negative weight $g(t, Z(\cdot))$ that will depend on the location t and on some extra random field $Z(\cdot)$. So we consider

$$G_u(X, B) = \sum_{t\in T:X(t)=u} g(t, Z(\cdot)). \tag{2.37}$$

This can be generalized to the case $D > d$ by considering a weighted measure of the level set:

$$G_u(X, B) = \int_{t \in T : X(t) = u} g(t, Z(\cdot)) \mathrm{d}\sigma_{D-d}(t). \tag{2.38}$$

What kind of weights are admissible? Let us give the most relevant examples:

- Up-crossings: Let $D = d = 1$, $Z(t) = X'(t)$, $g(t, Z(\cdot)) = \mathbb{1}_{Z(t)>0}$.
- Critical points above a level: Let $Y(t), t \in T \subset \mathbb{R}^d$ be a random field with real values and define $X(\cdot)$ as its gradient. The critical points of Y are the zeros of $X(\cdot)$. Their number is denoted by $\mathrm{Crt}(B)$. If we want to count only those above some level u we have to define $Z(\cdot) = Y(\cdot)$, $g(t, Z(\cdot)) = \mathbb{1}_{Z(t)>u}$.
- Critical points with a given index: We define the index $i(M)$ of a matrix as its number of negative eigenvalues. Among the critical values above we may want to count only those that are associated with a Hessian of index k. Their number is denoted by $\mathrm{Crt}^k(B)$. In such a case we have to define $Z(\cdot) = X'(\cdot) = Y''(\cdot)$, $g(t, Z(\cdot)) = \mathbb{1}_{i(Z(t))=k}$.
- Critical points with a given index and above a level: We can mix the two weights, we will skip the details. Their number is denoted by $\mathrm{Crt}^k_{u\uparrow}(B)$.
- Modified Euler Characteristic: let us define

$$\mathrm{EC} := \sum_{k=0}^{d} (-1)^{d-k} \mathrm{Crt}_{k,u}.$$

  If the random field is Morse, this quantity is equal, up to boundary problems, to the Euler Characteristic of the excursion set: $\{t \in T : X(t) > u\}$ (Adler and Taylor 2009).
- Global maxima: In the computation of the distribution of the maximum of a Random field, as in Azaïs and Wschebor (2008) we are interested in the number of global maxima above a certain level u. Note that this number takes the value 0 or 1. In that case $Z(\cdot) = Y(\cdot)$, $g(t, Z(\cdot)) = \mathbb{1}_{Y(t)>u, \forall s \in T : Y(s) \leq Y(t)}$.

We assume the following

Assumption (A.2.5)

(i) g is lower semi-continuous as a function of t;
(ii) g is lower semi-continuous as a function of $Z(\cdot)$ for the weak convergence of continuous functions;
(iii) For **every** $v \in N$, for every $t \in T$, the distribution of $\{Z(\cdot)\}$ conditional on $X(t) = v$ is well defined as a probability and is continuous, as a function of v with respect to the weak convergence of continuous functions.

Theorem 2.23 (Expected Integral on the Level Set) *Let $X : T \mapsto \mathbb{R}^d$ be a random field satisfying assumptions* **(A.2.3)***. To each point t of the parametric space, let us associate a weight $g(t, Z(\cdot))$, where the function g satisfies assumption*

(A.2.5)*. Then using the definition* (2.38),

$$\mathbb{E}\big(G_u(X, B)\big) = \int_B \mathbb{E}\left(\Delta(t) g(t, Z(\cdot)) \;\middle|\; X(t) = u\right) \; p_{X(t)}(u)\mathrm{d}t. \tag{2.39}$$

Remark 2.5 In the case when $X(\cdot)$, $Z(\cdot)$ are jointly Gaussian, (iii) is always met.

Proof Since the function $g(\cdot)$ is a monotone limit of a sequence of continuous function, by a monotone convergence argument, we can assume that $g(\cdot)$ is a continuous function. Hence the conditions above ensure the continuity of both sides of the Kac–Rice formula. □

2.7 Examples

2.7.1 Gaussian Random Fields

Theorem 2.24 *Let $X(\cdot)$ be a Gaussian random field from $T \subset \mathbb{R}^D$ to $\mathbb{R}^d$. Suppose that it has a.s. C^1 paths and that for every $t \in T$, $X(t)$ has a positive definite variance-covariance matrix. Then the conclusion of Theorem 2.17 holds true.*

In addition, if B is compact, both sides are finite and consequently the measure $\mathbb{E}(\sigma_{D-d}(\mathcal{L}_u(B)))$ is Radon.

Proof The random field clearly satisfies conditions (i) and (ii) of Assumption **(A.2.3)**.

It also satisfies the "difficult hypothesis" (iii). Indeed the conditional distribution of $X(\cdot)$ under the condition $X(t) = u$ is well defined by regression formulas for all u. In addition (ii) is equivalent to the fact that the determinant of the variance-covariance matrix of $X(t)$ is bounded-below on every compact set. This implies in turn the continuity in u of the conditional distribution: a variation δ_u of the conditioning causes only a translation of the expectation by a quantity which is $\delta_u f(t)$, where $f(\cdot)$ is a C^1 function. □

2.7.2 Stationary Gaussian Random Fields

This topic is addressed in detail in Sect. 3.4.

There, we establish, first and foremost, the result for every stationary Gaussian random field with C^1 paths and such that the distribution of $X(0)$ is N.D.

2.7.3 Functions of Gaussian Random Fields

Corollary 2.25 *Suppose now that $X(\cdot)$ is not Gaussian but a function of a Gaussian random field $Y(\cdot)$. More precisely*

$$X(t) = H\big(Y(t)\big) \quad t \in T,$$

where T is an open set of $\mathbb{R}^D$. We assume the following for v in a neighborhood of the considered level u.

Assumption (A.2.6)

(a) $Y(\cdot)$ is a Gaussian random field with parameter in $\mathbb{R}^D$, values in $\mathbb{R}^n$, $n \geq d$ with C^1 paths and N.D. distribution, i.e., for all $t \in T$: $\operatorname{Var}(Y(t)) \succ 0$.
(b) $H(\cdot) : \mathbb{R}^n \to \mathbb{R}^d$ is a C^1 function.
(c) For each $v \in R^d$, the irregular set

$$\{y \in H^{-1}(v) : \det(H'(y)H'(y)^\top) = 0\}$$

is empty.

Then the conclusion of Theorem 2.17 holds true.

Proof Without loss of generality, using additivity, we can assume that T is compact. Proposition 2.6 implies that, for all $t \in T$, the density of $X(t)$ at u is

$$p_{X(t)}(u) = \int_{y \in H^{-1}(u)} \frac{\phi_t(y)}{\sqrt{\det(H'(y)H'(y)^\top)}} \, \mathrm{d}\sigma_{n-d}(y).$$

Since $H^{-1}(u)$ is compact, then $\det(H'(\cdot)H'(\cdot)^\top)$ is bounded below on it. Therefore, $p_{X(t)}(v)$ is continuous and bounded. Here, $\phi_t(y)$ is the Gaussian density of $Y(t)$.

Let us check the conditions of Theorem 2.17. (i) and (ii) are clear.

Let us now consider the difficult hypothesis (iii). Let $\mathcal{F}$ be a bounded continuous non-negative functional on $C^1(T)$; our aim is to prove that

$$\mathbb{E}(\mathcal{F} \,|\, X(t) = v) \text{ is continuous as a function of } v.$$

We know that the conditional expectation $\mathbb{E}(\mathcal{F} \,|\, Y(t) = y)$ is well defined and that

$$\mathbb{E}(\mathcal{F}) = \int_{\mathbb{R}^n} \mathbb{E}(\mathcal{F} \,|\, Y(t) = y) p_{Y(t)}(y) \mathrm{d}y.$$

Using the generalized change of variable $y = H(x)$ (i.e. the co-area formula),

$$\mathbb{E}(\mathcal{F}) = \int_{\mathbb{R}^d} \mathrm{d}x \int_{H^{-1}(x)} \frac{\mathbb{E}(\mathcal{F} \,|\, Y(t) = y)}{\Delta(y)} p_{Y(t)}(y) \sigma_{n-d}(\mathrm{d}y),$$

with $\Delta(y) := \sqrt{\det(H'(y)H'(y)^\top)}$.

As a consequence, we choose as a version of the conditional expectation:

$$\mathbb{E}(\mathcal{F} \,|\, X(t) = x) := \frac{1}{p_{X(t)}(x)} \int_{H^{-1}(x)} \frac{\mathbb{E}(\mathcal{F} \,|\, Y(t) = y)}{\Delta(y)} p_{Y(t)}(y)\sigma_{n-d}(\mathrm{d}y).$$

Note that this integral is always well defined. According to the hypothesis, $\mathbb{E}(\mathcal{F} \,|\, Y(t) = y)$ is continuous. By (c), $\Delta(y)$ is continuous and almost everywhere non-zero. Applying Lemma 2.7,

$$y \mapsto \int_{H^{-1}(x)} \frac{\mathbb{E}(\mathcal{F} \,|\, Y(t) = y)}{\Delta(y)} p_{Y(t)}(y)\sigma_{n-d}(\mathrm{d}y)$$

is continuous. □

The classical example is, of course, the case of χ^2 processes:

$$X(t) = \|Y(t)\|^2$$

with $\mathrm{Var}(Y(t)) = I_n$. (a) and (b) are clear. As for (c),

$$H'(y) = 2(y_1, \dots, y_n); \quad \det(H'(y)H'(y)^\top) = 4\|y\|^2 = 4v$$

So if we exclude the non-interesting case $u = 0$, the hypothesis is met.

In the proof of Corollary 2.25, the Gaussianity of $Y(\cdot)$ is not needed: all that we need is the regularity of the density of $Y(t)$ and the regularity of the conditional expectation $\mathbb{E}(\mathcal{F} \,|\, Y(t) = y)$. The result will be generalized in the next section.

2.7.4 Sums of Parametrized Functions

In this section, we consider the following general framework:

$$X(t) = \sum_{i=1}^{n} L(Y_i, t) := H_t(\mathbf{Y}), \qquad \text{with } \mathbf{Y} := (Y_1, \dots, Y_n),$$

where

- $t \in T \subset \mathbb{R}^D$,
- the Y_i's are i.i.d. random variables in $\mathbb{R}^m$ with density $f(\cdot)$,
- $L(\cdot)$ is a function with value in $\mathbb{R}^d$, $d \leq D$ which is C^1 as a function of t and $\mathbf{Y}$,
- $nm \geq D$.

Proposition 2.26 *With the notation and assumptions above, assume that*

Assumption (A.2.7) *For every $t \in T$, $v \in \mathbb{R}^d$, the irregular set*

$$\{\mathbf{Y} \in H_t^{-1}(v) : H_t'(\mathbf{Y}) \textit{ is not of full rank}\}$$

is a.s. empty.

Then the conditions of Theorem 2.17 are fulfilled and the Kac–Rice formula holds true for $u \in \mathbb{R}^d$.

Proof Let us check that assumption **(A.2.3)** (i) and (ii) are implied by our hypotheses: (i) is clear, (ii) follows from the study of the density as in the proof of Theorem 2.25. We still have to establish (iii); to that end, we can follow the lines of the proof of Corollary 2.25. Fix a particular value of the parameter t_0 and consider a bounded continuous non-negative functional $\mathcal{F}$ on $C^1(T)$. We want to define $\mathbb{E}(\mathcal{F} \mid X(t_0) = v)$ as a continuous function of v.

It is clear that

$$\mathbb{E}(\mathcal{F} \mid \mathbf{Y}) = \mathcal{F}(\mathcal{Y}),$$

where $\mathcal{Y}$ is the function

$$t \mapsto H_t(\mathbf{Y}).$$

This functional is continuous as a function of $\mathbf{Y}$ because of the differentiability hypothesis. Now, we chose as a version of the conditional expectation

$$\mathbb{E}(\mathcal{F} \mid X(t_0) = v) := \frac{1}{p_{X(t_0)}(v)} \int_{H^{-1}(v)} \frac{\mathbb{E}(\mathcal{F} \mid \mathbf{Y} = y)}{\Delta(y)} p_{\mathbf{Y}}(y) \sigma_{n-d}(\mathrm{d}y),$$

where H stands for H_{t_0} and $\Delta(y)$ for its gradient. Since by hypothesis $\mathbb{E}(\mathcal{F} \mid \mathbf{Y} = y)$ is continuous, applying Lemma 2.7,

$$v \mapsto \int_{H^{-1}(v)} \frac{\mathbb{E}(\mathcal{F} \mid \mathbf{Y} = y)}{\Delta(y)} p_{\mathbf{Y}}(y) \sigma_{n-d}(\mathrm{d}y)$$

is continuous and consequently $\mathbb{E}(\mathcal{F} \mid X(t_0) = v)$ is a continuous function of v. This proves (iii) and completes the proof, giving:

$$\mathbb{E}(\sigma_{D-d}(\mathcal{L}_u)) = \int_T \mathbb{E}(\Delta_t \mid X(t) = u) p_{X(t)}(u) \, \mathrm{d}t.$$

□

2.7.5 The Mean Number of Critical Points of the Likelihood

An example of a situation corresponding to the above framework is given by the case of the number of critical points of the likelihood. We consider a parametric statistical model. The observation $\mathbf{Y} := (Y_1, \dots, Y_n)$ consists of n i.i.d. observations Y_i that lie in $\mathbb{R}^m$.

$L(Y, \theta)$ is the log-likelihood associated with the observation of one $Y \in \mathbb{R}^m$. We define

$$L(\theta) := \sum_{i=1}^{n} L(Y_i, \theta).$$

We assume that the model is parametric. That is, θ varies in an open set Θ of $\mathbb{R}^d$. Let us denote by $L'(\theta) \in \mathbb{R}^d$ the gradient of $L(\theta)$:

$$L'(\theta) = \frac{\partial}{\partial \theta} L(\theta).$$

We have

$$L'(\theta) := \sum_{i=1}^{n} L'(Y_i, \theta).$$

Setting $\mathcal{L} := L'$, under convenient conditions, we can apply Proposition 2.26 to compute the expectation of the number N of critical points of the likelihood

$$N := \#\{\theta \in \theta : L'(\theta) = 0\}.$$

2.7.6 Gravitational Stochastic Microlensing

This subsection is mainly based on the paper of Petters et al. (2009). Let us consider a gravitational potential $\psi : T \subset \mathbb{R}^2 \mapsto \mathbb{R}$, and define the transformation:

$$\eta(x) = x - \nabla\psi(x). \tag{2.40}$$

A lensed image is defined as a point solution of the equation $\eta(x) = y$.

From now on, we will focus on a type of lens system known as stochastic microlensing. For the definition, let us consider a number N of stars, each located at a random position ξ_j, $j = 1, \dots, N$. In this case, the above quantities are defined as follows:

$$\psi(x) = \kappa_c \frac{|x|^2}{2} - \frac{\gamma}{2}(x_1^2 - x_2^2) + m \sum_{j=1}^{N} \log|x - \xi_j|,$$

$$\eta(x) = (1 - \kappa_c + \gamma)x - 2m \sum_{i=1}^{N} \frac{x - \xi_i}{||x - \xi_i||^2},$$

where $x = (x_1, x_2) \in \mathbb{R}^2 - \{\xi_1, \ldots, x_N\}$, γ is the external shear, in continuous matter with constant density $\kappa_c \geq 0$, m the mass of the stars and N the number of stars, that remain fixed. We also use the condition $(1 - \kappa_c + \gamma) < 0$, which is known as the supercritical case in the physics literature, see Petters et al. (2012, pp. 434). We assume that the positions of the stars are independent random vectors $\xi_j = (U_j, V_j)$ with a density f supported on the ball $D = B_2(0, R)$. The lensed image with y corresponding to the light source is given by the equation

$$\eta(x^*) = y \text{ that can be expressed as } 2m \sum_{i=1}^{N} \frac{x^* - \xi_i}{||x^* - \xi_i||^2} - (1 - \kappa_c + \gamma)x^* = y.$$

We will consider the expectation of the number of images produced by the source y in a region $D \subset \mathbb{R}^2$, which we will denote by $N(D, y)$.

To place this problem within the framework of the previous development, we introduce $L(\Xi, x) : D \subset \mathbb{R}^2 \mapsto \mathbb{R}^2$ through the expression

$$L(\Xi, x) = \sum_{i=1}^{N} \left[\frac{(1 - \kappa_c + \gamma)x}{N} - 2m \frac{x - \xi_i}{||x - \xi_i||^2} \right] \text{ where } \Xi = (\xi_1, \ldots, \xi_m).$$

The function $L(\Xi, x)$ has m singularities at the points $x = \xi_j$, but is a C^∞ function in $\mathbb{R}^2 - \{\xi_1, \ldots, \xi_N\}$. Let $y \in \mathbb{R}^2$, and let $\mathbf{V}$ be a neighborhood of y. Since we have $\lim_{x \to \xi_j} ||\eta(x)|| = +\infty$, all the roots $\eta(x) = y'$ for $y' \in \mathbf{V}$ are located in $\mathbb{R}^2 - \{\xi_1, \ldots, \xi_N\}$. Therefore, in the set $\eta^{-1}(\mathbf{V})$, the function η is C^∞, which confirms condition (i) in the localized version of Theorem 2.17.

Let us consider the following push-forward formula:

$$p_{L(\Xi,x)}(y) = \int_{L^{-1}(\Xi,x)(y)} \frac{1}{\Delta(\Xi, x)} \prod_{j=1}^{N} f(\xi_j) d\xi_j,$$

where

$$\Delta(\Xi, x) = |\det \frac{\partial L(\Xi, x)}{\partial x}|.$$

Hence, to check (ii) and (iii) of Theorem 2.17, we will utilize Proposition 2.26. Let us examine the condition of this proposition. Specifically, we need to analyze the behavior of $\Delta(\Xi, x)$. Hence, we first compute

$$\frac{\partial L(\Xi, x)}{\partial x} =$$

$$(1-\kappa_c+\gamma)I + \begin{bmatrix} -2m\sum_{i=1}^N(\frac{1}{||x-\xi_i||^2} - 2\frac{(x_1-\xi_i^1)^2}{||x-\xi_i||^4}) & 2m\sum_{i=1}^N \frac{(x_1-\xi_i^1)(x_2-\xi_i^2)}{||x-\xi_i||^4} \\ 2m\sum_{i=1}^N \frac{(x_1-\xi_i^1)(x_2-\xi_i^2)}{||x-\xi_i||^4} & -2m\sum_{i=1}^N(\frac{1}{||x-\xi_i||^2} - \frac{(x_2-\xi_i^2)^2}{|||x-\xi_i||^4}) \end{bmatrix}.$$

In this form, we obtain

$$\langle\frac{\partial L(\Xi, x)}{\partial x}v, v\rangle$$

$$= \Big[(1-\kappa_c+\gamma) - 2m\sum_{i=1}^N \frac{1}{||x-\xi_i||^2}]||v||^2 + 2m\sum_{i=1}^N \frac{1}{||x-\xi_i||^4}\langle x-\xi_i, v\rangle^2$$

$$= [(1-\kappa_c+\gamma) - 2m\sum_{i=1}^N \frac{1}{||x-\xi_i||^2}(1-\langle\frac{x-\xi_i}{||x-\xi_i||}, \frac{v}{||v||}\rangle^2)\Big]||v||^2.$$

When $||v|| = 1$, we have

$$\langle\frac{\partial L(\Xi, x)}{\partial x}v, v, \rangle = [(1-\kappa_c+\gamma) - 2m\sum_{i=1}^N \frac{1}{||x-\xi_i||^2}(1-\langle\frac{x-\xi_i}{||x-\xi_i||}, v\rangle^2)].$$

Since $(1-\kappa_c+\gamma) < 0$, for any vector v, it holds that $\langle\frac{\partial L(\Xi,x)}{\partial x}v, v\rangle < 0$. Thus, in this case, we get $\Delta(\Xi, x) > 0$. Therefore, conditions (ii) and (iii) of Theorem 2.17 hold true.

The Kac–Rice formula in this case is written as

$$\mathbb{E}(N(D, y)) = \int_D \mathbb{E}[|\det \frac{\partial L(\Xi, x)}{\partial x}| \,|\eta(x) = y] f_{\eta(x)}(y)\mathrm{d}x, \quad \forall y.$$

Remark 2.6 In Petters et al. (2009, Theorem 4) a version of a similar formula for almost every y has been proved. Our method allows us to obtain, in the supercritical case, the formula **for all** y. We believe that the formula is true for all y also in the subcritical case; however,the preceding procedure cannot be applied.

2.7.7 Shot-Noise Processes

Following Biermé and Desolneux (2012), the shot-noise process $X(\cdot) = \{X(t)\colon t \in \mathbb{R}^d\}$ is defined by

$$X(t) = \sum_i \beta_i g((t-\tau_i)), \tag{2.41}$$

where the β_i's form a sequence of i.i.d. random real variables, the "impulse", and $(\tau_i)_i$ is a Poisson field with intensity (say) 1.

We assume the following

Assumption (A.2.8)

(i) The kernel $g(\cdot)$ is of class C^1 with compact support included in, say, $(-\eta, \eta)^d$
(ii) The impulse β has a finite first moment.
(iii) $\beta_1 g(\mathcal{T})$ has a bounded density, where $\mathcal{T}$ has a uniform distribution over $(-\eta, \eta)^d$.

See Chap. 7 for more details. Define for some Borel set T included in $\mathbb{R}^d$

$$\mathcal{M}_u := \sigma_{d-1}\{t \in T : X(t) = u\}.$$

Let us consider a fixed instant t. Without loss of generality, we can assume that $t = 0$. The number $\mathcal{P}$ of (i)'s in (2.41) follows a Poisson distribution with parameter $(2\eta)^d$ and thus is almost surely finite. Let $u \neq 0$. We can use the law of total probability to write

$$\mathbb{E}(\mathcal{M}_u) = \mathbb{E}(\mathcal{M}_u \mid \mathcal{P} = 1)\mathbb{P}\{\mathcal{P} = 1\} + \cdots + \mathbb{E}(\mathcal{M}_u \mid \mathcal{P} = 1)\mathbb{P}\{\mathcal{P} = p\} + \cdots \tag{2.42}$$

By definition of the Poisson process, conditional on $\mathcal{P} = p$ the τ_i's are i.i.d. with uniform distribution on $(-\eta, \eta)^d$. Each term of the sum can be computed using Proposition 2.26.

Let us check the condition **(A.2.7)** of this proposition. Under the condition $\{\mathcal{P} = p\}$, H_t' is a $(2p \times 1)$ matrix which is of full rank unless all its elements are zero. A sufficient condition to avoid this is, for example, that:

$$\mathbb{P}\{g((\varpi)) \neq 0\} = 1,$$

where ϖ is uniformly distributed on $(-\eta, \eta)^d$. As a conclusion, Proposition 2.26 can be applied conditionally to $\mathcal{P}$.

Happy end: finally, by de-conditioning, we get the usual expression: (2.28) holds true for every $u \neq 0$ as soon as we have defined the conditional expectation by all the processes above.

2.7.8 *The Mean Number of Critical Points of a Gaussian Field Defined on a Manifold*

To avoid heavy notation, we will limit our attention to Gaussian random fields. Let $X(\cdot)$ be a Gaussian field, with C^2 paths, defined on a C^2 manifold M of dimension d as in the preceding section and with real values. Using a convenient parameterization of the different tangent spaces $\mathbb{T}_t M$ (see the details of the proof of the next theorem) we can view the Riemannian derivative $\nabla X(t)$ as a random field $M \mapsto \mathbb{R}^d$. For B a Borel set of M we recall that we have defined the number of critical points as:

$$\text{Crt}(B) := \#\{t \in B : \nabla X(t) = 0\},$$

and in that sense the Kac–Rice formula is a direct consequence of Theorem 2.20. We also present another proof (version 2) that avoids the delicate problem of parameterization of the tangent space $\mathbb{T}_t$.

Theorem 2.27 *Let $X(\cdot)$ be a C^2 Gaussian field defined on a C^2 manifold M of dimension d and taking real values. We define* Crt(B)) *as above. Suppose that for all $t \in M$, $\nabla X(t)$ has an N.D. distribution, then, for every Borel set B in M:*

$$\mathbb{E}(\text{Crt}(B)) = \int_B \mathbb{E}\left(|\det \nabla^2 X(t)| \mid \nabla X(t) = 0\right) p_{\nabla X(t)}(0) \mathrm{d}\sigma_t, \tag{2.43}$$

where σ is the surface measure on M.

This theorem can be generalized by defining $\text{Crt}_{u\uparrow}(B)$, $\text{Crt}_{u\downarrow}(B)$ $\text{Crt}^k(B)$) $\text{Crt}^k_{u\uparrow}(B)$, $\text{Crt}^k_{u\downarrow}(B)$ as the number of critical points above u (resp. beneath u, with index k, with index k above u, with index k beneath u) in B.

By a symmetry argument, the study of the maxima above u and that of the minima under $-u$ are of course equivalent, but we need to consider both of them in order to engage with the existing literature. When the level u is considered we need $X(t)$ and $\nabla X(t)$ to have an N.D. distribution.

We will give one formula: the reader is invited to guess the other statements,

$$\mathbb{E}(\text{Crt}^k_{u\uparrow}(B)) = \tag{2.44}$$
$$\int_B \int_u^\infty \mathbb{E}\left(|\det \nabla^2 X(t) \mathbb{1}_{i(\nabla^2 X(t))=k}| \mid \nabla X(t) = 0, X(t) = x\right) p_{\nabla X(t), X(t)} \mathrm{d}\sigma_t \mathrm{d}x.$$

Proof We will limit our attention to the case of the computation of $\mathbb{E}(\text{Crt}(B))$. Since we are on a manifold some details are needed.

Version 1 The first method is very short but non-elementary. The Riemannian gradient $\nabla X(t)$ as defined in Sect. 2.2 belongs to the tangent space $\mathbb{T}_t$, which is a moving space (tangent bundle). To be able to view it as an element of $\mathbb{R}^d$ we need a "good" parameterization of the different tangent spaces. This can be carried out using the Levi-Civita connection. Once this is done, $\nabla X(t)$ can be viewed as belonging to $\mathbb{R}^d$ and the result is a direct application of Theorem 2.20 to $\nabla X(t)$ in the simplest case. For the other cases we have to add some weights depending on the value of $X(t)$ or on the value of the index and apply Theorem 2.23. Since the considered random field is Gaussian, the only point condition is the non-degeneracy of the distribution of $\nabla X(t)$ or $X(t)$, $\nabla X(t)$ depending on the formula considered.

Version 2 Let $\phi : \mathbb{R}^d \mapsto M$ be a local parametrization onto an open set $\phi(U) \subset M$. We can assume that $B \subset \phi(U)$.

Then $X \circ \phi : \mathbb{R}^d \mapsto \mathbb{R}$, and $(X \circ \phi)' : \mathbb{R}^d \mapsto \mathbb{R}^d$ are random fields on "flat spaces".

Note that the number of critical points of ∇X in $\phi(U) \subset M$ coincides with the number of solutions of $(X \circ \phi)'(s) = 0,\ s \in \mathbb{R}^d$.

Then if the distribution of $\nabla X(t)$ (or equivalently the distribution of $(X \circ \phi)'$) is N.D. Theorem 2.17 implies that

$$\mathbb{E}(\mathrm{Crt}(\phi(U)) = \int_{\mathbb{R}^d} \mathbb{E}\left(|\det(X \circ \phi)''(s)| \mid (X \circ \phi)'(s) = 0\right) \cdot p_{(X\circ\phi)'(s)}(0)\, ds.$$

By the change of variable formula it is easy to check that

$$p_{\nabla X(\phi(s))}(0) = \sqrt{\det\left(\left(\langle \frac{\partial \phi(s)}{\partial s_i}, \frac{\partial \phi(s)}{\partial s_j} \rangle\right)\right)}\, p_{(X\circ\phi)'(s)}(0) \tag{2.45}$$

and that

$$|\det(\nabla^2 F(t_0))| = \det(F \circ \phi)''(t_0)| \cdot \det\left(\left(\langle \frac{\partial \phi(s_0)}{\partial s_i}, \frac{\partial \phi(s_0)}{\partial s_j} \rangle\right)\right)^{-1}$$

where $\phi(s_0) = t_0$ and $\nabla^2 F(t_0)$ is the Riemannian Hessian of F at $t_0 \in M$.

Then

$$\begin{aligned}
\mathbb{E}(\mathrm{Crt}(\phi(U)) &= \int_{\mathbb{R}^d} \mathbb{E}\left(|\det\left((X \circ \phi)''(s)\right)| \mid (X \circ \phi)'(s) = 0\right) \cdot p_{(X\circ\phi)'(s)}(0)\, ds \\
&= \int_{\mathbb{R}^d} \mathbb{E}\left(|\det \nabla^2 X(\phi(s))| \mid \nabla X(\phi(s)) = 0\right) \cdot p_{\nabla X(\phi(s))}(0) \cdot \\
&\qquad \mathrm{vol}\left(\frac{\partial \phi(s)}{\partial s_1}, \dots, \frac{\partial \phi(s)}{\partial s_d}\right) ds \\
&= \int_{t\in\phi(U)} \mathbb{E}\left(|\det \nabla^2 X(t)| \mid \nabla X(t) = 0\right) \cdot p_{\nabla X(t)}(0)\, d\sigma(t),
\end{aligned}$$

where the last equality follows from the fact that the pullback of the volume form $d\sigma(\phi(s))$ in local coordinates is $\sqrt{\det\left(\left(\langle \frac{\partial \phi(s)}{\partial s_i}, \frac{\partial \phi(s)}{\partial s_j} \rangle\right)\right)}\, ds$. □

Examples of the application of this formula are given by papers discussing the complexity of random fields on high-dimension spheres (Auffinger et al. 2013; Auffinger and Ben Arous 2013) and even in a non-Gaussian framework (Maillard et al. 2020). More details are given in Chaps. 3 and 5.

2.8 Finiteness of the Second Moment for Gaussian Stationary Random Fields

This section is a continuation of Sect. 1.9.1, and we therefore assume that $D > 1$. It naturally finds its place in this chapter, though it does use some small results from the next chapter.

Here we will present, with some improvements, certain results of Estrade and Fournier (2016) and Azaïs and León (2020a). As in Sect. 1.9.1, the main message is that under weak conditions (and sometimes under no conditions) the second moment of the measure of the level set is finite. This has some implications for the Hermite representation used in Chap. 4.

2.8.1 *Statement of the Problem*

Let us consider a random field $X(\cdot)$ satisfying the following assumption:

Assumption (A.2.9) $X(\cdot)$ is a C^1 stationary Gaussian field $\mathbb{R}^D \mapsto \mathbb{R}^d, D > 1\ D \geq d$. The distributions of $X(0)$ and $X'(0)$ are N.D.

By a rescaling in space, we can assume without loss of generality that

$$\mathbb{E}[X(t)] = 0 \quad \text{and} \quad \operatorname{Var}(X(t)) = \mathrm{Id}_d,$$

and this will be done below.

The purpose of this section is to give conditions under which the second moment of the measure of the level set

$$\mathbb{E}\Big(\big(\sigma_{D-d}(\mathcal{L}_u(B))\big)^2\Big) \tag{2.46}$$

is finite and to address the problem of its continuity as a function of u. The answer depends on whether $D = d$ or not. This has several applications, in particular for establishing Central limit theorems or Hermite representations, see Chap. 4.

2.8.2 *Case d < D*

In this particular case no additional hypotheses are needed, but for the second assertion we need the set to be sufficiently small:

Theorem 2.28 *Let $X(\cdot)$ be a stationary random field, $\mathbb{R}^D \mapsto \mathbb{R}^d$, $d < D$, satisfying Assumption* **(A.2.9)**. *Then, for every u and every compact set B,*

$$\mathbb{E}\Big(\big(\sigma_{D-d}(\mathcal{L}_u(B))\big)^2\Big) < +\infty.$$

$$\textit{For } B \textit{ sufficiently small, } \mathbb{E}\Big(\big(\sigma_{D-d}(\mathcal{L}_u(B))\big)^2\Big) \textit{ is a continuous function of } u. \tag{2.47}$$

The proof of the theorem uses the easy lemma below.

Lemma 2.29 *Let S and $(Z_n)_n$ be in the same Gaussian space. Assume that $Z_n \to Z$ a.s. or in probability or in $\mathbb{L}^2(\Omega)$ and the random variable Z is N.D. Then*

$$\forall z, \mathbb{E}(S \mid Z_n = z) \to \mathbb{E}(S \mid Z = z),$$
$$\mathrm{Var}(S \mid Z_n) \to \mathrm{Var}(S \mid Z).$$

Proof of Theorem 2.28 By use of the triangle inequality, we see that it is sufficient to establish the finiteness of the second moment for a sufficiently small set B. Set

$$C = \{X(0) = X(t) = u\}, \tag{2.48}$$

and let $\mathbb{E}_C$ denote the expectation conditional on C. We consider the following quantity

$$\mathcal{A}(t, u) = \mathbb{E}_C\big(\Delta(0)\Delta(t)\big). \tag{2.49}$$

To prove the finiteness, using the triangle inequality, we can assume that B and thus t are sufficiently small. We define the Jacobian matrix $X'(0)$ by $X'_{ij} = \frac{\partial X_i}{\partial t_j}(0)$. By applying the Cauchy–Schwarz inequality and symmetry of the roles of 0 and t:

$$\mathcal{A}(t, u) \le \mathbb{E}_C\Big(\det((X'(0))^\top X'(0))\Big).$$

Using Hadamard's inequality for the non-negative definite matrix $M = (M_{i,j})_{ij}$, we know that

$$\det(M) \le M_{1,1} \dots M_{d,d}. \tag{2.50}$$

This gives

$$\mathcal{A}(t, u) \le \mathbb{E}_C\big(\|X'_{:1}\|^2 \dots \|X'_{:d}\|^2\big) = \sum_{1 \le i_1, \dots, i_d \le d} \mathbb{E}_C((X'_{i_1,1})^2 \dots (X'_{i_d,d})^2), \tag{2.51}$$

where $X'_{:i}$ denotes the ith column of the matrix X', see the basic notation at the beginning of the book. We study the expectation of X'. Let $t_i, i \in \mathbb{N}$ such that

$$\frac{t_i}{\|t_i\|} \text{ converges to (say) } \mu \in S^{D-1}.$$

Then using Lemma 2.29

$$\mathbb{E}\big(X'(0) \mid X(0) = X(t_i) = u\big) = \mathbb{E}\Big(X'(0) \mid X(0) = u, \frac{1}{\|t_i\|}\big(X(t_i) - X(0)\big) = 0\Big)$$
$$\to \mathbb{E}\Big(X'(0) \mid X(0) = u, X'(0)\mu = 0\Big) \qquad (2.52)$$

which is bounded because $X(0)$ and $X'(0)$ are independent. By compactness of the unit sphere this implies that $\mathbb{E}_C(X'(0))$ is bounded. Since the conditional expectation is contractive, the conditional variance is bounded and this implies for all i, j,

$$\mathbb{E}_C\big((X'_{ij})^2\big) \leq (\texttt{const}). \qquad (2.53)$$

As a consequence $\mathcal{A}(t, u)$ is bounded.

We can now study the validity of the Kac–Rice formula of order 2. To this end, we will study the joint density

$$p_{X(0),X(t)}(0, 0) = (\texttt{const})\big(\det \mathrm{Var}(X(0), X(t))\big)^{-\frac{1}{2}}.$$

Again here, we use a sequence $t_i, i \in \mathbb{N}$ such that

$$\frac{t_i}{\|t_i\|} \to \mu \in S^{D-1}.$$

Using the fact that a determinant is invariant under adding to some row (or column) a linear combination of the others rows (or columns), (Belyaev 1966), we get

$$\det\Big(\mathrm{Var}(X(0), X(t_i))\Big) = \|t_i\|^{2d} \det\Big(\mathrm{Var}(X(0), \frac{1}{\|t_i\|}\big(X(t_i) - X(0)\big) = 0\Big).$$

Note that we have used Assumption **(A.2.9)**. By the same argument of compactness as above this implies that for t sufficiently small:

$$p_{X(0),X(t)}(0, 0) \asymp \|t\|^{-d}. \qquad (2.54)$$

This implies that the joint density is N.D. and that we can apply the Kac–Rice formula of order 2 (Theorem 2.21) as soon as the considered set B is sufficiently small:

$$\mathbb{E}\Big(\big(\sigma_{D-d}(\mathcal{L}_u(B))\big)^2\Big)$$
$$= \int_{B^2} \mathcal{A}(t-s,u)p_{X(s),X(t)}(u,u)\mathrm{d}s\mathrm{d}t \leq (\texttt{const})|B| \int_B \|t\|^{-d}\mathrm{d}t, \tag{2.55}$$

where $|B|$ is the Lebesgue measure of B. The integrability of $||t||^{-d}$ in $\mathbb{R}^D$ gives the result. The continuity comes from a dominated convergence argument. □

2.8.3 Case $D = d > 1$

In this case, under Assumption **(A.2.9)** and thanks to Proposition 2.11 the level set $\mathcal{L}_u$ is a set of isolated points and its σ_0 measure is just the number of these points. We denote it by $N(u, B)$.

Choose any norm in the space of matrices, trivariant tensors and quadrivariant tensors.

Theorem 2.30 *Suppose that $X(\cdot)$ is a random field, $\mathbb{R}^d \mapsto \mathbb{R}^d$, $d > 1$, satisfying*

- *Assumption* ***(A.2.9)****.*
- *For all $\mu \in S^{d-1}$, $\mu \neq 0$, the vector $X'(t)\mu$ is N.D.*
- $$\int \frac{\|r''(t) - r''(0)\|}{\|t\|^d} \mathrm{d}t \text{ converges at } 0.$$

Then:

- *For all compact $B \subset \mathbb{R}^d$, the second moment of $N_u(B)$ is finite.*
- *For B sufficiently small, the second moment of $N_u(B)$ is continuous as a function of u.*

Note that the covariance $r(\cdot)$ is a $d \times d$ matrix so that $r''(\cdot)$ is a quadrivariant tensor of size d.

This theorem has a direct corollary in the particular case of critical points:

Let $X(\cdot)$ be a stationary Gaussian random field defined on $\mathbb{R}^d$, $d > 1$, with C^2 paths and real values. The critical points are the zeros of the gradient $X'(t)$, which is $\mathbb{R}^d$-valued. This situation is exactly that of Theorem 2.30. We get

Corollary 2.31 *Suppose that*

- *$X(\cdot)$ is Gaussian stationary, centered and has C^2 sample paths.*
- *$X'(t)$ is N.D. For all $\mu \in S^{d-1}$, $\mu \neq 0$, the vector $X'(t)\mu$ has a non-degenerate distribution.*

•

$$\int \frac{\|r^{(4)}(t) - r^{(4)}(0)\|}{\|t\|^d} \mathrm{d}t \textit{ converges at } 0.$$

Then:

- *For all compact $B \subset \mathbb{R}^d$, the second moment of the number of critical points of $X(\cdot)$ included in B is finite.*
- *For B sufficiently small, the second moment of the number of points included in B such that $X'(t) = u$ is continuous as a function of u.*

Proof of Theorem 2.30 We retake the computations of Theorem 2.28. We consider a set B which is the ball B_T with center 0 and radius T, for T sufficiently small. Using the Cauchy–Schwarz inequality, we have to give bounds to

$$\int_{(B_T)^2} \mathbb{E}_C(\det(X'(s))\det(X'(t)))p_{X(0),X(t)}(u,u)\mathrm{d}s\mathrm{d}t$$
$$\leq (\texttt{const}) \int_{B_{2T}} \mathbb{E}_C(\det(X'(0))^2)p_{X(0),X(t)}(u,u)\mathrm{d}t. \tag{2.56}$$

Now we can write the integral in (2.56) in polar coordinates, setting $t = \rho\mu$, $\rho > 0$, $\mu \in \mathbb{S}^{d-1}$. We perform the integral in that order:

$$\int_{\mathbb{S}^{d-1}} \mathrm{d}\sigma(\mu) \int_0^{2T} \mathrm{d}\rho\, \rho^d \big|\det(X''(0))\det(X''(\rho\mu))\big| p_{X'(0),X'(\rho\mu)}(u,u).$$

We see that it is sufficient to establish a uniform bound (in μ) of the radial limits as $\rho \to 0$.

We consider the conditional expectation. Note that ρ and μ appear in the condition C. This condition can be written as $C := \{X(0) = u, \frac{X(\mu\rho)-X(0)}{\rho} = 0\}$.

The $2d \times 2d$ variance-covariance matrix of the conditioners defining C is now

$$\mathrm{Var} := \begin{pmatrix} \Lambda_2 & -\rho K(\rho,\mu) \\ -\rho K(\rho,\mu) & 2K(\rho,\mu) \end{pmatrix}, \tag{2.57}$$

where Λ_2 is the variance-covariance matrix of $X(0)$, and

$$2K(\rho,\mu) = 2\rho^{-2}\big(\mathcal{M} + r'(\mu\rho)\big) = r^{(2)}(0)_{:,:,\mu,\mu} + o(1), \tag{2.58}$$

where $r^{(2)}(0)_{:,:,\mu,\mu}$ is the matrix with (i,j) entry, $i,j = 1,\ldots,d$, equal to $r^{(2)}(0)[e_i, e_j, \mu, \mu]$, e_i being the i-th basis vector.

The covariance between $X'(0)$ and $X(t)$ is given by the tri-variant d^3 tensor $r'(t)$. This implies that the covariance between $X'(0)$ and the conditioners takes the value

$$\mathrm{Cov} := \mathrm{Cov}\Big(X'(0), \big(X(0), \frac{X(\mu\rho) - X(0)}{\rho}\big)\Big) = \Big(0_{d^3}, \frac{1}{\rho}\big(r'(\mu\rho)\big)\Big).$$

We now consider another conditioning, namely with respect to

$$C' := \{X(0) = u, X'(0)\mu = 0\}.$$

It is trivial to see that

$$\mathbb{E}_{C'}(\det^2(X'(0))) = 0.$$

The variance-covariance matrix on the conditioners is now

$$\mathrm{Var}' := \begin{pmatrix} \Lambda_2 & 0 \\ 0 & r^{(2)}(0)_{:,:,\mu,\mu} \end{pmatrix}.$$

Because of the last assumption, this matrix is N.D. The covariance is now

$$\mathrm{Cov}' := \mathrm{Cov}\big(X'(0), \big(X(0), X'(0)\mu\big)\big) = \big(0_{d^3}, r^{(2)}(0)_{:,:,:,\mu}\big),$$

where $r^{(2)}(0)_{:,:,:,\mu}$ is the 3- tensor with i, j, k entry, $i, j, k = 1, \ldots, d$, equal to $r^{(2)}(0)[e_i, e_j, e_k, \mu]$.

The regression formulas imply that

$$\mathbb{E}_C(X'(0)) = \mathrm{Cov}\ \mathrm{Var}^{-1}\begin{pmatrix} u\mathrm{J}_d \\ 0_d \end{pmatrix},$$

$$\mathrm{Var}_C(X'(0)) = r^{(2)}(0) - \mathrm{Cov}\ \mathrm{Var}^{-1}\mathrm{Cov}^\top,$$

$$\mathbb{E}_{C'}(X'(0)) = \mathrm{Cov}'(\mathrm{Var}')^{-1}\begin{pmatrix} u\mathrm{J}_d \\ 0_d \end{pmatrix},$$

$$\mathrm{Var}_{C'}(X'(0)) = r^{(2)}(0) - \mathrm{Cov}'(\mathrm{Var}')^{-1}(\mathrm{Cov}')^\top,$$

where $\mathrm{J}_{1,d}$ is the "all one vector" of size d. The formulas above need some additional details: for the first one, for example, let C_{ijk} be the entries of Cov and $V_{k,k'}$ those of Var^{-1}, then the ij entry of $\mathbb{E}_C(X'(0))$ is

$$u\sum_{k=0}^{2d}\sum_{k'=0}^{d} C_{ijk}V_{kk'}.$$

For the second one, the i_1, i_2, i_3, i_4 entry, $1 \leq i_1, i_2, i_3, i_4 \leq d$ of Cov $\text{Var}^{-1}\text{Cov}^\top$ is

$$\sum_{k,k'} C_{i_1,i_2,k} V_{k,k'} C_{i_3,i_4,k'}.$$

In these formulas the only point that could cause divergence is the inversion of the variance covariance matrices. For Var′, it depends on μ only and, because of the second assumption, it is uniformly, in μ, N.D.

By (2.57) and (2.58) Var converges to Var′ again uniformly by a compactness argument. As a consequence the inversion of variance covariance matrices does not present problems.

We use the following lemma. Its proof is given at the end of this section.

Lemma 2.32 *Let M be some $d \times d$ Gaussian matrix with variance-covariance 4-tensor T and expectation matrix E, then $\mathbb{E}\big(\det^2(M)\big)$ is a locally Lipschitz function of T, E.*

The remarks above imply that $\mathbb{E}_C(\det^2(X'(0)))$, $\mathbb{E}_{C'}(\det^2(X'(0)))$, $\text{Var}_C(X'(0))$, and $\text{Var}_{C'}(X'(0))$ are bounded uniformly in μ. This implies that the locally Lipschitz function $\mathbb{E}\big(\det^2(M)\big)$ is in fact globally Lipschitz, giving

$$\begin{aligned}&\mathbb{E}_C(\det^2(X'(0))) = \mathbb{E}_C(\det^2(X'(0))) - \mathbb{E}_{C'}(\det^2(X'(0)))\\ &\leq (\texttt{const})\|\text{Var}_C(X'(0)) - \text{Var}_{C'}(X'(0))\| + (\texttt{const})\|\mathbb{E}_C(X'(0)) - \mathbb{E}_{C'}(X'(0))\|.\end{aligned}$$

This implies that

$$\mathbb{E}_C(\det^2(X'(0))) \leq (\texttt{const})\|\text{Var} - \text{Var}'\| + (\texttt{const})\|\text{Cov} - \text{Cov}'\|. \tag{2.59}$$

Let us consider the first term on the r.h.s. of (2.59). It is bounded by

$$\Big[2\rho^{-2}\big(r(\mu\rho) - r(0)\big)\Big] - r^{(2)}(0)_{:,:,\mu,\mu}.$$

By application of the Taylor formula with integral remainder, this last term is equal to

$$\int_0^1 W(\xi)\big[r^{(2)}(\xi\mu\rho)_{:,:,\mu,\mu} - r^{(2)}(0)_{:,:,\mu,\mu}\big]\mathrm{d}\xi.$$

where $W(\xi)$ is some weight function.

Let us consider the second term. It is bounded by

$$(\texttt{const})\big(\frac{1}{\rho}\big(r'(\mu\rho)\big) - r^{(2)}(0)_{:,:,:,\mu}\big).$$

By direct integration,

$$\frac{1}{\rho}\big(r'(\mu\rho)\big) - r^{(2)}(0)_{:,:,:,\mu} = \frac{1}{\rho}\big(r'(\mu\rho) - r'(0)\big) - r^{(2)}(0)_{:,:,:,\mu}$$
$$= \int_0^1 r^{(2)}(\xi\mu\rho)_{:,:,:,\mu} - r^{(2)}(0)_{:,:,:,\mu}\, d\xi,$$

giving that the r.h.s. of (2.59) is bounded by

$$\texttt{(const)} \int_0^1 W'(\xi)\|r^{(2)}(\xi\mu\rho)_{:,:,:,\mu} - r^{(2)}(0)_{:,:,:,\mu}\| d\xi$$
$$\leq \texttt{(const)} \int_0^1 \|r^{(2)}(\xi\mu\rho)_{:,:,:,\mu} - r^{(2)}(0)_{:,:,:,\mu}\| d\xi,$$

since $W'(\xi)$ is bounded.

Gathering together the different bounds obtained so far, we get that the l.h.s. of (2.56) is bounded by

$$\texttt{(const)} \int_B \int_0^1 \frac{\|r^{(2)}(\xi t) - r^{(2)}(0)\|}{\|t\|^d} d\xi dt + O(\|t\|^{d-1})$$
$$= \texttt{(const)} \int_0^1 \int_B \frac{\|r^{(2)}(\xi t) - r^{(2)}(0)\|}{\|t\|^d} dt d\xi + O(\|t\|^{d-1}).$$

This concludes the proof of the first assertion.

The same calculation proves the uniform domination of the integrand in the Kac–Rice formula of order 2, which implies the continuity. □

2.8.3.1 Proof of Lemma 2.32

Note that det(M) is the expectation of a polynomial of degree $2d$ in the entries of M. To compute it, we can decompose each entry as the sum of its expectation and a centered random variable. It is well known that the expectation of the product of m Gaussian centered variables (possibly equal) is zero if m is odd, and, if m is even, is given by the Wick formula as a sum over all pairings of products of the $m/2$ covariances associated with each pairing. In conclusion, $\mathbb{E}\big(\det^2(M)\big)$ is a degree $2d$ polynomial in the $d(d+1)/2$ entries of the expectation and in the $d(d+1)(d(d+1)+2)/8$ entries of the variance-covariance matrix. But these details are not really needed: all that we need to know is that this function is a polynomial and hence, a local Lipschitz function.

2.8.4 Necessary and Sufficient Condition

The preceding section gives a nice sufficient condition but says nothing about its necessity. To obtain such a result we are obliged to reinforce our conditions assuming isotropy. More precisely we assume in addition

Assumption (A.2.10) The coordinates $X_i(\cdot)$ of $X(\cdot)$ are independent and isotropic. We set

$$\operatorname{Cov}(X_i(s), X_i(t)) = \mathbf{r}_i(\|s-t\|^2),$$

where $\mathbf{r}_i$ is some function of class C^1 because of Assumption **(A.2.9)**

We set:

$$\sigma^2_{max}(r) := \max_{i=1,\dots,d} \operatorname{Var}\big((X'_{i\lambda}(0)) \bigm| X(0), X(\lambda r)\big).$$

We obtain the following result, extracted from Azaïs and León (2020a).

Theorem 2.33 *Under Assumptions* **(A.2.9)** *and* **(A.2.10)** *the following are equivalent*

(a) For all compact $B \subset \mathbb{R}^d$ *and all* $u \in \mathbb{R}^d$, $\mathbb{E}\big((N(u,B))^2\big)$ *is finite.*
(b) $\mathbb{E}\big((N(u,B))^2\big)$ *is finite for some* u *and some compact* B *with non-empty interior.*

Then

(c) $\displaystyle\int \frac{\sigma^2_{max}(\rho)}{\rho} \mathrm{d}\rho$ *converges at* 0.

2.9 Finiteness of Higher Moments for Gaussian Fields

In a recent paper (Gass and Stecconi 2024) the problem of higher moments in the Gaussian case is addressed using the Kac–Rice formula. Their main results take various forms, and we have chosen to present below their Theorems 1.1 and 1.2.

Theorem 2.34 *Let k,d be positive integers,* T *be an open subset of* $\mathbb{R}^d$ *and* $X(\cdot)$ *be a Gaussian random field from* T *to* $\mathbb{R}^d$ *with a.s.* C^k *sample paths. Assume that for every* $t \in T$, *the random Gaussian vector* $X(t), X'(t), \dots, X^{(k-1)}(t)$ *is non-degenerate. Then, for every compact subset* K *of* T,

$$\mathbb{E}\Big(\big(N_0(X,K)\big)^k\Big) < +\infty.$$

Note that considering only the case $u = 0$ is not restrictive. The extension of this theorem to random fields defined on a manifold is easy, because the statement is local, and we can apply the preceding theorems in a chart. The result can be extended to the case of random fields from $\mathbb{R}^D$ to $\mathbb{R}^d$, $D > d$ (Gass and Stecconi 2024, Theorem 1.5).

This result can be extended to critical points. The non-degeneracy of the gradient random field $X'(\cdot)$ does not hold, by symmetry of the second partial derivatives. This implies that the case of critical points requires a separate proof. Actually, the proof of Gass and Stecconi (2024) relies on Kergin projection and the Kergin projection of a gradient field is the gradient of a polynomial. The non-degeneracy of the Gaussian random field $X(\cdot)$ implies that the Kergin interpolant is non-degenerate on the space of gradients of polynomials, hence the proof on the zeros can be used verbatim for critical points. Kergin interpolation is a natural multivariate extension of the Newton form of classical Hermite polynomial interpolation, introduced by Kergin (1980).

Theorem 2.35 *Let k,d be positive integers, T be an open subset of $\mathbb{R}^d$ and $X(\cdot)$ be a Gaussian random field from T to $\mathbb{R}$ with a.s. C^{k+1} sample paths. Assume that for every $t \in T$, the random Gaussian vector $X(t), X'(t), \dots, X^{(k)}(t)$ is non-degenerate. Then, for every compact subset K of T,*

$$\mathbb{E}\Big(\big(\mathrm{Crt}(X, K)\big)^k\Big) < +\infty.$$

Proof We give a sketch of the proof of Theorem 2.34. First we can consider an arbitrarily small set K and the finiteness of moments is equivalent to that of factorial moments. The non-degeneracy of

$$X(t), X'(t), \dots, X^{(k-1)}(t)$$

implies that locally the distribution of $X(t_1), \dots, X(t_k)$ is non-degenerate (in fact this point is not trivial and is done using polynomial interpolation). So we can write the Kac–Rice formula of order k (Theorem 2.21) to get

$$\begin{aligned} &\mathbb{E}\left[\big(N_0(X, K)\big)\big(N_0(X, K) - 1\big) \dots \big(N_0(X, K) - k + 1\big)\right] \\ &\quad = \int_{K^k} \mathbb{E}_C\left(\Delta(t_1) \dots \Delta(t_k)\right)\ p_{(X(t_1),\dots,X(t_k))}(0, \dots, 0)\mathrm{d}t_1 \dots \mathrm{d}t_k, \end{aligned} \tag{2.60}$$

where $\mathbb{E}_C$ is the expectation conditional on the condition:

$$C := \big\{(X(t_1), \dots, X(t_k)) = (0, \dots, 0)\big\}.$$

The main problem is to study the behavior of the integrand in the neighborhood of the large diagonal

$$\Delta := \big\{t_1, \dots, t_k \in K : \exists i, j \text{ with } i \neq j \text{ s.t. } t_i = t_j\big\}.$$

There exist billions of different ways to tend to this diagonal and a direct study seems very complicated. The construction of some compactification of $K^k \setminus \Delta$ is performed in Ancona and Letendre (2025) but we prefer to follow the route of Gass and Stecconi (2024), which is more elementary. Both approaches are detailed in the recent survey (Nicolaescu 2023b).

Our main tool to give a bound to the integrand in (2.60) is polynomial interpolation spaces. First, when $Y(\cdot)$ lies in a polynomial space of fixed degree, Bezout's theorem implies that the moments of the number of solutions to $Y(t) = 0$ are uniformly bounded.

The second point is the construction of an interpolation $Y(\cdot)$ such that

$$Y_{t_1,\dots,t_n}(t_i) = X(t_i) \text{ and } Y'(t_i) = X'(t_i), \quad \text{for } i = 1, \dots, k,$$

which is continuous as $t := (t_1, \dots, t_k)$ converges to Δ in the space of C^{k-1} functions. This is done using Kergin's interpolation. The last point is a factorization of the integrand based on a Gram–Schmidt orthogonalization of the evaluation linear forms at points t_i

$$\begin{aligned}\rho(t_1, \dots, t_k) : &= \mathbb{E}_C\left(\Delta(t_1) \dots \Delta(t_k)\right)\ p_{(X(t_1),\dots,X(t_k))}(0, \dots, 0)\\ &= R(t_1, \dots, t_k)\sigma_X(t_1, \dots, t_k).\end{aligned}$$

As the notation indicates, $\sigma_X(\cdot)$ depends on the distribution of the random field but $R(\cdot)$ depends only on the linear space $V \subset C^k(T, \mathbb{R}^d)$ in which the random field lies. Moreover, $\sigma_X(\cdot)$ is uniformly bounded above and below in $(t_1, \dots, t_k) \in K^k$.

Putting together these elements, we get the boundedness of the moments. □

2.10 Finiteness of Moments in the Co-dimension 1 Case

In this section, we consider the particular case $d = 1$ and $D > 1$. The level set $\mathcal{L}_u$ is, under general conditions, of co-dimension 1 in $\mathbb{R}^D$. The first two results concern moments of order 1 and 2 for Gaussian stationary random fields while the last one is for a general situation.

2.10.1 A Necessary and Sufficient Condition for Finiteness of the Expectation

Let $X : T \mapsto \mathbb{R}$ be a Gaussian stationary random field, T an open subset of $\mathbb{R}^D$, $D > 1$. We assume that it is non-constant in the sense that $\text{Var}X(0) > 0$. Without loss of generality, we can assume that it is normalized, i.e. $\mathbb{E}(X(t)) = 0$, $\text{Var}(X(t)) = 1$.

We know from Theorem 2.24 that, if the paths are C^1, the expectation of $\sigma_{D-1}(\mathcal{L}_u(B))$ is finite as soon as B is compact, and given by formula (2.63). But in fact, we want to be more general: recall that in dimension 1, ($D = 1$) Proposition 1.15 states that, when the spectral measure is well defined (i.e. the covariance function is continuous), the Kac–Rice formula is always true since both sides of the equation:

$$\mathbb{E}\big(N_u(X, T)\big) = |T|\sqrt{\lambda_2}\frac{\mathrm{e}^{-u^2/2}}{\pi}$$

are simultaneously finite or infinite.

The goal of this section is to generalize the result to $D > 1$ and to prove that, in a broad sense, this formula is always true.

Let Λ_2 be the second spectral moment matrix defined by

$$(\Lambda_2)_{ij} := \int_{\mathbb{R}^D} \lambda_i \lambda_j \mathrm{d}F(\lambda).$$

This matrix may be finite or infinite, infinite meaning by convention that at least one entry is infinite. Define

$$\mathcal{F}(\Lambda_2) = \frac{1}{(2\pi)^{(D+1)/2}} \int_{z\in\mathbb{R}^D} (z^\top \Lambda_2 z)^{1/2} \mathrm{e}^{-\|z\|^2/2} \mathrm{d}\lambda_D(z). \tag{2.61}$$

We want to prove, in all cases, that

$$\mathbb{E}(\sigma_{D-1}(\mathcal{L}_u(B))) = |T|\mathcal{F}(\Lambda_2)\mathrm{e}^{-u^2/2}. \tag{2.62}$$

More precisely:

- Whenever Λ_2 is finite, the r.h.s. of (2.62) is of course finite, and we have to prove that the l.h.s. is equal to the r.h.s.
- If Λ_2 is infinite then we have to prove that both sides of (2.62) are infinite.

When Λ_2 is finite, it is easy to prove that $X(\cdot)$ is differentiable in quadratic mean. If, in addition, the sample paths are almost surely differentiable (which is a little stronger) we have, by Theorem 2.17:

$$\begin{aligned}\mathbb{E}(\sigma_{D-1}(\mathcal{L}_u(B))) &= \lambda_D(T)\mathbb{E}(\|X'(0)\|)\frac{\mathrm{e}^{-u^2/2}}{\sqrt{2\pi}} \\ &= \lambda_D(T)\mathcal{F}(\Lambda_2)\mathrm{e}^{-u^2/2},\end{aligned} \tag{2.63}$$

where $\mathcal{F}(\Lambda_2)$ is given by (2.61) The first equality in (2.63) is the true Kac–Rice formula, the second is due to classical integration.

When Λ_2 is infinite, i.e. when at least one of its entries is infinite, we need to extend the definition of $\mathcal{F}(\Lambda_2)$ by setting it to $+\infty$. If we consider (2.62), both sides are now always well defined, finite, or infinite.

This kind of property has been studied in Wschebor (1982), Wschebor (1985), for $D > 1$ using a particular definition of the measure of the level set though the de Giorgi perimeter. Our proof uses Proposition 1.15 and Crofton's formula. Since the Bulinskaya lemma (Proposition 2.11) cannot be applied we must measure the level set by its D-1-integral geometric measure, setting

$$\bar{m}_u := \mathcal{I}_{D,D-1}(\mathcal{L}_u). \tag{2.64}$$

Theorem 2.36 *Let $X(t)$ be a centered, Gaussian stationary random field $X : \mathbb{R}^D \mapsto \mathbb{R}$. We assume that it is normalized as above. Then, for $u \in \mathbb{R}$, T a compact set of $\mathbb{R}^d$ and $\bar{m}_u$ defined by* (2.64) *we have the equivalence between:*

- $\mathbb{E}(\bar{m}_u) < \infty$.
- *The second spectral moment matrix Λ_2 is finite.*

In such a case we have

$$\mathbb{E}(\bar{m}_u) = |T|\mathcal{F}(\Lambda_2)\mathrm{e}^{-u^2/2}.$$

Proof Since $X(\cdot)$ is almost surely continuous, $\mathcal{L}_u$ is a Borel set on $\mathbb{R}^d$ a.s., and therefore its D-1-integral geometric measure is well defined. By Fubini's theorem and Crofton's formula we obtain that

$$\begin{aligned}&\mathbb{E}(\mathcal{I}_{D,D-1}(\mathcal{L}_u))\\&= c_{D,D-1}\int_{v\in S^{D-1}}\left(\int_{y\in v^\perp}\mathbb{E}(\#\left\{\mathcal{L}_u\cap \ell_{v,y}\right\})\,d\sigma_{D-1}(y)\right)\,\mathrm{d}S^{D-1}(v).\end{aligned}$$

By Proposition 1.15 applied to the process $t \mapsto X(y+tv)$, we obtain

$$\mathbb{E}(\#\left\{\mathcal{L}_u\cap \ell_{v,y}\right\}) = \sigma_1(T\cap \ell_{v,y})\sqrt{v^\top\Lambda_2 v}\,\frac{1}{\pi}\mathrm{e}^{-u^2/2}.$$

Then, $\mathbb{E}(\mathcal{I}_{D,D-1}(\mathcal{L}_u))$ is equal to

$$\mathrm{e}^{-u^2/2}\frac{c_{D,D-1}}{\pi}\cdot\int_{v\in S^{D-1}}\sqrt{v^\top\Lambda_2 v}\left(\int_{y\in v^\perp}\sigma_1(T\cap \ell_{v,y})\,\mathrm{d}\sigma_{D-1}(y)\right)\,\mathrm{d}S^{D-1}(v)$$

and by Fubini's theorem, we obtain that

$$\mathbb{E}(\mathcal{I}_{D-1}(\mathcal{L}_u)) = \mathrm{Leb}_D(T)\mathrm{e}^{-u^2/2}\frac{c_{D,D-1}}{\pi}\int_{S^{D-1}}\sqrt{v^\top\Lambda_2 v}\,\mathrm{d}S^{D-1}(v). \tag{2.65}$$

Integrating in polar coordinates the expression $\mathcal{F}(\Lambda_2)$, given in (2.61), we obtain

$$\mathcal{F}(\Lambda_2) = \frac{1}{(2\pi)^{(D+1)/2}} \int_0^{+\infty} \rho^D \mathrm{e}^{-\rho^2/2}\, \mathrm{d}\rho \int_{v \in S^{D-1}} (v^\top \Lambda_2 v)^{1/2} \mathrm{d}S^{D-1}(v).$$

Furthermore, making the change of variable $u = \rho^2/2$, it is straightforward to show that

$$\mathcal{F}(\Lambda_2) = \frac{c_{D,D-1}}{\pi} \int_{S^{D-1}} \sqrt{v^\top \Lambda_2 v}\, \mathrm{d}S^{D-1}(v), \tag{2.66}$$

and therefore from (2.65) we have

$$\mathbb{E}(\mathcal{I}_{D,D-1}(\mathcal{L}_u)) = \lambda_D(T)\mathrm{e}^{-u^2/2}\mathcal{F}(\Lambda_2). \tag{2.67}$$

We can now consider the two following cases.

- When Λ_2 is finite, the integral on the r.h.s. of (2.67) is finite and therefore we get the desired result in this case.
- When Λ_2 is infinite, we define the linear subspace

$$G(\Lambda_2) := \{v \in \mathbb{R}^D : v^\top \Lambda_2 v < +\infty\}.$$

We prove that $G(\Lambda_2)$ is of dimension strictly smaller than D. Let $v_1, \dots, v_{d_0}$ be a maximal set of linearly independent vectors of $G(\Lambda_2)$. Then by standard linear algebra:

- The space $\mathrm{Span}(v_1, \dots, v_{d_0})$ generated by $v_1, \dots, v_{d_0}$ is in $G(\Lambda_2)$. This implies that $d_0 < D$.
- For every $v \notin \mathrm{Span}(v_1, \dots, v_{d_0})$: $v^\top \Lambda_2 v = +\infty$ (unless $v_1, \dots, v_{d_0}$ is not maximal).
- This implies that $G(\Lambda_2) = \mathrm{Span}(v_1, \dots, v_{d_0})$.

In conclusion the integrand in (2.67) is almost everywhere infinite so the integral is infinite and as a consequence the expectation of the integral geometric measure is infinite.

□

2.10.2 *Higher Moments, Non-stationary and Non-Gaussian Results*

This section is a generalization of the results of Sect. 1.9.2 to random fields with real values. It builds on the paper Armentano et al. (2023).

Let us assume that the parametric space T is of dimension D, $D > 1$, and that the random field $X(\cdot)$ defined on T is real-valued. We study the moments of the $D-1$-dimensional Hausdorff measure m_u of the level set $\mathcal{L} = \mathcal{L}_u$ restricted to $B_D(0,a)$ (the closed ball with radius a centered at the origin)

$$\mathcal{L}_u = \mathcal{L}_u(B_D(0,a)) := \{t \in B_D(0,a) \colon X(t) = u\}.$$

We assume that the random field $X(\cdot)$ has C^1 paths and uniformly bounded density so that the Bulinskaya lemma (Proposition 2.11) applies, implying that the level set $\mathcal{L}_u$ is rectifiable. The Crofton formula takes the form:

$$\sigma_{D-1}(B) = \frac{\pi^{D/2}}{\Gamma(D/2)} a^{D-1} \cdot \int_{v \in S^{D-1}} \int_{y \in v^{\perp} \cap B_D(0,a)} \# \left\{B \cap \ell_{v,y}\right\} \mathrm{d}v_a^{\perp}(y) \, \mathrm{d}S^{D-1}(v). \tag{2.68}$$

Here $\mathrm{d}S^{D-1}$ is the uniform probability on the sphere S^{D-1}, $\mathrm{d}v_a^{\perp}$ is the uniform probability on $v^{\perp} \cap B_D(0,a)$, and $\ell_{v,y}$ is the affine linear space $\{y + tv : t \in \mathbb{R}\}$.

In view of (2.68), to obtain the finiteness of the moments of $\sigma_{D-1}(\mathcal{L}_u)$, the idea is to give conditions on $X(\cdot)$ that ensure that its restriction to an arbitrary line in $\mathbb{R}^D$ satisfies the hypothesis of Theorem 1.22 for some values of m, h and k, as stated in the following result.

Theorem 2.37 *Consider a real-valued random field $X(\cdot) = \{X(t) \colon t \in B_D(0,a)\}$. Assume that $X(\cdot)$ satisfies*

Assumption (A.2.11) *The sample paths of $X(\cdot)$ are $C^k(B_D(0,a))$, $k \geq 2$.*

Assumption (A.2.12) *For some $m = 1, 2, \ldots$ there exists a constant D_m such that*

$$\max_{|v|=1} \mathbb{E}\left(\left\|\frac{\partial^k X}{\partial v^k}\right\|_{\infty}^{m}\right) \leq D_m,$$

where $\frac{\partial^k}{\partial v^k}$ denotes the k-th directional derivative w.r.t. the vector v.

Assumption (A.2.13) *For some $0 \leq h \leq k$ there exists a constant $C > 0$ such that the joint density of*

$$X(t), \frac{\partial X(t)}{\partial v}, \ldots, \frac{\partial^h X(t)}{\partial v^h}$$

is bounded by C uniformly in $t \in B_D(0,a)$, $v \in S^{d-1}$, and $(u_1, \ldots, u_{h+1}) \in \mathbb{R}^{h+1}$. When $h > 0$, we assume in addition that the density of $X(t)$ is bounded uniformly in t.

Then, for $p \geq 1$ satisfying (1.22)*:*

$$p < \left(k - \frac{h}{2} - \frac{1}{1+h}\right)\left(\frac{1}{m} + \frac{1}{1+h}\right)^{-1},$$

the pth moment of the measure $\sigma_{D-1}(\mathcal{L}_u)$ of the level set is finite and is bounded by

$$(c_{D,a})^p \left((k-1)^p + D_m \cdot E_{\alpha,k,p} + C|2a|^{(h+1)(k-h/2)} \cdot D_{\alpha,k,h,p}\right),$$

where $c_{D,a} = \frac{\pi^{D/2}}{\Gamma(D/2)} a^{D-1}$ is the coefficient in (2.68)*, and α and the coefficients $E_{\alpha,k,p}$ and $D_{\alpha,k,h,p}$ are as in Theorem 1.22.*

Proof of Theorem 2.37 We can first apply Jensen's inequality in (2.68):

$$\begin{aligned}(\sigma_{D-1}(\mathcal{L}_u))^p &= \left(c_{D,a} \int_{v \in S^{d-1}} \int_{y \in v^{\perp} \cap B_D(0,a)} \# \left\{C_u \cap \ell_{v,y}\right\} \, \mathrm{d}S^{d-1}(v) \, \mathrm{d}v_a^{\perp}(y)\right)^p \\ &\leq (c_{D,a})^p \int_{v \in S^{d-1}} \int_{y \in v^{\perp} \cap B_D(0,a)} \left(\# \left\{C_u \cap \ell_{v,y}\right\}\right)^p \, \mathrm{d}S^{d-1}(v) \, \mathrm{d}v_a^{\perp}(y).\end{aligned}$$

Now, take expectation and apply Fubini's Theorem,

$$\begin{aligned}&\mathbb{E}\left(\sigma_{D-1}(\mathcal{L}_u)\right)^p \\ &\quad \leq (c_{D,a})^p \int_{v \in S^{d-1}} \int_{y \in v^{\perp} \cap B_D(0,a)} \mathbb{E}\left(\# \left\{C_u \cap \ell_{v,y}\right\}\right)^p \, \mathrm{d}S^{d-1}(v) \, \mathrm{d}v_a^{\perp}(y).\end{aligned}$$

Let us apply Theorem 1.22 to the expectation inside the integral. It is clear that the bound is maximal when the interval is maximal, and it corresponds to $y = 0$. So,

$$\begin{aligned}&\mathbb{E}\left(\sigma_{D-1}(\mathcal{L}_u)\right)^p \\ &\quad \leq (c_{D,a})^p \left((k-1)^p + D_m \cdot E_{\alpha,k,p} + C|2a|^{(h+1)(k-h/2)} \cdot D_{\alpha,k,h,p}\right).\end{aligned}$$

The expectation in the r.h.s. above is finite due to Theorem 1.22. □

An easy consequence is

Corollary 2.38 *Assume that the random field in Theorem 2.37 is Gaussian stationary with C^k paths and a spectral measure with a support admitting an accumulation point. Using the same arguments as in Sect. 1.9.3.2, we establish that the moment of order p of the integral geometric measure of $\mathcal{L}_u$ is finite with $p = \frac{k(k+1)}{2} - 2$.*

Let us recall that the case $p = 2$ and the case of arbitrary p were considered in Azaïs and León (2020a) and in Armentano et al. (2019) respectively.

An extension to the sphere is given in Armentano et al. (2023) and an extension to critical points is given in Sect. 5.6. Examples with shot noise random fields are given in Chap. 7.

2.11 More Results

The paper by Stecconi (2022) considers a more general situation. Consider a random field $X(\cdot)$ between two C^1 manifolds M and N of dimension m and n respectively, and let $W \subset N$ be a codimension m manifold. Consider now

$$N_W(X, B) := \#\{t \in B : X(t) \in W\}.$$

Suppose that $X(t)$ admits a density $p_{X(t)}$ in a certain sub-manifold S_t of N. Then under some condition (KROK) we have

$$\mathbb{E}\big(N_W(X, B)\big) = \int_B \mathrm{d}\sigma(t) \int_{W \cap S_t} \mathbb{E}\big(|\nabla X(t)| \frac{\sigma_u(X, W)}{\sigma_u(S_t, W)} \,\big|X(t) = u\big) P_X(t)(u) \mathrm{d}\sigma(u),$$

where $\sigma_u(X, W)$ and $\sigma_u(S_t, W)$ are angles to be defined, and σ denotes the surface measure in the relevant space.

Note that our result Theorem 2.17 corresponds to the particular case where W is just the singleton $\{u\}$.

Chapter 3
Distributions, Expectations and Moments for Gaussian Processes and Fields

Designed as a comprehensive tool box, this chapter aggregates the technical results required to apply Kac–Rice formulas to Gaussian models. It details the joint distributions, expectations, and higher moments for stationary and non-stationary Gaussian processes and fields. Particular attention is given to isotropic Gaussian fields, providing technical results for random fields defined on both flat spaces and spheres.

3.1 Introduction

As we have shown above, there is a rich body of literature concerning the geometry of Gaussian processes. A bibliographic review has already been given in the preface to this book, so we will not repeat it here. Each of the papers cited contains numerous calculations, some of which are at times a little tedious. This chapter brings together the most common calculations that appear across the papers, and on occasion recalculates them. The proofs for some of these calculations will be given here, but for the most part, we refer our readers to the original papers.

Note that this chapter has not been imagined for linear reading: the specific cases can be consulted as needed and in any order. Likewise, there is almost nothing new in this chapter: its purpose is to bring together and present clearly key calculations.

This chapter is of course geared towards the computation of the elements of the Kac–Rice formula and the use of inequalities such as Arcones' inequality, see Lemma 4.10.

It is divided into four sections: one on stationary Gaussian processes ($d = 1$); one on the non-stationary case; one devoted to stationary Gaussian fields ($d > 1$) and one on the particular case of isotropic Gaussian fields.

D. Armentano et al., *Geometry of Level Sets of Random Fields, Kac–Rice Formulas, Hermite Expansions and Applications*, Probability Theory and Stochastic Modelling 111, https://doi.org/10.1007/978-3-032-29144-8_3

Each of the main sections opens with a passage on the description of distributions, and each concludes by looking at the computation of the elements of the Kac–Rice formula and the computation of the moments of the number of roots or of the moments of the measure of the level set.

3.2 Stationary Gaussian Processes

Let $\{X(t) : t \in \mathbb{R}\}$ be a zero-mean stationary Gaussian process. It satisfies

$$\mathbb{E}\left(X(s)X(t)\right) = c(s,t) = r(s-t). \tag{3.1}$$

Let us recall the Bochner theorem (Theorem 1.10): suppose that $r(\cdot)$ is continuous, then a necessary and sufficient condition for (3.1) to define a valid covariance is for $r(\cdot)$ to be the Fourier transform of a positive measure: **the spectral measure** $\mathrm{d}F$. This measure is symmetric and we can write

$$r(t) = \int_{-\infty}^{+\infty} \mathrm{e}^{\imath\lambda t}\mathrm{d}F(\lambda) = 2\int_{0}^{+\infty} \cos(\lambda t)\mathrm{d}F(\lambda).$$

When this measure admits a density, it is called the spectral density. The moments of the spectral measure, when they exist, are called the spectral moments and are traditionally denoted by $\lambda_0, \lambda_2, \ldots$. **In this section, unlike in other sections and mainly because we will interpret critical points as zeros of the derivative, we don't assume that the variance λ_0 is equal to one.** Note that, by definition, $\lambda_0 < \infty$. We assume, to avoid uninteresting cases, that $\lambda_0 > 0$ and $\lambda_2 > 0$. Note also that the odd moments vanish.

3.2.1 *Differentiability*

Using classical results Feller (1991) see also Azaïs and Wschebor (2009, Exercise 1.4) we have the following properties:

- $X(\cdot)$ is ℓ times differentiable in quadratic mean if and only if the spectral moment of order 2ℓ, $\lambda_{2\ell}$, is finite.
- In such a case we have the following Taylor expansion:

$$r(\tau) = 1 + \sum_{i=1}^{\ell}(-1)^i\lambda_{2i}\frac{\tau^{2i}}{(2i)!} + o(\tau^{2\ell}). \tag{3.2}$$

- If, in addition, the remainder in (3.2) is

$$O\left(\tau^{2\ell}\frac{1}{\log(|\tau|^{1+\delta})}\right), \qquad \delta > 0$$

then using, for example, Adler and Taylor (2009, Th.1.4.2), $X(\cdot)$ admits a version with C^ℓ sample paths.

3.2.2 Distributions

In the sequel we assume, except when explicitly specified, that the process $X(\cdot)$ is of class C^ℓ, $\ell \geq 1$. This is equivalent to the existence of a finite 2ℓth spectral moment. We have for $k = 1, \ldots, \ell$

$$\lambda_0 := \mathrm{Var}(X(t)) = r(0),$$
$$\mathbb{E}\left(X^{(k_1)}(t)X^{(k_2)}(t)\right) = (-1)^{k_2} r^{(2k)}(0) = (-1)^{k+k_2}\lambda_{2k} \text{ for } k_1 + k_2 = 2k \leq 2\ell.$$

The last quantity is null when $k_1 + k_2$ is odd.

Using the Cauchy–Schwarz inequality, it is easy to check that the spectral moments satisfy, for n and m two integers such that $n + m \leq \ell$,

$$\lambda_{2n}\lambda_{2n+2m} \geq \lambda_{2n+m}^2. \tag{3.3}$$

3.2.2.1 One-Point Distribution

The main result is that even derivatives are independent of odd derivatives.

Recall that we have assumed that $\lambda_{2\ell}$ is finite, $\left(X(t), X'(t), \ldots, X^{(\ell)}(t)\right)$ is a zero-mean Gaussian vector with variance-covariance matrix

$$\begin{bmatrix} \lambda_0 & 0 & -\lambda_2 & 0 & \lambda_4 & 0 & \ldots & v_\ell \\ 0 & \lambda_2 & 0 & -\lambda_4 & 0 & \lambda_6 & \ldots & v_{\ell+1} \\ -\lambda_2 & 0 & \lambda_4 & 0 & -\lambda_6 & 0 & \ddots & \vdots \\ 0 & -\lambda_4 & 0 & \lambda_6 & 0 & -\lambda_8 & \ddots & \vdots \\ \lambda_4 & 0 & -\lambda_6 & 0 & \lambda_8 & 0 & \ddots & \vdots \\ 0 & \lambda_6 & 0 & -\lambda_8 & 0 & \lambda_{10} & \ddots & \vdots \\ \vdots & \ddots & \ddots & \ddots & \ddots & \ddots & \ddots & \vdots \\ v_\ell & v_{\ell+1} & \ldots & \ldots & \ldots & \ldots & \ldots & \lambda_{2\ell} \end{bmatrix},$$

with

$$v'_\ell = 0 \text{ when } \ell' \text{ is odd, } v'_\ell = -\lambda'_\ell \text{ when } \ell' = 4k+2 \text{ and } v'_\ell = \lambda'_\ell \text{ when } \ell' = 4k.$$

Put the even derivatives on one side and the odd derivatives on the other side to get

$$\mathrm{Var}(X(t), X''(t), \ldots, X'(t), X^{(3)}(t), \ldots) = V_\ell$$

$$= \begin{bmatrix} \lambda_0 & -\lambda_2 & \lambda_4 & -\lambda_6 & \ldots & 0 & 0 & 0 & \ldots \\ -\lambda_2 & \lambda_4 & -\lambda_6 & \lambda_8 & \ldots & 0 & 0 & 0 & \ldots \\ \lambda_4 & -\lambda_6 & \lambda_8 & -\lambda_{10} & \ldots & 0 & 0 & 0 & \ldots \\ -\lambda_6 & \lambda_8 & -\lambda_{10} & \lambda_{12} & \ldots & 0 & 0 & 0 & \ldots \\ \vdots & \vdots & \vdots & \vdots & \ddots & \vdots & \vdots & \vdots & \vdots \\ 0 & 0 & 0 & 0 & \ldots & \lambda_2 & -\lambda_4 & \lambda_6 & \ldots \\ 0 & 0 & 0 & 0 & \ldots & -\lambda_4 & \lambda_6 & -\lambda_8 & \ldots \\ 0 & 0 & 0 & 0 & \ldots & \lambda_6 & -\lambda_8 & \lambda_{10} & \ldots \\ \vdots & \vdots & \vdots & \vdots & \vdots & \vdots & \vdots & \vdots & \ddots \end{bmatrix}$$

$$= \begin{bmatrix} \mathrm{Hank}(\lambda_0, -\lambda_2, \lambda_4, \ldots) & 0 \\ 0 & \mathrm{Hank}(\lambda_2, -\lambda_4, \lambda_6, \ldots) \end{bmatrix},$$

where $\mathrm{Hank}(c_0, c_1, c_2, \ldots, c_{2p})$ is the $(p+1) \times (p+1)$ Hankel matrix defined by

$$\mathrm{Hank}(c_0, c_1, c_2, \ldots, c_{2p})_{i,j} = c_{i+j-2}, \; 1 \le i, j \le p+1.$$

Generally speaking, there is no easy way to compute the determinant of a Hankel matrix. However, for small values of ℓ we have an explicit expression of $\det(V_\ell)$. For example

$$\det(V_1) = \lambda_0\lambda_2, \quad \det(V_2) = (\lambda_0\lambda_4 - \lambda_2^2)\lambda_2, \quad \det(V_3) = (\lambda_0\lambda_4 - \lambda_2^2)(\lambda_2\lambda_6 - \lambda_4^2),$$

$$\det(V_4) = \frac{1}{\lambda_0}((\lambda_0\lambda_8 - \lambda_4^2)(\lambda_0\lambda_4 - \lambda_2^2) - (\lambda_0\lambda_6 - \lambda_2\lambda_4)^2) \times (\lambda_2\lambda_6 - \lambda_4^2),$$

$$\det(V_5) = \frac{1}{\lambda_0\lambda_2}((\lambda_0\lambda_8 - \lambda_4^2)(\lambda_0\lambda_4 - \lambda_2^2) - (\lambda_0\lambda_6 - \lambda_2\lambda_4)^2)$$
$$\times ((\lambda_2\lambda_{10} - \lambda_6^2)(\lambda_2\lambda_6 - \lambda_4^2) - (\lambda_2\lambda_8 - \lambda_4\lambda_6)^2).$$

3.2.2.2 Two-Point Distribution

We assume that $\lambda_{2\ell}$ is finite with $\ell \ge 2$. We recall that this is equivalent to the fact that $r(\cdot)$ is of class $C^{2\ell}$. Let $s \neq t$ two points and set: $\tau = s - t$. We have for k_1 and

k_2 in $\{1, \ldots, \ell\}$,

$$\mathbb{E}\left(X^{(k_1)}(s)X^{(k_2)}(t)\right) = (-1)^{k_2} r^{(k_1+k_2)}(\tau).$$

The joint distribution of $(X(s), X(t), X'(s), X'(t), X''(s), X''(t))$ is zero-mean Gaussian with variance-covariance matrix given by

$$\begin{pmatrix} \lambda_0 & r(\tau) & 0 & -r'(\tau) & -\lambda_2 & r''(\tau) \\ r(\tau) & \lambda_0 & r'(\tau) & 0 & r''(\tau) & -\lambda_2 \\ 0 & r'(\tau) & \lambda_2 & -r''(\tau) & 0 & r^{(3)}(\tau) \\ -r'(\tau) & 0 & -r''(\tau) & \lambda_2 & -r^{(3)}(\tau) & 0 \\ -\lambda_2 & r''(\tau) & 0 & -r^{(3)}(\tau) & \lambda_4 & r^{(4)}(\tau) \\ r''(\tau) & -\lambda_2 & r^{(3)}(\tau) & 0 & r^{(4)}(\tau) & \lambda_4 \end{pmatrix}.$$

The Taylor expansion of $r(\cdot)$ at zero: $r(\tau) = \lambda_0 - \frac{\tau^2}{2}\lambda_2 + O(\tau^4)$ implies that for τ sufficiently small $|r(\tau)| \neq \lambda_0$, implying that (i) the distribution of $(X(s), X(t))$ is N.D. and (ii) the Kac–Rice formula of order 2 holds true on a sufficiently small set (Theorem 1.12). The conditional distribution of $(X'(s), X'(t))$ given $C := (X(s) = X(t) = u)$ is Gaussian with mean vector

$$\mathbb{E}_C\left(X'(s), X'(t)\right) = \left(\frac{ur'(\tau)}{\lambda_0 + r(\tau)}, \frac{-ur'(\tau)}{\lambda_0 + r(\tau)}\right), \tag{3.4}$$

and variance-covariance matrix

$$\mathrm{Var}_C\left(X'(s), X'(t)\right) := \begin{pmatrix} \lambda_2 - \dfrac{\lambda_0(r'(\tau))^2}{\lambda_0^2 - r(\tau)^2} & -r''(\tau) - \dfrac{r(\tau)(r'(\tau))^2}{\lambda_0^2 - r(\tau)^2} \\ -r''(\tau) - \dfrac{r(\tau)(r'(\tau))^2}{\lambda_0^2 - r(\tau)^2} & \lambda_2 - \dfrac{\lambda_0(r'(\tau))^2}{\lambda_0^2 - r(\tau)^2} \end{pmatrix}. \tag{3.5}$$

The conditional distribution of $(X''(s), X''(t))$ given $X'(s) = X'(t) = 0$ is zero-mean Gaussian with variance-covariance matrix

$$\Sigma := \begin{pmatrix} \lambda_4 - \dfrac{\lambda_2(r^{(3)}(\tau))^2}{\lambda_2^2 - (r''(\tau))^2} & r^{(4)}(\tau) + \dfrac{r''(\tau)(r^{(3)}(\tau))^2}{\lambda_2^2 - (r''(\tau))^2} \\ r^{(4)}(\tau) + \dfrac{r''(\tau)(r^{(3)}(\tau))^2}{\lambda_2^2 - (r''(\tau))^2} & \lambda_4 - \dfrac{\lambda_2(r^{(3)}(\tau))^2}{\lambda_2^2 - (r''(\tau))^2} \end{pmatrix}. \tag{3.6}$$

Again, since by (3.3) $\lambda_4 > 0$, it is well defined if τ is sufficiently small.

3.2.3 Moments

3.2.3.1 First Moment

We recall that when Z is a Gaussian random variable with mean μ and variance σ^2, then

$$\mathbb{E}(Z^+) = \sigma\phi\left(\frac{\mu}{\sigma}\right) + \mu\Phi\left(\frac{\mu}{\sigma}\right), \tag{3.7}$$

$$\mathbb{E}(|Z|) = 2\sigma\phi\left(\frac{\mu}{\sigma}\right) + \mu\left(1 - 2\Phi\left(-\frac{\mu}{\sigma}\right)\right). \tag{3.8}$$

We obtain the following simple expressions for the mean numbers, on the Borel set B, of u-crossings $N_u(B)$, u-upcrossings $U_u(B)$, critical points $\mathrm{Crt}(B)$, critical points with index i, $\mathrm{Crt}^i(B)$, and critical points above the level u with index i, $\mathrm{Crt}^i_{u\uparrow}(B)$.

The first two items of the next proposition simply rephrase Proposition 1.15.

Proposition 3.1

- *Whether λ_2 is finite or not:*

$$\mathbb{E}\left(N_u(B)\right) = 2\mathbb{E}\left(U_u(B)\right) = \frac{|B|}{\pi}\sqrt{\frac{\lambda_2}{\lambda_0}}\exp\left(-\frac{u^2}{2\lambda_0}\right).$$

- *Assume that $\lambda_2 < +\infty$, then whether λ_4 is finite or not:*

$$\mathbb{E}(\mathrm{Crt}(B)) = 2\mathbb{E}(\mathrm{Crt}^1(B)) = 2\mathbb{E}(\mathrm{Crt}^0(B)) = \frac{|B|}{\pi}\sqrt{\frac{\lambda_4}{\lambda_2}}.$$

- *Assume in addition that $\lambda_4 < +\infty$, $\lambda_0\lambda_4 > \lambda_2^2$, then, concerning the number of maxima:*

$$\mathbb{E}(\mathrm{Crt}^1_{u\uparrow}(B))$$
$$= \frac{|B|}{2\pi}\sqrt{\frac{\lambda_4}{\lambda_2}}\bar{\Phi}\left(u\sqrt{\frac{\lambda_4}{\lambda_0\lambda_4 - \lambda_2^2}}\right) + \frac{|B|}{\sqrt{2\pi}}\sqrt{\frac{\lambda_2}{\lambda_0}}\phi\left(\frac{u}{\sqrt{\lambda_0}}\right)\Phi\left(\frac{\lambda_2 u}{\sqrt{\lambda_0(\lambda_0\lambda_4 - \lambda_2^2)}}\right).$$

For the number of minima:

$$\mathbb{E}(\mathrm{Crt}^0_{u\uparrow}(B))$$
$$= \frac{|B|}{2\pi}\sqrt{\frac{\lambda_4}{\lambda_2}}\bar{\Phi}\left(u\sqrt{\frac{\lambda_4}{\lambda_0\lambda_4 - \lambda_2^2}}\right) - \frac{|B|}{\sqrt{2\pi}}\sqrt{\frac{\lambda_2}{\lambda_0}}\phi\left(\frac{u}{\sqrt{\lambda_0}}\right)\bar{\Phi}\left(\frac{\lambda_2 u}{\sqrt{\lambda_0(\lambda_0\lambda_4 - \lambda_2^2)}}\right).$$

Proof The main argument for the last item is that, conditional on $X'(t) = 0$ and $X(t) = x$, $X''(t)$ is distributed as

$$\mathcal{N}\left(-\frac{\lambda_2}{\lambda_0}x, \frac{\lambda_0\lambda_4 - \lambda_2^2}{\lambda_0}\right),$$

where $\mathcal{N}(\mu, \sigma^2)$ is the Gaussian distribution with mean μ and variance σ^2. Then, we deduce

$$\mathbb{E}\left(|X''(t)| \mid X(t) = x, X'(t) = 0\right) = \mathbb{E}\left(|X''(t)| \mid X(t) = x\right)$$

$$= \frac{2\sqrt{\lambda_0\lambda_4 - \lambda_2^2}}{\sqrt{\lambda_0}}\phi\left(\frac{x\lambda_2}{\sqrt{\lambda_0(\lambda_0\lambda_4 - \lambda_2^2)}}\right) + \frac{x\lambda_2}{\lambda_0}\left(2\Phi\left(\frac{x\lambda_2}{\sqrt{\lambda_0(\lambda_0\lambda_4 - \lambda_2^2)}}\right) - 1\right),$$

and

$$\mathbb{E}\left(|X''(t)|\mathbb{1}_{\{X''(t)<0\}} \mid X(t) = x, X'(t) = 0\right) = \mathbb{E}\left(|X''(t)|\mathbb{1}_{\{X''(t)<0\}} \mid X(t) = x\right)$$

$$= \frac{\sqrt{\lambda_0\lambda_4 - \lambda_2^2}}{\sqrt{\lambda_0}}\phi\left(\frac{x\lambda_2}{\sqrt{\lambda_0(\lambda_0\lambda_4 - \lambda_2^2)}}\right) + \frac{x\lambda_2}{\lambda_0}\Phi\left(\frac{x\lambda_2}{\sqrt{\lambda_0(\lambda_0\lambda_4 - \lambda_2^2)}}\right).$$

□

3.2.3.2 Moment of Order 2

The study of the moment of order two of the number of u-crossings presents some numerical difficulties. If for all $s \neq t$ the distribution of $X(s), X(t)$ is N.D., then Theorem 1.12 implies the validity of the Kac–Rice formula of order 2, giving

$$\mathbb{E}\left(N_u([0,T])(N_u([0,T]) - 1)\right) = \int_{[0,T]^2} A_u(s,t)\,\mathrm{d}s\,\mathrm{d}t = 2\int_0^T (T-\tau)A_u(0,\tau)\mathrm{d}\tau, \tag{3.9}$$

with the integrand

$$A_u(s,t) = \mathbb{E}\left(|X'(s)X'(t)| \mid X(s) = X(t) = u\right) p_{X(s),X(t)}(u,u).$$

When s and t are close, the above density tends to infinity, but in most of the cases, this singularity is completely compensated by the expectation term, as shown in Proposition 1.18. Let us recall that it is proven in Proposition 1.16 that the non-degeneracy condition is true for all non-constant stationary random processes when the considered interval is sufficiently small.

We have equivalent formulas for the number of u-upcrossings

$$\begin{aligned}\mathbb{E}\left(U_u([0,T])(U_u([0,T])-1)\right) &= \int_{[0,T]^2} A_u^+(s,t)\,\mathrm{d}s\,\mathrm{d}t \\ &= 2\int_0^T (T-\tau)A_u^+(0,\tau)\mathrm{d}\tau,\end{aligned}$$

with the integrand

$$A_u^+(s,t) = \mathbb{E}\left((X'(s))^+(X'(t))^+ \mid X(s)=X(t)=u\right) p_{X(s),X(t)}(u,u).$$

Before giving Propositions 3.2 and 3.4, we will introduce some notation.

Let m_1, m_2, c be three reals with $|c|<1$. We define

$$J_{11}(m_1,m_2,c) = \int_0^{+\infty}\int_0^{+\infty}$$
$$\frac{xy}{2\pi\sqrt{1-c^2}}\exp\left(-\frac{(x-m_1)^2-2c(x-m_1)(y-m_2)+(y-m_2)^2}{2(1-c^2)}\right)\mathrm{d}y\mathrm{d}x.$$

In Azaïs and Wschebor (2009, pp. 103–104) it is proven that

$$\begin{aligned}J_{11}(m_1,-m_1,c) &= \sqrt{1-c^2}\phi(km_1)\phi(m_1) + 2m_1\phi(m_1)\left(\Phi(km_1)-\frac{1}{2}\right) \\ &\quad + 2\left(c-m_1^2\right)\int_{m_1}^{+\infty}\left[\Phi(kx)-\frac{1}{2}\right]\phi(x)\mathrm{d}x, \qquad (3.10)\end{aligned}$$

where $k := \sqrt{\dfrac{1+c}{1-c}}$. In a general way we can prove that

$$\begin{aligned}J_{11}(m_1,m_2,c) &= \sqrt{1-c^2}\phi(m_1)\phi\left(\frac{cm_1-m_2}{\sqrt{1-c^2}}\right) + m_2\phi(m_1)\bar{\Phi}\left(\frac{cm_1-m_2}{\sqrt{1-c^2}}\right) \\ &\quad + m_1\phi(m_2)\bar{\Phi}\left(\frac{cm_2-m_1}{\sqrt{1-c^2}}\right) + (c+m_1m_2)\int_{-m_2}^{+\infty}\Phi\left(\frac{m_1+cx}{\sqrt{1-c^2}}\right)\phi(x)\mathrm{d}x. \\ &\qquad (3.11)\end{aligned}$$

Proposition 3.2 *If for all $s \neq t \in T = [0,T]$ the distribution of $X(s), X(t)$ is N.D.:*

$$\mathbb{E}\left(U_u([0,T])(U_u([0,T])-1)\right) = \int_0^T 2(T-\tau)A_u^+(0,\tau)\mathrm{d}\tau,$$

where

$$A_u^+(0,\tau) = \Lambda_{11} J_{11}(b(\tau), -b(\tau), c(\tau)) p_{X(0),X(\tau)}(u,u), \tag{3.12}$$

with

$$b(\tau) := \frac{1}{\sqrt{\Lambda_{11}}} \frac{u r'(\tau)}{\lambda_0 + r(\tau)}, \qquad c(\tau) := \frac{\Lambda_{12}}{\Lambda_{11}},$$

and Λ_{ij} *the* (i, j) *entry of the conditional variance-covariance matrix defined in* (3.5). *Moreover, we have*

$$\begin{aligned}\mathbb{E}\left(N_u([0,T])(N_u([0,T])-1)\right) &= \int_{[0,T]^2} A_u(s,t)\mathrm{d}s\mathrm{d}t \\ &= 2\int_0^T (T-\tau) A_u(0,\tau)\mathrm{d}\tau,\end{aligned}$$

where

$$\begin{aligned}A_u(0,\tau) = p_{X(0),X(\tau)}(u,u)\,&[2\Lambda_{11} J_{11}(b(\tau), -b(\tau), c(\tau)) \\ &+ \Lambda_{11} J_{11}(b(\tau), b(\tau), -c(\tau)) + \Lambda_{11} J_{11}(-b(\tau), -b(\tau), -c(\tau))].\end{aligned}$$

In the following lemma we give some equivalents, as $|\tau| \to 0$, of the main quantities appearing in the preceding proposition. They are easily obtained using the same techniques as those used in the proof of Azaïs and Delmas (2022, Lemma 9).

Lemma 3.3 *As* $\tau \to 0$,

$$\Lambda_{11} = \Lambda_{22} \simeq \frac{\tau^2}{4} \frac{(\lambda_0\lambda_4 - \lambda_2^2)}{\lambda_0}, \qquad \Lambda_{12} = \Lambda_{21} \simeq -\frac{\tau^2}{4} \frac{(\lambda_0\lambda_4 - \lambda_2^2)}{\lambda_0},$$

$$\det \Lambda \simeq \frac{\tau^6}{144} \frac{(\lambda_0\lambda_4 - \lambda_2^2)(\lambda_2\lambda_6 - \lambda_4^2)}{\lambda_0\lambda_2}, \qquad b(\tau) \simeq -\frac{u\lambda_2}{\sqrt{\lambda_0(\lambda_0\lambda_4 - \lambda_2^2)}},$$

$$c(\tau) \simeq -1, \qquad 1 - c(\tau)^2 = \frac{\det \Lambda}{\Lambda_{11}^2} \simeq \frac{\tau^2}{9} \frac{\lambda_0(\lambda_2\lambda_6 - \lambda_4^2)}{\lambda_2(\lambda_0\lambda_4 - \lambda_2^2)},$$

$$k(\tau) = \sqrt{\frac{1+c(\tau)}{1-c(\tau)}} = \sqrt{\frac{\det \Lambda}{(\Lambda_{11} - \Lambda_{12})^2}} \simeq \frac{|\tau|}{6} \sqrt{\frac{\lambda_0(\lambda_2\lambda_6 - \lambda_4^2)}{\lambda_2(\lambda_0\lambda_4 - \lambda_2^2)}},$$

$$p_{X(0),X(\tau)}(u,u) \simeq \frac{1}{|\tau| 2\pi \sqrt{\lambda_0\lambda_2}} \exp\left(-\frac{u^2}{2\lambda_0}\right).$$

The following proposition describes the behavior of $A_u^+(0,\tau)$ and $A_u(0,\tau)$ when τ is small.

Proposition 3.4 *Let $\mathcal{X}$ be a real-valued, zero-mean, stationary Gaussian process with a C^8 covariance function r. Then as $\tau \to 0$,*

$$A_u^+(0,\tau) \simeq \frac{\tau^4}{1296} \frac{(\lambda_2\lambda_6 - \lambda_4^2)^{3/2}}{(\lambda_0\lambda_4 - \lambda_2^2)^{1/2}\pi^2\lambda_2^2} \exp\left(-\frac{u^2}{2}\frac{\lambda_4}{(\lambda_0\lambda_4 - \lambda_2^2)}\right). \tag{3.13}$$

$$A_u(0,\tau) \simeq \frac{|\tau|(\lambda_0\lambda_4 - \lambda_2^2)}{8\pi\lambda_0^{3/2}\lambda_2^{1/2}}\left(1 + \frac{u^2\lambda_2^2}{\lambda_0(\lambda_0\lambda_4 - \lambda_2^2)}\right)\exp\left(-\frac{u^2}{2\lambda_0}\right). \tag{3.14}$$

Note that (3.13) is due to Azaïs et al. (1999). It can be found in Azaïs and Wschebor (2009, p. 104) with a proof based on a Maple computation. We give a proof of this result below and we extend it to obtain an equivalent of $A_u(0,\tau)$.

Proof of Proposition 3.4 Here, we will give the proof of (3.13); (3.14) is obtained in the same way. $A_u^+(0,\tau)$ is given by (3.12). It depends on $J_{11}(b(\tau), -b(\tau), c(\tau))$ given by (3.10). Here we denote $b(\tau)$, $c(\tau)$ and $k(\tau)$ by b, c and k for simplicity. We have to obtain the equivalents of the three following terms appearing in (3.10) as $\tau \to 0$,

$$T_1 := \sqrt{1-c^2}\phi(kb)\phi(b),$$

$$T_2 := 2b\phi(b)\left(\Phi(kb) - \frac{1}{2}\right),$$

$$T_3 := 2\left(c - b^2\right)\int_b^{+\infty}\left[\Phi(kx) - \frac{1}{2}\right]\phi(x)\mathrm{d}x.$$

We expand $\phi(kb)$, $\Phi(kb)$ and $\Phi(kx)$ around 0 respectively in T_1, T_2 and T_3,

$$T_1 \simeq \frac{\phi(b)}{\sqrt{2\pi}}\left[\sqrt{1-c^2} - \frac{k^2b^2}{2}\sqrt{1-c^2}\right],$$

$$T_2 \simeq \frac{\phi(b)}{\sqrt{2\pi}}\left(2kb^2 - \frac{2k^3b^4}{3}\right),$$

$$T_3 \simeq \frac{\phi(b)}{\sqrt{2\pi}}\left[2ck - 2kb^2 - \frac{ck^3}{3}(2+b^2) + \frac{b^2k^3}{3}(2+b^2)\right].$$

Summing T_1, T_2 and T_3 we get

$$J_{11}(b,-b,c) \simeq \frac{\phi(b)}{\sqrt{2\pi}}\left[\sqrt{1-c^2}+2ck-\frac{k^2b^2}{2}\sqrt{1-c^2}-\frac{2ck^3}{3}-\frac{cb^2k^3}{3}+\frac{2b^2k^3}{3}\right].$$

Using Lemma 3.3, it is easy to check that

$$-\frac{k^2b^2}{2}\sqrt{1-c^2}-\frac{2ck^3}{3}-\frac{cb^2k^3}{3}+\frac{2b^2k^3}{3}\simeq\frac{2k^3}{3}.$$

Moreover

$$\sqrt{1-c^2}+2ck=\sqrt{\frac{1+c}{1-c}}(1+c)=k^3(1-c)\simeq 2k^3.$$

We thus obtain

$$J_{11}(b,-b,c)\simeq\frac{\phi(b)}{\sqrt{2\pi}}\frac{8k^3}{3}. \tag{3.15}$$

Lemma 3.3 gives

$$p_{X(s),X(t)}(u,u)\simeq\frac{1}{|\tau|2\pi\sqrt{\lambda_0\lambda_2}}\exp\left(-\frac{u^2}{2\lambda_0}\right) \tag{3.16}$$

and

$$\Lambda_{11}\simeq\frac{\tau^2}{4}\frac{(\lambda_0\lambda_4-\lambda_2^2)}{\lambda_0}. \tag{3.17}$$

Using (3.12), (3.15), (3.16), and (3.17), we obtain (3.13). □

In Chap. 5 the functions $A_u^+(s,t)$ and $A_u(s,t)$ will be understood as the correlation functions between up-crossings and crossings of the level u of $X(\cdot)$. Considering the derivative $X'(\cdot)$ and the level $u=0$, Propositions 3.2 and 3.4 can be applied to measure correlation function of critical points in the case $d=1$.

3.3 Non-stationary Gaussian Processes

In this section $\{X(t) : t \in \mathbb{R}\}$ is a real-valued centered Gaussian process with continuously differentiable sample paths and covariance function $c(s,t)$.

3.3.1 Distribution of the Maximum

Here, we will simply give a nice result for non-stationary Gaussian processes that have a constant variance. The variance-covariance matrix of $X(s), X(t), X'(s), X'(t)$ takes the value

$$\begin{pmatrix} c(s,s) & c(s,t) & c_{0,1}(s,s) & c_{0,1}(s,t) \\ c(s,t) & c(t,t) & c_{0,1}(t,s) & c_{0,1}(t,t) \\ c_{1,0}(s,s) & c_{1,0}(s,t) & c_{1,1}(s,s) & c_{1,1}(s,t) \\ c_{1,0}(t,s) & c_{1,0}(t,t) & c_{1,1}(t,s) & c_{1,1}(t,t) \end{pmatrix},$$

where $c_{i,j}(s,t) := \dfrac{\partial^{i+j} c(s,t)}{\partial s^i \partial t^j}$.

If we assume that the variance is constant equal to 1, $c(t,t) \equiv 1$, the distribution of $X'(s), X'(t)$ conditional on $C := \{X(s) = X(t) = u\}$ has for expectation (Azaïs and Wschebor 2009, p. 96).

$$\mathbb{E}_C(X'(s), X'(t)) = u\left(\frac{c_{0,1}(s,t)}{1+c(s,t)}, \frac{c_{0,1}(t,s)}{1+c(s,t)}\right).$$

and

$$\mathrm{Var}_C(X'(s)) = c_{1,1}(s,s) - \frac{c_{1,0}^2(s,t)}{1-c^2(s,t),}$$

$$\mathrm{Var}_C(X'(T)) = c_{1,1}(t,t) - \frac{c_{1,0}^2(t,s)}{1-c^2(s,t)}.$$

Using this result we have Proposition 4.1 of Azaïs and Wschebor (2009):

Proposition 3.5 *For $T > 0$ let $M_T = \sup_{t\in[0,T]} X(t)$. Then,*

$$\mathbb{P}\{M_T > u\} \leq \frac{e^{-u^2/2}}{2\pi} \int_0^T \sqrt{c_{11}(t,t)}\mathrm{d}t + 1 - \Phi(u). \tag{3.18}$$

Assume that the covariance is C^4, that $c(s,t) \neq 1$ for $s, t \in [0,T]$, $s \neq t$, and $c_{11}(s,s) > 0$ for all $s \in [0,T]$. Then Azaïs and Wschebor (2009, p. 94), if $u > 0$:

$$\mathbb{P}\{M_T > u\} = \frac{e^{-u^2/2}}{2\pi} \int_0^T \sqrt{c_{11}(t,t)}\mathrm{d}t + 1 - \Phi(u) + O\big(\phi(u(1+\delta))\big)$$

for some positive real number δ.

3.3.2 Higher Moments

We present here Azaïs and Wschebor (2009, Proposition 5.10). We consider k points $t_1, \ldots, t_k$. We set, when it exists,

$$A_0^+(t_1, \ldots, t_k) := p_{X(t_1),\ldots,X(t_k)}(0, \ldots, 0)$$
$$\mathbb{E}\left(X'(t_1)\mathbb{1}_{X'(t_1)>0} \ldots X'(t_k)\mathbb{1}_{X'(t_k)>0} \,\middle|X(t_1) = \cdots = X(t_k)=0\right).$$

Proposition 3.6 *Suppose that $X(\cdot)$ is a real-valued, zero-mean Gaussian process with C^{2k-1} paths and that for each pairwise distinct values of the parameters $t_1, t_2, \ldots, t_k$, the joint distribution of $(X(t_i), X'(t_i), \ldots, X^{(2k-1)}(t_i), i = 1, 2, \ldots, k)$ is non-degenerate. Then, as $t_1, t_2, \ldots, t_k \to t^\star$,*

$$A_0^+(t_1, \ldots, t_k) \simeq J_k(t^\star) \prod_{1\le i<j\le k} (t_j - t_i)^4,$$

where

$$J_k(t) = \frac{2! \ldots (2k-2)!}{(2\pi)^k[(2k-1)!]^{2k-1}} \frac{I_k(1)}{[D_{2k-1}(t)]^{1/2}} \left(\frac{D_{2k-1}(t)}{D_{2k-2}(t)}\right)^k,$$

with $D_\ell(t) = \det\left(\text{Var}\left(X(t), X'(t), \ldots, X^{(l)}(t)\right)\right)$ and $I_k(1)$ given by

$$I_k(1) = \int_{(\mathbb{R}^+)^k} y_1 \ldots y_k \exp\left(-\frac{1}{2}\left(\sum_{i=1}^k y_i\right)^2\right) \mathrm{d}y_1 \ldots \mathrm{d}y_k.$$

Remark 3.1 In the stationary case, when $k = 2$ and $u = 0$, the preceding proposition allows us to prove (3.13).

3.4 Stationary Gaussian Fields

In this section we assume that $\{X(t) : t \in \mathbb{R}^d\}$ is a real-valued stationary Gaussian field with covariance function

$$\mathbb{E}(X(s)X(t)) = c(s, t) = r(s - t).$$

Except when we explicitly specify otherwise, we assume until the end of the chapter that the **considered random field is normalized** in the sense that for all t

$$\mathbb{E}(X(t)) = 0, \quad \text{Var}(X(t)) = 1.$$

3.4.1 Bochner's Theorem

As in the case $d = 1$, we assume that r is continuous, which is a minimal hypothesis.

We say that a function $k : \mathbb{R}^d \to \mathbb{R}$ is positive definite if

$$\sum_{i=1}^{n}\sum_{j=1}^{n} k(x_i - x_j)\gamma_i\overline{\gamma}_j \geq 0, \tag{3.19}$$

for any collection of vectors $\{x_i\}_{i=1}^n$, $x_i \in \mathbb{R}^d$ and scalars $\{\gamma_i\}_{i=1}^n$, $\gamma_i \in \mathbb{C}$. This is obviously the case for the function $r(\cdot)$. Below we will prove Bochner's theorem, which states that every continuous function with $k(0) = 1$ is the characteristic function of a probability measure defined on $\mathbb{R}^d$. Our proof adapts the proof given in Feller (1991, pp. 621–623) for the multidimensional case.

We recall that the characteristic function $\hat{\mu}$ of a probability measure μ is given by

$$\hat{\mu}(t) = \int_{\mathbb{R}^d} \mathrm{e}^{\imath\langle t,x\rangle} \mathrm{d}\mu(x).$$

To prove the theorem we need some lemmas.

Lemma 3.7 *Let $k(\cdot)$ be an integrable and bounded continuous function. In this case, we can use a Fourier inversion formula, defining*

$$f(x) = \frac{1}{(2\pi)^d}\int_{\mathbb{R}^d} \mathrm{e}^{-\imath\langle x,t\rangle} k(t)\mathrm{d}t.$$

It holds that $k(\cdot)$ is the characteristic function of a probability measure if and only if $k(0) = 1$ and $\forall x : f(x) \geq 0$.

Proof Let us first prove that the conditions are necessary. If $k = \hat{\mu}$, with $\mu(\mathbb{R}^d) = 1$, then $k(0) = 1$. Moreover, as k is integrable, the inversion formula shows that the function f is the density of μ, implying $f(x) \geq 0$.

In order to prove sufficiency, let g be a probability density defined on $\mathbb{R}^d$ such that $\hat{g}(x) \geq 0$, for instance the density of the standard Gaussian vector, thus

$$\begin{aligned}
\int_{\mathbb{R}^d} \hat{g}(hx)\mathrm{e}^{\imath\langle x,a\rangle} f(x)\mathrm{d}x &= \frac{1}{(2\pi)^d}\int_{\mathbb{R}^d}\int_{\mathbb{R}^d} \hat{g}(hx)\mathrm{e}^{\imath\langle x,a-t\rangle} k(t)\mathrm{d}t\mathrm{d}x \\
&= \frac{1}{(2\pi)^d}\int_{\mathbb{R}^d}\int_{\mathbb{R}^d} \hat{g}(y)\mathrm{e}^{\imath\langle y,\frac{a-t}{h}\rangle} k(t)\mathrm{d}y\frac{\mathrm{d}t}{h^d} \\
&= \int_{\mathbb{R}^d} g\left(\frac{t-a}{h}\right) k(t)\frac{\mathrm{d}t}{h^d}.
\end{aligned}$$

By using that g is a density we obtain

$$|\int_{\mathbb{R}^d} \hat{g}(hx)e^{\imath\langle x,a\rangle} f(x)dx| \leq ||k||_\infty.$$

Furthermore,

$$\int_{\mathbb{R}^d} \hat{g}(hx)f(x)dx = \int_{\mathbb{R}^d} g\left(\frac{t}{h}\right)k(t)\frac{dt}{h^d}.$$

The integrand on the left-hand side is positive and by Fatou's lemma we get

$$\int_{\mathbb{R}^d} f(x)dx \leq \limsup_{h\to 0}\int_{\mathbb{R}^d} g\left(\frac{t}{h}\right)k(t)\frac{dt}{h^d} \leq ||k||_\infty.$$

As a consequence, as $k(\cdot)$ is a continuous function,

$$\int_{\mathbb{R}^d} e^{\imath\langle x,a\rangle} f(x)dx = \lim_{h\to 0}\int_{\mathbb{R}^d} g\left(\frac{t-a}{h}\right)k(t)\frac{dt}{h^d} = k(a).$$

□

Now we do not assume that $k(\cdot)$ is integrable anymore. Consider the function $k(t)e^{-\frac{\varepsilon^2||t||^2}{2}}$. We have

$$0 \leq f_\varepsilon(x) = \int_{\mathbb{R}^d} f(x-y)\frac{e^{-\frac{||y||^2}{2\varepsilon^2}}}{(2\pi\varepsilon^2)^{\frac{d}{2}}}dy = \frac{1}{(2\pi)^d}\int_{\mathbb{R}^d} e^{-i\langle x,t\rangle}k(t)e^{-\frac{\varepsilon^2||t||^2}{2}}dt.$$

Then, applying the preceding result we get that $k(t)e^{-\frac{\varepsilon^2||t||^2}{2}}$ is the characteristic function of a probability measure. Making $\varepsilon \to 0$, the Levy continuity theorem gives that k is the characteristic function of a probability measure.

Theorem 3.8 *A continuous function k is positive definite if and only if it is the characteristic function of a probability measure.*

Proof Necessity:

$$\sum_{i=1}^{n}\sum_{j=1}^{n} k(x_i - x_j)\gamma_i\overline{\gamma}_j = \int_{\mathbb{R}^d} |\sum_{j=1}^{n} e^{\imath\langle x_j,x\rangle}\gamma_j|^2 d\mu(x) \geq 0.$$

Sufficiency: let us introduce the discrete probability measure $\nu_n = \sum_{j=1}^n p_j\delta_{x_j}(t)$ and its symmetrized measure $\nu_n^{sim} = \nu_n * \tilde{\nu}_n$, where $\tilde{\nu}_n(t) = \nu(-t)$. Computing the convolution we get

$$\nu_n^{sim}(t) = \sum_{i=1}^{n}\sum_{j=1}^{n} p_i p_j \delta_{x_i - x_j}(t).$$

Thus, computing the characteristic function, we obtain

$$\sum_{i=1}^{n}\sum_{j=1}^{n} k(x_i - x_j) \mathrm{e}^{-i\langle x, x_i - x_j\rangle} p_i p_j = \int_{\mathbb{R}^d} \mathrm{e}^{-\imath\langle x,t\rangle} k(t) \mathrm{d}\nu_n^{sim}(t) \geq 0.$$

The density of discrete measures in the space of measures implies that the above relation holds for any symmetric measure, in particular for the centered Gaussians. This form yields the following:

$$\int_{\mathbb{R}^d} \mathrm{e}^{-\imath\langle y,t\rangle} k(t) \mathrm{e}^{-\frac{\varepsilon^2 \|t\|^2}{2}} \mathrm{d}t \geq 0.$$

The preceding lemma gives the result whenever k is bounded. In the next lemma we prove that this is always the case. □

Lemma 3.9 *For any positive definite function k it holds that*

$$k(0) \geq 0, \qquad |k(x)| \leq k(0),$$

and $k(-x) = \overline{k(-x)}$.

Proof We use (3.19) with $n = 2$ and $x_2 = 0$ and $\gamma_2 = 1$. Hence we get

$$k(0)(1 + |\gamma_1|^2) + k(x_1)\gamma_1 + k(-x_1)\overline{\gamma_1} \geq 0.$$

Letting $\gamma_1 = 0$ we obtain $k(0) \geq 0$. If $\gamma_1 > 0$ the above inequality turns into

$$k(0)(1 + |\gamma_1|^2) + (k(x_1) + k(-x_1))\gamma_1 \geq 0,$$

then $k(x_1) = \overline{k(-x_1)}$. Finally, if $k(0) \neq 0$ and by taking $\gamma_1 = -\frac{\overline{k(x_1)}}{k(0)}$ we get

$$k(0) - \frac{|k(x_1)|^2}{k(0)} \geq 0,$$

and the announced result follows, because $k(0) = 0$ implies $k \equiv 0$. □

By way of conclusion, Bochner's theorem implies that $r(\cdot)$ is the Fourier transform of a symmetric positive measure, the spectral measure that will be denoted F,

$$r(s - t) = \int_{\mathbb{R}^d} \mathrm{e}^{\imath\langle s-t,\lambda\rangle} \mathrm{d}F(\lambda).$$

3.4.2 Distributions

We begin with a lemma on the computation of the joint law of the derivatives.

Lemma 3.10 *Let $\{X(t) : t \in \mathbb{R}^d\}$ be a real-valued zero-mean stationary Gaussian field defined on $\mathbb{R}^d$. Let $(i_1, \ldots, i_d)$ and $(j_1, \ldots, j_d)$ be two multi-indices. We set $|i| := i_1 + \ldots + i_d$. When $|i| + |j|$ is even and when all the quantities hereunder are well defined*

$$\mathbb{E}\left(\frac{\partial^{|i|}X(t)}{\partial t_1^{i_1}, \ldots, \partial t_d^{i_d}} \frac{\partial^{|j|}X(t)}{\partial t_1^{j_1}, \ldots, \partial t_d^{j_d}}\right) = \left.\frac{\partial^{|i|+|j|}c(s,t)}{\partial s_1^{i_1}, \ldots, \partial s_d^{i_d} \partial t_1^{j_1}, \ldots, \partial t_d^{j_d}}\right|_{s=t}$$

$$= (-1)^{|j|} \left.\frac{\partial^{|i|+|j|}r(x)}{\partial x_1^{i_1+j_1} \ldots \partial x_d^{i_d+j_d}}\right|_{x=0} = (-1)^{|j|}(-1)^{\frac{|i|+|j|}{2}} \int_{\mathbb{R}^d} \lambda_1^{i_1+j_1} \ldots \lambda_d^{i_d+j_d} \mathrm{d}F(\lambda)$$

$$=: (-1)^{|j|}(-1)^{\frac{|i|+|j|}{2}} \lambda_{(i_1+j_1),\ldots,(i_d+j_d)} \cdot \tag{3.20}$$

When $|i| + |j|$ is odd, the covariances vanish.

Remark 3.2 Here, as in Sect. 3.2.1 for $d = 1$, we have some relations between the existence of moments and differentiability. Consider the 2ℓ tensor $\Lambda_{2\ell}$ of moments whose entries are all the spectral moments of order 2ℓ. The random field $X(\cdot)$ is ℓ times differentiable in quadratic mean if and only if this tensor is finite. Sample paths' differentiability demands a slightly stronger control of the Taylor expansion of the function $r(\cdot)$ at zero.

From Lemma 3.10 we deduce the conditional distribution of $X''(t)$ given $X(t) = x$ and $X'(t) = 0$.

Lemma 3.11 *Let $\{X(t) : t \in \mathbb{R}^d\}$ be a real-valued zero-mean stationary Gaussian field defined on $\mathbb{R}^d$. We define $\Lambda_2 := \mathrm{Var}\left(X'(t)\right)$, the matrix of second-order spectral moments*

$$\lambda_{ij} = \int_{\mathbb{R}^d} \lambda_i \lambda_j \mathrm{d}F(\lambda),$$

and

$$\lambda_{ijkl} := \mathbb{E}\left(X''_{ij}(t)X''_{kl}(t)\right) = \int_{\mathbb{R}^d} \lambda_i \lambda_j \lambda_k \lambda_l \mathrm{d}F(\lambda).$$

Under the condition $X(t) = x$ and $X'(t) = 0$, $X''(t)$ is Gaussian with mean

$$\mathbb{E}_C(X''(t)) = -x\Lambda_2$$

and variance-covariance matrix

$$\mathbb{E}_C(X''_{ij}(t)X''_{kl}(t)) = \lambda_{ijkl} - \lambda_{ij}\lambda_{kl}. \tag{3.21}$$

Let us set $\Gamma := X''(t) + x\Lambda_2$. Conditionally on $X(t) = x$, Γ is Gaussian centered with variance-covariance matrix given by (3.21). To compute the mean number of local maxima above a level u on a subset T, given in Theorem 3.15 below, we will have to calculate $\mathbb{E}\left(\det(\Gamma - x\Lambda_2) \mid X(t) = x\right)$. The result is given without proof in Lemma 3.14. To obtain this result we need to compute the expectation of determinants and sub-determinants of some particular Gaussian matrices given in Definition 3.1. These expectations are obtained below in Lemma 3.12. These results can be interesting in their own right. For more details about the proofs we refer to Delmas (2001, pages 47–48) or Azaïs and Delmas (2002).

Definition 3.1 Let Λ be a symmetric $d \times d$ matrix. A random $d \times d$ matrix Γ is said to belong to the class $M(\Lambda)$ if:

- Γ is a Gaussian matrix,
- $\mathbb{E}(\Gamma) = 0$,
- $\exists (S_{ijkl})_{i,j,k,l}$ symmetric in i, j, k, l (that is to say, $S_{ijkl} = S_{p(i)p(j)p(k)p(l)}$ for all permutations p of i, j, k, l) such that

$$\mathbb{E}(\Gamma_{ij}\Gamma_{kl}) = S_{ijkl} - \lambda_{ij}\lambda_{kl},$$

where $\Lambda = (\lambda_{ij})_{1 \le i,j \le d}$.

Then the following is true.

Lemma 3.12 *Let I and J be two subsets of $\{1, 2, \ldots, d\}$. For $\Gamma \in M(\Lambda)$ and m an integer*

- $B^\top \Gamma B \in M(B^\top \Lambda B)$*, for all fixed $d \times d$ matrices B,*
- $\mathbb{E}(\Delta_{I,J}(\Gamma)) = 0$ *when* $\#I = \#J = 2m + 1$,
- $\mathbb{E}(\Delta_{I,J}(\Gamma)) = \dfrac{(-1)^m (2m)!}{m! 2^m} \Delta_{I,J}(\Lambda)$ *when* $\#I = \#J = 2m$,

where $\Delta_{I,J}(\Lambda) = \det(\Lambda_{I,J})$ and $\Lambda_{I,J}$ is the $(\#I) \times (\#J)$ submatrix of Λ composed of its rows I and columns J.

Remark 3.3 The proof of this lemma can be found in Delmas (2001, page 45) and Azaïs and Delmas (2002). It is a generalization of Adler (1981, Lemma 5.3.2).

3.4.3 Moments

We consider here a random field $\{X(t) : t \in \mathbb{R}^D\}$ taking values in $\mathbb{R}^d$, such that

$$\mathbb{E}(X(t)) = 0, \quad \mathrm{Var}(X(t)) = \mathrm{Id}_d.$$

Lemma 3.13 *Suppose that $X(\cdot) : B \subset \mathbb{R}^D \to \mathbb{R}^d$ with $D > 1$ is a zero-mean Gaussian stationary random field with a.s. C^1 paths and non-degenerate variance-covariance matrix. Then a.s. there are no irregular roots.*

(a) *In addition the Kac–Rice formula of order 1 holds true in the sense of* (2.28)*:*

$$\begin{aligned}\mathbb{E}\left(\sigma_{D-d}\left(\mathcal{L}_u(B)\right)\right) &= \int_B \mathbb{E}\left(\Delta(t) \mid X(t) = u\right) p_{X(t)}(u)\mathrm{d}t \\ &= |B|\,\mathbb{E}\left(|\det(X'(0))|\right) p_{X(0)}(u).\end{aligned} \tag{3.22}$$

In addition if B is compact, the expectation is finite.

(b) *Moreover if $(X(0), X(t))$ is non-degenerate for all $t \neq 0$, then we have the Kac–Rice formula of order 2:*
When $D > d$

$$\begin{aligned}\mathbb{E}\left(\sigma^2_{D-d}\left(\mathcal{L}_u(B)\right)\right) &= \int_B \mathrm{d}s \int_B \mathrm{d}t \\ \mathbb{E}\left(|\det(X'(s))\det(X'(t))| \mid X(s) = X(t) = u\right) &p_{X(s),X(t)}(u,u).\end{aligned} \tag{3.23}$$

In the particular case when $B = [0, T]^D$,

$$\begin{aligned}\mathbb{E}\left(\sigma^2_{D-d}\left(\mathcal{L}_u(B)\right)\right) &= 2^D \int_{[0,T]^D} \mathrm{d}v_1 \ldots \mathrm{d}v_D \prod_{i=1}^{D}(T - v_i) \\ \mathbb{E}\left(|\det(X'(0))\det(X'(v))| \mid X(0) = X(v) = u\right) &p_{X(0),X(v)}(u,u).\end{aligned} \tag{3.24}$$

When $D = d$ the l.h.s. above must be replaced by the factorial moment:

$$\mathbb{E}\left(N_u(X, B)(N_u(X, B) - 1)\right).$$

(c) *Suppose in addition to (b) that the d coordinates $X_1(\cdot), \ldots, X_d(\cdot)$ of $X(\cdot)$ are independent and that the spectral moments matrices Λ_2^k defined by*

$$\left(\Lambda_2^k\right)_{i,j} := \mathbb{E}\Big(\frac{\partial X_k(t)}{\partial t_i}\frac{\partial X_k(t)}{\partial t_j}\Big) \text{ for } i, j = 1, \ldots, D,\ k = 1, \ldots, d \tag{3.25}$$

are non-degenerate, then the Kac–Rice formulas of order 2 (3.23) *and* (3.24) *hold true for a sufficiently small set B.*

Remark 3.4 When $D = d = 1$ we have already seen in Sect. 1.9.1 that the Kac–Rice formula of order 1 always holds true as soon as the variance is non-zero. When $0 < \lambda_2 < +\infty$ we have seen in Proposition 1.16 that the Kac–Rice formula of order 2 always holds true for a sufficient small interval T. This is generalized to higher dimension in Lemma 3.13.

Proof Since the random field is stationary, $X(t)$ and $X'(t)$ are independent. This directly gives the first result in (b). The second one follows from the change of variables $v_i = s_i - t_i$ for $i = 1, \ldots, d$. For (c), we need to prove that the distribution of $X(0), X(t)$ is N.D. for t sufficiently small. Since the coordinates are independent, it is sufficient to prove that for every k, $X_k(0), X_k(t)$ is N.D. Since the random field has been assumed to be normalized, a Taylor expansion gives

$$\mathbb{E}\big(X_k(0)X_k(t)\big) = 1 - \frac{1}{2}t^\top \Lambda_2^k t + O(|t|^2).$$

This implies that for $t \neq 0$, sufficiently small, the quantity above differs from ± 1. As a consequence $X(0), X(t)$ is N.D. □

3.4.4 Local Maxima Above a High Level and Modified Euler Characteristic

Let $\{X(t) : t \in \mathbb{R}^d\}$ be a real-valued zero-mean stationary Gaussian field with C^2-paths and unit variance. We denote by $M_u^X(T)$ the number of local maxima above the level u belonging to the subset T of $\mathbb{R}^d$

$$M_u^X(T) := \#\{t \in T \subset \mathbb{R}^d : X(t) \geq u, X'(t) = 0, i\left(X''(t)\right) = d\},$$

where $X'(\cdot)$ and $X''(\cdot)$ denote respectively the gradient and the Hessian of $X(\cdot)$ and $i(X''(t))$ is the index (the number of negative eigenvalues) of $X''(t)$.

We recall the notation $\Lambda_2 := \operatorname{Var}\left(X'(t)\right)$ and $H_i(\cdot)$ is the ith Hermite polynomial (see Sect. 4.4 for more details).

Lemma 3.14 *Let $\{X(t) : t \in \mathbb{R}^d\}$ be a real-valued zero-mean stationary Gaussian field with C^2-paths and unit variance. We assume $(X(t), X'(t), X''(t))$ has a non-degenerate distribution. Using the notation above*

$$\mathbb{E}\left(\det(X''(t)) \,\middle|\, X(t) = x, X'(t) = 0\right) = (-1)^d \det(\Lambda_2) H_d(x)\,.$$

Proof The proof of this lemma can be found in Delmas (2001, page 48) and Azaïs and Delmas (2002). It is an extension of Adler (1981, Corollary p. 110). □

Theorem 3.15 *Let $\{X(t) : t \in \mathbb{R}^d\}$ be a real-valued zero-mean stationary Gaussian field defined on $\mathbb{R}^d$ with C^2-paths and unit variance. We assume that $\forall t \neq s$ $\left(X(s), X(t), X'(s), X'(t), X"(s), X"(t)\right)$ has a non-degenerate distribution. Using the notation above for $\delta > 0$ and $u \to +\infty$*

$$\mathbb{E}\left(M_u^X(T)\right) = \frac{|T|\sqrt{\det(\Lambda_2)}}{(2\pi)^{(d+1)/2}} H_{d-1}(u) \exp\left(-\frac{u^2}{2}\right) + O\big(\phi\left((1+\delta)u\right)\big). \quad (3.26)$$

Moreover

$$\mathbb{E}\left(M_u^X(T)(M_u^X(T)-1)\right) = O\big(\phi\left((1+\delta)u\right)\big). \tag{3.27}$$

In the particular case $T = [0, L]^d$, *we have for* $\delta > 0$ *and* $u \to +\infty$*:*

$$\mathbb{P}\left(\max_{t\in T} X(t) \geq u\right) = \sum_{n=1}^{d}\left(\sum_{I\in\mathcal{F}_n(d)} \Delta_{I,I}^{1/2}(\Lambda_2)\right)$$
$$\frac{L^n}{(2\pi)^{(n+1)/2}}\exp\left(-\frac{u^2}{2}\right)H_{n-1}(u) + \bar{\Phi}(u) + O\big(\phi\left((1+\delta)u\right)\big), \tag{3.28}$$

where $\mathcal{F}_n(d)$ *is the set of all subsets of* $\{1,\ldots,d\}$ *of size n and* $\Delta_{I,I}(\Lambda_2)$, *as defined in Lemma 3.12.*

Proof See Delmas (2001) and Azaïs and Delmas (2002). Our assumptions here are weaker because the assumptions for the Kac–Rice formula are weaker. □

Remark 3.5 Hasofer (1976a) and Adler (1981) give a one-term approximation for $\mathbb{E}\left(M_u^X(S)\right)$ when $u \to +\infty$ for stationary Gaussian fields. The more accurate asymptotic expansion (3.26) is due to Delmas (1998) and the expansion (3.28) is due to Delmas (2001) and Azaïs and Delmas (2002). Note that (3.26) and (3.28) are written in a slightly different way in Delmas (1998), Delmas (2001), and Azaïs and Delmas (2002) but it is easy to check that the expressions are the same as those given in the previous theorem.

Remark 3.6 An evaluation of δ has been carried out in Azaïs and Wschebor (2009, Theorem 8.15).

Recall that in Sect. 2.6.3.3 we defined $\mathrm{EC}(T)$, the modified Euler characteristic of the excursion set $E(T,u) := \{t \in T : X(t) \geq u\}$, as

$$\mathrm{EC}(T) := \sum_{k=0}^{d}(-1)^k \mathrm{Crt}_{u\uparrow}^{d-k}(T) \qquad \text{where}$$

$$\mathrm{Crt}_{u\uparrow}^{d-k}(T) := \#\{t \in T : X(t) \geq u, X'(t) = 0, i\left(X''(t)\right) = d-k\}.$$

The following lemma can be deduced from Lemma 3.14.

Lemma 3.16 *Let* $\{X(t) : t \in \mathbb{R}^d\}$ *be a real-valued zero-mean stationary Gaussian field with* C^2*-paths and unit variance. We assume that* $\forall t$, $(X(t), X'(t), X''(t))$ *has a non-degenerate distribution. Using the notation above, we have*

$$\mathbb{E}\left(\mathrm{EC}(T)\right) = \frac{|T|\sqrt{\det(\Lambda_2)}}{(2\pi)^{(d+1)/2}} H_{d-1}(u)\exp\left(-\frac{u^2}{2}\right).$$

3.4.5 Mean Length of Level Curves

We assume that $\{X(t) : t \in \mathbb{R}^2\}$ is a real-valued, zero-mean stationary Gaussian field defined on $\mathbb{R}^2$. We define $\mathcal{L}_u(X, B)$, the u-level curve contained in a Borel set $B \in \mathbb{R}^2$ as

$$\mathcal{L}_u(X, B) := \{t \in B : X(t) = u\}.$$

Proposition 3.17 *Assume that the moment matrix* $\Lambda_2 = \text{Var}(X'(t))$ *is positive definite. Then, with the above notation,*

$$\begin{aligned}\mathbb{E}\left(\sigma_1(\mathcal{L}_u(X, B))\right) &= |B| p_X(u) \mathbb{E}\left(\|X'(0)\|\right) \\ &= |B| \sqrt{\frac{2}{\pi}} \phi(u) \sqrt{\mathbf{R}_2} \varepsilon(k),\end{aligned}$$

where $k^2 := (1 - \mathbf{R}_1/\mathbf{R}_2)$ *with* $\mathbf{R}_2 > \mathbf{R}_1$ *the eigenvalues of* Λ_2 *and*

$$\varepsilon(k) := \int_0^{\pi/2} \left(1 - k^2 \sin^2 \theta\right)^{1/2} \mathrm{d}\theta$$

is the complete elliptic integral of the second kind.

The proof of this proposition can be found in Azaïs and Wschebor (2009, Theorem 11.3 p. 289).

This theorem has a direct generalization to random fields $\mathbb{R}^d \to \mathbb{R}$, $d > 2$. In that case we find the $d-1$-dimensional measure of the level set and the result is less explicit:

$$\mathbb{E}\left(\sigma_{d-1}(\mathcal{L}_u(X, B))\right) = |B| \phi(u) \mathbb{E}\left(\|X'(0)\|\right),$$

where $X'(0)$ is a Gaussian vector with variance-covariance matrix Λ_2.

3.5 Isotropic Gaussian Fields

3.5.1 Distributions in the "Flat Case"

The term "flat case" is used here in contrast to "spherical case". In the spherical case we consider random fields on the sphere whereas in the flat case we consider random fields defined on $\mathbb{R}^d$.

In this section $X(\cdot) = \{X(t) : t \in \mathbb{R}^d\}$ is a real-valued centered stationary and isotropic Gaussian field. Its covariance function satisfies

$$\mathbb{E}\big(X(s)X(t)\big) = c(s,t) = \mathbf{r}(\|s-t\|^2) = \rho(\|s-t\|). \tag{3.29}$$

The spectral moments defined by (3.20) are related when they have the same degree. For example, when all the considered quantities are finite, when $d = 2$

$$\lambda_{2,0} = \lambda_{0,2} =: \lambda_2 = -2\mathbf{r}'(0) = -\rho''(0).$$

In the general case, we define

$$\lambda_{2k} := \lambda_{2k,0,\cdots,0} = \lambda_{0,2k,0,\cdots,0} = \lambda_{0,\cdots,0,2k,0,\cdots,0}$$
$$= (-1)^k \frac{(2k)!}{k!} \mathbf{r}^{(k)}(0) = (-1)^k \rho^{(2k)}(0)$$

and we have, see Azaïs and Delmas (2022, Lemme 1)

$$\lambda_{2k_1,\ldots,2k_d} = \lambda_{2k} \frac{k!}{(2k)!} \prod_{\ell=1}^{d} \frac{(2k_\ell)!}{k_\ell!},$$

with $k = k_1 + \cdots + k_d$. Note that, due to isotropy, all the spectral moments with an odd index are zero. A necessary and sufficient condition for differentiability in quadratic mean of order ℓ is $\lambda_{2\ell} < +\infty$.

When λ_{2k} is finite, the spectral moments satisfy for $n = 2, \ldots, k$

$$\lambda_{2n}\lambda_{2n-4} \geq \frac{(2n-1)}{(2n-3)} \frac{(2n-4+d)}{(2n-2+d)} \lambda_{2n-2}^2. \tag{3.30}$$

Remark 3.7 Note that the coefficients

$$K(n,d) := \frac{(2n-1)}{(2n-3)} \frac{(2n-4+d)}{(2n-2+d)}$$

are greater than 1 when $d > 1$ and take the value 1 when $d = 1$.

A particular class of covariances merits particular consideration, namely Schoenberg covariances.

Proposition 3.18 *A real function* $\mathbf{r}(\cdot)$ *is a Schoenberg covariance if* (3.29) *defines a valid covariance in any dimension. A characterization of such covariance was given by Schoenberg (1938):* $\mathbf{r}$ *with* $\mathbf{r}(0) = 1$ *is a Schoenberg covariance if and only if there exists a probability measure G on* $[0, +\infty)$ *such that*

$$\mathbf{r}(x) = \int_0^{+\infty} \mathrm{e}^{-xw} G(\mathrm{d}w) \quad \textit{for all } x \geq 0. \tag{3.31}$$

Definition 3.2 A covariance function satisfying Proposition 3.18 is called a Schoenberg covariance.

Lemma 3.19 *A Schoenberg covariance satisfies $\lambda_4 \geq 3\lambda_2^2$ when these quantities are well defined. Equality holds if and only if $\mathbf{r}(t) = \mathrm{e}^{-a^2 t}$. In that case, $\mathbb{E}(X(0)X(t)) = \mathrm{e}^{-a^2\|t\|^2}$.*

3.5.1.1 One-Point Distribution

Assume $\lambda_4 < \infty$. This implies that $X(\cdot)$ is twice differentiable in quadratic mean. Below, we give a description of the distribution of $X(t)$, $X'(t)$, $X''(t)$.

Definition 3.3 We say that a symmetric $d \times d$ matrix G has a d-GOE distribution if

(i) it is centered Gaussian,
(ii) all its $d(d+1)/2$ free elements are independent,
(iii) its diagonal elements have variance 1 and its off-diagonal elements have variance 1/2.

Proposition 3.20 (Azaïs and Wschebor 2008, Azaïs and Delmas 2022) *Let $X(\cdot)$ be a Gaussian isotropic random field with $\lambda_4 < \infty$, then*

- *$X''(t)$ follows a $\sqrt{\frac{2\lambda_4}{3}}(G - Z\mathrm{Id}_d)$ distribution where G is a d-GOE matrix and Z is a $\mathcal{N}(0, 1/2)$ random variable independent of G.*
- $\mathrm{Cov}(X(t), X''(t)) = -\lambda_2 \mathrm{Id}_d$.
- *When $\lambda_4 \geq 3\lambda_2^2$, the distribution of $X''(t)$ conditional on $X(t) = x$ is*

$$\sqrt{\frac{2\lambda_4}{3}}\left(G - \tilde{Z}\mathrm{Id}_d\right) - \lambda_2 x \mathrm{Id}_d,$$

where $\tilde{Z}$ is an $\mathcal{N}\left(0, \frac{\lambda_4 - 3\lambda_2^2}{2\lambda_4}\right)$ random variable independent of G.

- *$X'(t)$ is independent of $(X(t), X''(t))$ and has for variance-covariance $\lambda_2 \mathrm{Id}_d$.*

3.5.1.2 Two-Point Distribution

Assume again $\lambda_4 < \infty$ and set $\tau := \|t - s\|$. For the sake of brevity, in the sequel, $\mathbf{r}(\tau^2)$, $\mathbf{r}'(\tau^2)$, $\mathbf{r}''(\tau^2)$, $\mathbf{r}'''(\tau^2)$ and $\mathbf{r}^{(4)}(\tau^2)$ are denoted respectively by $\mathbf{r}$, $\mathbf{r}'$, $\mathbf{r}''$, $\mathbf{r}'''$ and $\mathbf{r}^{(4)}$. We first give a lemma that describes the covariances between $X(0)$, $X'(0)$, $X''(0)$ and $X(t)$, $X'(t)$, $X''(t)$ for $t = \tau e_1$.

Lemma 3.21 *$\forall i \neq j \neq k \in \{1, \dots, d\}$:*

$$\mathbb{E}\left(X(0)X_i'(t)\right) = -\mathbb{E}\left(X(t)X_i'(0)\right) = 2\tau\mathbf{r}'\delta_{1i},$$

$$\mathbb{E}\left(X_i'(0)X_j'(t)\right) = 0, \quad \mathbb{E}\left(X_i'(0)X_i'(t)\right) = -2\mathbf{r}' - 4\tau^2\mathbf{r}''\delta_{1i},$$

$$\mathbb{E}\left(X_{ij}''(0)X_i'(t)\right) = -\mathbb{E}\left(X_{ij}''(t)X_i'(0)\right) = \mathbb{E}\left(X_{ii}''(0)X_j'(t)\right) = -\mathbb{E}\left(X_{ii}''(t)X_j'(0)\right)$$

$$= 4\tau\mathbf{r}''\delta_{1j},$$

$$\mathbb{E}\left(X_{ij}''(0)X_k'(t)\right) = 0, \quad \mathbb{E}\left(X_{ii}''(0)X_i'(t)\right) = -\mathbb{E}\left(X_{ii}''(t)X_i'(0)\right)$$

$$= \left(12\tau\mathbf{r}'' + 8\tau^3\mathbf{r}'''\right)\delta_{1i},$$

$$\mathbb{E}\left(X_{ii}''(0)X_{ik}''(t)\right) = \mathbb{E}\left(X_{ij}''(0)X_{ik}''(t)\right) = 0,$$

$$\mathbb{E}\left(X_{ii}''(0)X_{ii}''(t)\right) = 12\mathbf{r}' + \left(48\tau^2\mathbf{r}''' + 16\tau^4\mathbf{r}^{(4)}\right)\delta_{1i},$$

$$\mathbb{E}\left(X_{ii}''(0)X_{jj}''(t)\right) = \mathbb{E}\left(X_{ij}''(0)X_{ij}''(t)\right) = 4\mathbf{r}'' + 8\tau^2[\delta_{1i} + \delta_{1j}]\mathbf{r}'''.$$

We deduce the conditional distribution of $X'(0), X'(t)$ given $X(0) = X(t) = 0$.

Lemma 3.22 *Let $(A(0), A(t))$ be a representation of the distribution of $(X'(0), X'(t))$ given $X(0) = X(t) = 0$ with $t = \tau e_1$. We assume $(X(0), X(t))$ is non-degenerate. The joint distribution of $A(0), A(t)$ is centered Gaussian with variance-covariance matrix*

$$\mathbb{E}\left(A_i(t)A_i(t)\right) = \mathbb{E}\left(A_i(0)A_i(0)\right) = \lambda_2 - \frac{4\tau^2\mathbf{r}'^2}{1 - \mathbf{r}^2}\delta_{i1},$$

$$\mathbb{E}\left(A_i(0)A_i(t)\right) = -2\mathbf{r}' - \left(4\tau^2\mathbf{r}'' + \frac{4\tau^2\mathbf{r}\mathbf{r}'^2}{1 - \mathbf{r}^2}\right)\delta_{i1},$$

$$\mathbb{E}\left(A_i(t)A_j(t)\right) = 0 \text{ and } \mathbb{E}\left(A_i(0)A_j(t)\right) = 0 \forall i \neq j.$$

Below we give some results about the conditional distribution of $X''(0), X''(t)$ given $X'(0) = X'(t) = 0$ (from Azaïs and Delmas 2022, Lemma 8)

Lemma 3.23 *Let $\xi(0)$ and $\xi(t)$ be a representation of the distribution of $X''(0)$ and $X''(t)$ given $X'(0) = X'(t) = 0$. We assume $\left(X'(0), X'(t)\right)$ is non-degenerate. We denote by $\xi_d(t)$ the vector*

$$(\xi_{11}(t), \dots, \xi_{dd}(t))$$

and by $\xi_u(t)$ *the vector*

$$\left(\xi_{12}(t), \xi_{13}(t), \dots, \xi_{(d-1)d}(t)\right).$$

The joint distribution of $(\xi_d(0), \xi_u(0), \xi_d(t), \xi_u(t))$ *is zero-mean Gaussian with variance-covariance matrix:*

$$\begin{bmatrix} \Gamma_1 & 0 & \Gamma_3 & 0 \\ 0 & \Gamma_2 & 0 & \Gamma_4 \\ \Gamma_3 & 0 & \Gamma_1 & 0 \\ 0 & \Gamma_4 & 0 & \Gamma_2 \end{bmatrix}.$$

Γ_1 *is the* $d \times d$ *matrix:*

$$\Gamma_1 = \begin{bmatrix} 12\mathbf{r}''(0) & 4\mathbf{r}''(0) & \cdots & 4\mathbf{r}''(0) \\ 4\mathbf{r}''(0) & \ddots & 4\mathbf{r}''(0) & \vdots \\ \vdots & 4\mathbf{r}''(0) & \ddots & 4\mathbf{r}''(0) \\ 4\mathbf{r}''(0) & \cdots & 4\mathbf{r}''(0) & 12\mathbf{r}''(0) \end{bmatrix} + \frac{\tau^2\mathbf{r}'(0)}{2[\mathbf{r}'(0)^2 - (\mathbf{r}' + 2\mathbf{r}''\tau^2)^2]} \times M,$$

with

$$M = \begin{bmatrix} (12\mathbf{r}'' + 8\tau^2\mathbf{r}''')^2 & 4\mathbf{r}''(12\mathbf{r}'' + 8\tau^2\mathbf{r}''') & \cdots & 4\mathbf{r}''(12\mathbf{r}'' + 8\tau^2\mathbf{r}''') \\ 4\mathbf{r}''(12\mathbf{r}'' + 8\tau^2\mathbf{r}''') & 16\mathbf{r}''^2 & \cdots & 16\mathbf{r}''^2 \\ \vdots & \vdots & \ddots & \vdots \\ 4\mathbf{r}''(12\mathbf{r}'' + 8\tau^2\mathbf{r}''') & 16\mathbf{r}''^2 & \cdots & 16\mathbf{r}''^2 \end{bmatrix}.$$

Γ_2 *is the* $\frac{d(d-1)}{2} \times \frac{d(d-1)}{2}$ *diagonal matrix:*

$$\Gamma_2 = \begin{bmatrix} M_1 & 0 \\ 0 & M_2 \end{bmatrix} \text{ with } M_1 = \left(4\mathbf{r}''(0) + \frac{8\tau^2(\mathbf{r}'')^2\mathbf{r}'(0)}{\mathbf{r}'(0)^2 - (\mathbf{r}')^2}\right)\mathrm{Id}_{d-1}$$

$$\text{and } M_2 = \left(4\mathbf{r}''(0)\right)\mathrm{Id}_{\frac{(d-1)(d-2)}{2}}.$$

We set $a := 4\mathbf{r}'' + 8\tau^2\mathbf{r}'''$ *and* $b := 12\mathbf{r}'' + 48\tau^2\mathbf{r}''' + 16\tau^4\mathbf{r}^{(4)}$.

Γ_3 *is the* $d \times d$ *matrix:*

$$\Gamma_3 = \begin{bmatrix} b & a & \cdots & \cdots & a \\ a & 12\mathbf{r}'' & 4\mathbf{r}'' & \cdots & 4\mathbf{r}'' \\ \vdots & 4\mathbf{r}'' & \ddots & 4\mathbf{r}'' & \vdots \\ \vdots & \vdots & 4\mathbf{r}'' & \ddots & 4\mathbf{r}'' \\ a & 4\mathbf{r}'' & \cdots & 4\mathbf{r}'' & 12\mathbf{r}'' \end{bmatrix} + \frac{\tau^2(\mathbf{r}' + 2\mathbf{r}''\tau^2)}{2\left[\mathbf{r}'(0)^2 - (\mathbf{r}' + 2\mathbf{r}''\tau^2)^2\right]} \times M.$$

Γ_4 *is the* $\frac{d(d-1)}{2} \times \frac{d(d-1)}{2}$ *diagonal matrix:*

$$\Gamma_4 = \begin{bmatrix} \tilde{M}_1 & 0 \\ 0 & \tilde{M}_2 \end{bmatrix} \text{ with } \tilde{M}_1 = \left(a + \frac{8\tau^2(\mathbf{r}'')^2\mathbf{r}'}{\mathbf{r}'(0)^2 - \mathbf{r}'^2} \right) \mathrm{Id}_{d-1}$$

$$\text{and } \tilde{M}_2 = 4\mathbf{r}'' \, \mathrm{Id}_{\frac{(d-1)(d-2)}{2}}.$$

We can choose the other parameterization, using $\rho(\cdot)$ given in (3.29) to express Lemma 3.23. It simplifies some expressions but makes others more complex. We can use the following relations between ρ and $\mathbf{r}$,

$$\begin{aligned}
\rho(\tau) &= \mathbf{r}(\tau^2), \\
\rho'(\tau) &= 2\tau\mathbf{r}'(\tau^2), \\
\rho''(\tau) &= 2\mathbf{r}'(\tau^2) + 4\tau^2\mathbf{r}''(\tau^2), \\
\rho^{(3)}(\tau) &= 12\tau\mathbf{r}''(\tau^2) + 8\tau^3\mathbf{r}^{(3)}(\tau^2), \\
\rho^{(4)}(\tau) &= 12\mathbf{r}''(\tau^2) + 48\tau^2\mathbf{r}^{(3)}(\tau^2) + 16\tau^4\mathbf{r}^{(4)}(\tau^2).
\end{aligned}$$

In particular, the joint distribution of $X'(s)$, $X'(t)$ can be written

$$p_{X'(s),X'(t)}(0,0) = (2\pi)^{-d}\left(\lambda_2^2 - \left(2\mathbf{r}'(\tau^2) + 4\tau^2\mathbf{r}''(\tau^2)\right)^2\right)^{-1/2} \left(\lambda_2^2 - 4\mathbf{r}'(\tau^2)^2\right)^{(1-d)/2}.$$

Setting $\tilde{a} := 4\tau\mathbf{r}''(\tau^2)$ and $\tilde{b} := 2\mathbf{r}'(\tau^2)$, $\Gamma_1, \Gamma_2, \Gamma_3, \Gamma_4$ can be rewritten as

$$\Gamma_1 = \begin{bmatrix} \lambda_4 & \lambda_4/3 & \cdots & \lambda_4/3 \\ \lambda_4/3 & \ddots & \lambda_4/3 & \vdots \\ \vdots & \lambda_4/3 & \ddots & \lambda_4/3 \\ \lambda_4/3 & \cdots & \lambda_4/3 & \lambda_4 \end{bmatrix} - \frac{\lambda_2}{\lambda_2^2 - \rho''(\tau)^2} \times M,$$

where

$$M := \begin{bmatrix} \rho^{(3)}(\tau)^2 & \tilde{a}\rho^{(3)}(\tau) & \cdots & \tilde{a}\rho^{(3)}(\tau) \\ \tilde{a}\rho^{(3)}(\tau) & \tilde{a}^2 & \cdots & \tilde{a}^2 \\ \vdots & \vdots & \ddots & \vdots \\ \tilde{a}\rho^{(3)}(\tau) & \tilde{a}^2 & \cdots & \tilde{a}^2 \end{bmatrix}.$$

Γ_2 is the $\frac{d(d-1)}{2} \times \frac{d(d-1)}{2}$ diagonal matrix $\Gamma_2 = \begin{bmatrix} M_1 & 0 \\ 0 & M_2 \end{bmatrix}$ with

$$M_1 = \left(\frac{\lambda_4}{3} - \frac{\lambda_2 \tilde{a}^2}{\lambda_2^2 - \tilde{b}^2} \right) \mathrm{Id}_{d-1} \text{ and } M_2 = \left(\frac{\lambda_4}{3} \right) \mathrm{Id}_{\frac{(d-1)(d-2)}{2}}.$$

Γ_3 is the $d \times d$ matrix:

$$\Gamma_3 = \frac{1}{\tau} \begin{bmatrix} \tau\rho^{(4)}(\tau) & \rho^{(3)}(\tau) - 2\tilde{a} & \cdots & \cdots & \rho^{(3)}(\tau) - 2\tilde{a} \\ \rho^{(3)}(\tau) - 2\tilde{a} & 3\tilde{a} & \tilde{a} & \cdots & \tilde{a} \\ \vdots & \tilde{a} & \ddots & \tilde{a} & \vdots \\ \vdots & \vdots & \tilde{a} & \ddots & \tilde{a} \\ \rho^{(3)}(\tau) - 2\tilde{a} & \tilde{a} & \cdots & \tilde{a} & 3\tilde{a} \end{bmatrix} + \frac{\rho''(\tau)}{\lambda_2^2 - \rho''(\tau)^2} \times M.$$

Γ_4 is the $\frac{d(d-1)}{2} \times \frac{d(d-1)}{2}$ diagonal matrix:

$$\Gamma_4 = \frac{1}{\tau} \begin{bmatrix} \tilde{M}_1 & 0 \\ 0 & \tilde{M}_2 \end{bmatrix} \text{ with } \tilde{M}_1 = \left(\rho^{(3)}(\tau) - 2\tilde{a} + \tau \frac{\tilde{a}^2 \tilde{b}}{\lambda_2^2 - \tilde{b}^2} \right) \mathrm{Id}_{d-1}$$

$$\text{and } \tilde{M}_2 = \tilde{a}\, \mathrm{Id}_{\frac{(d-1)(d-2)}{2}}.$$

In the following lemma, we give the equivalent of the variance-covariance matrix of $X''(0), X''(t)$ given $X'(0) = X'(t) = 0$ as $\tau \to 0$, see Azaïs and Delmas (2022, lemme 9).

Lemma 3.24 *Using the same notation as in Lemma 3.23, for $j \neq k$ and $j, k \neq 1$, as $\tau \to 0$, when the considered quantities are finite,*

$$\det\left(\mathrm{Var}\left(X'(0), X'(t)\right)\right) \simeq \tau^{2d} \frac{\lambda_2^d \lambda_4^d}{3^{d-1}}, \tag{3.32}$$

$$\mathrm{Var}\left(\xi_{11}(t)\right) \simeq \frac{\tau^2}{4} \frac{(\lambda_2\lambda_6 - \lambda_4^2)}{\lambda_2}, \quad \mathrm{Var}\left(\xi_{1j}(t)\right) \simeq \frac{\tau^2}{4} \frac{(9\lambda_2\lambda_6 - 5\lambda_4^2)}{45\lambda_2}, \tag{3.33}$$

$$\mathrm{Var}\left(\xi_{jj}(t)\right) \simeq \frac{8\lambda_4}{9}, \quad \mathrm{Var}\left(\xi_{jk}(t)\right) \simeq \frac{\lambda_4}{3}, \tag{3.34}$$

$$\mathrm{Cov}\left(\xi_{11}(t), \xi_{jj}(t)\right) = \tau^2 \frac{11\lambda_2\lambda_6 - 15\lambda_4^2}{180\lambda_2} + o(\tau^2), \quad \mathrm{Cov}\left(\xi_{jj}(t), \xi_{kk}(t)\right) \simeq \frac{2\lambda_4}{9}. \tag{3.35}$$

All the other covariances $\mathrm{Cov}(\xi_{il}(t), \xi_{mn}(t))$ *are zero*, $\forall i, l, m, n \in \{1, \ldots, d\}$. (3.36)

We have of course the same relations for $\xi(0)$.

Moreover, we have $\forall j,k \in \{2,\ldots,d\}$ *and* $\forall i \in \{1,\ldots,d\}$*; as* $\tau \to 0$,

$$\operatorname{Cov}\left(\xi_{jk}(0), \xi_{jk}(t)\right) \simeq \operatorname{Var}\left(\xi_{jk}(t)\right), \ \operatorname{Cov}\left(\xi_{1i}(0), \xi_{1i}(t)\right) \simeq -\operatorname{Var}\left(\xi_{1i}(t)\right), \tag{3.37}$$

$$\operatorname{Cov}\left(\xi_{11}(0), \xi_{jj}(t)\right) = \tau^2 \frac{15\lambda_4^2 - 7\lambda_2\lambda_6}{180\lambda_2} + o(\tau^2),$$

$$\operatorname{Cov}\left(\xi_{jj}(0), \xi_{kk}(t)\right) \simeq \frac{2\lambda_4}{9} \textit{ for } j \neq k, \tag{3.38}$$

All the other covariances $\operatorname{Cov}(\xi_{il}(0), \xi_{mn}(t))$ *are zero,* $\forall i,l,m,n \in \{1,\ldots,d\}$. (3.39)

Finally, we also have

$$\det\left(\operatorname{Var}\left(\xi_{11}(t), \xi_{11}(0)\right)\right) \simeq \frac{\tau^6}{144} \frac{(\lambda_4\lambda_8 - \lambda_6^2)(\lambda_2\lambda_6 - \lambda_4^2)}{\lambda_2\lambda_4}. \tag{3.40}$$

The proof of Lemma 3.24 is given in Azaïs and Delmas (2022). Though it is based on some lengthy computations, it uses a trick that drastically simplifies them. We present it here in the example of the proof of (3.32).

Because a determinant is invariant under adding a linear combination of other rows (resp. columns) to a given row (resp. a given column), as $\tau \to 0$,

$$\begin{aligned}
\det\left(\operatorname{Var}\left(X'(0), X'(t)\right)\right) &= \det(\operatorname{Var}(X_1'(0), \ldots, X_d'(0), X_1'(t), \ldots, X_d'(t))) \\
&= \det(\operatorname{Var}(X_1'(0), \ldots, X_d'(0), X_1'(t) - X_1'(0), \ldots, X_d'(t) - X_d'(0))) \\
&\simeq \tau^{2d} \det(\operatorname{Var}(X_1'(0), \ldots, X_d'(0), X_{11}''(0), \ldots, X_{1d}''(0))).
\end{aligned}$$

Using Proposition 3.20 we obtain (3.32).

3.5.2 Independent Copies of Isotropic Gaussian Random Fields

This section is borrowed from Azaïs and Wschebor (2009, Section 12.4). We consider d independent zero-mean stationary and isotropic random fields $X_i(\cdot) = \{X_i(t) : t \in \mathbb{R}^d\}$ with C^1 paths. We assume that

$$\mathbb{E}\left(X_i(s), X_i(t)\right) = \mathbf{r}_i(\|t - s\|^2), \quad (i = 1, \ldots, d), \tag{3.41}$$

where $\mathbf{r}_i$ is of class C^2. and, with no loss of generality, that $\mathbf{r}_i(0) = 1$.

The random field $\mathbf{X}(\cdot)$ that consists of $\left(X_1(\cdot), \ldots, X_d(\cdot)\right)$ is a random field from $\mathbb{R}^d$ to $\mathbb{R}^d$. We are interested in the number of roots of $\mathbf{X}(t) = 0$.

3.5.2.1 Computation of the Expectation of the Number of Roots

The computation of the expectation of the number of roots of $\mathbf{X}(t) = 0$ belonging to a Borel set $B \in \mathbb{R}^d$ can be carried out using the Rice formula and the result (3.22) of Sect. 3.4, obtaining

$$\mathbb{E}\big(N_0(\mathbf{X}, B)\big) = |B|(2\pi)^{-d/2}\mathbb{E}\big(|\det(\mathbf{X}'(0))|\big)$$

$$= \frac{|B|}{\sqrt{\pi}}\left(\frac{2}{\pi}\right)^{d/2}\Gamma\left(\frac{d+1}{2}\right)\left[\prod_{i=1}^{d}|\mathbf{r}_i'(0)|^{1/2}\right]. \tag{3.42}$$

The proof of this result is essentially based on the computation of the expectation of a Gaussian determinant given by Lemma 2.18.

3.5.2.2 Computation of the Variance

To go further, we need two additional hypotheses

(1) All the $\mathbf{r}_i$ coincide, $\mathbf{r}_i = \mathbf{r}$, $i = 1, \ldots, d$.
(2) The function $\mathbf{r}$ is a "Schoenberg covariance" in the sense of Definition 3.2.

We define the normalized number of roots:

$$\mathcal{N}(B) = \frac{N_0(\mathbf{X}, B))}{\mathbb{E}\,(N_0(\mathbf{X}, B))}.$$

Theorem 3.25 (Azaïs and Wschebor 2005b) *Under the notation and assumptions above and assuming further that*

1. *the distribution G in* (3.31) *is not concentrated at a single point and*

$$\int_0^{+\infty} x^2 G(\mathrm{d}x) < \infty,$$

2. *$\{B_d\}_{d=1,2\ldots}$ is a sequence of Borel sets, $B_d \subset \mathbb{R}^d$, and there exist two positive constants δ, Δ such that for each d, B_d contains a ball with radius δ and is contained in a ball with radius Δ.*

Then,

$$\mathrm{Var}\big(\mathcal{N}(B_d)\big) \to +\infty, \tag{3.43}$$

exponentially fast as $d \to +\infty$.

Proof As in many other situations it suffices to show that

$$\frac{\mathbb{E}\Big(N_0(\mathbf{X}, B)\big(N_0(\mathbf{X}, B) - 1\big)\Big)}{\big[\mathbb{E}\big(N_0(\mathbf{X}, B)\big)\big]^2} \to +\infty \tag{3.44}$$

exponentially fast as $d \to +\infty$. The denominator in (3.44) is given by Formula (3.42). For the numerator, we can apply the Kac–Rice formula for the second-order factorial moment Theorem 2.21:

$$\begin{aligned}&\mathbb{E}\big(N_0(\mathbf{X}, B)\,(N_0(\mathbf{X}, B) - 1)\big)\\ &= \iint_{B\times B} \mathbb{E}\big(\,\big|\det(\mathbf{X}'(s))\det(\mathbf{X}'(t))\big|\;\big|\,\mathbf{X}(s) = \mathbf{X}(t) = 0\big)\; p_{\mathbf{X}(s),\mathbf{X}(t)}(0, 0)\; \mathrm{d}s\; \mathrm{d}t.\end{aligned} \tag{3.45}$$

Next, we compute the ingredients of the integrand in (3.45). Because of invariance under translations, the integrand is a function of $\tau = t - s$. The Gaussian density is given by

$$p_{\mathbf{X}(s),\mathbf{X}(t)}(0, 0) = \frac{1}{(2\pi)^d(1 - \mathbf{r}^2(\|\tau\|^2))^{d/2}}.$$

Note that (3.31) implies that for $\tau \neq 0$, $\mathbf{r}^2 < 1$ so the density is N.D. and the validity of the Kac–Rice formula is proven. Let us turn to the conditional expectation in (3.45). Let $C := \{\mathbf{X}(s) = \mathbf{X}(t) = 0\}$ and let $A^s = (A^s_{i\alpha})$, $A^t = (A^t_{i\alpha})$ be the $d \times d$ random matrices having as joint distribution the conditional Gaussian distribution of the pair $\mathbf{X}'(s)$, $\mathbf{X}'(t)$ given C. We have

$$\mathbb{E}_C\big(\,\big|\det(\mathbf{X}'(s))\det(\mathbf{X}'(t))\big|\,\big) = \mathbb{E}\big(\,\big|\det(A^s)\det(A^t)\big|\,\big).$$

Take now an orthonormal basis of $\mathbb{R}^d$ having the unit vector $\frac{\tau}{\|\tau\|}$ as first element. By Lemma 3.22, the $(2d) \times (2d)$ variance matrix of the pair A^s_i, A^t_i—the ith rows of A^s and A^t respectively—takes the following form

$$\left[\begin{array}{cccc|cccc} U_0 & 0 & \dots & 0 & U_1 & 0 & \dots & 0\\ 0 & V_0 & 0 & \vdots & 0 & V_1 & 0 & \vdots\\ \vdots & 0 & \ddots & 0 & \vdots & 0 & \ddots & 0\\ 0 & \dots & 0 & V_0 & 0 & \dots & 0 & V_1\\ \hline U_1 & 0 & \dots & 0 & U_0 & 0 & \dots & 0\\ 0 & V_1 & 0 & \vdots & 0 & V_0 & 0 & \vdots\\ \vdots & 0 & \ddots & 0 & \vdots & 0 & \ddots & 0\\ 0 & \dots & 0 & V_1 & 0 & \dots & 0 & V_0 \end{array}\right],$$

where

$$U_0 = U_0(\|\tau\|^2) = -2\mathbf{r}'(0) - 4\frac{\mathbf{r}'^2 \|\tau\|^2}{1-\mathbf{r}^2} ;$$

$$V_0 = -2\mathbf{r}'(0) ;$$

$$U_1 = U_1(\|\tau\|^2) = -4\mathbf{r}'' \|\tau\|^2 - 2\mathbf{r}' - 4\frac{\mathbf{r}\mathbf{r}'^2\|\tau\|^2}{1-\mathbf{r}^2};$$

$$V_1 = V_1(\|\tau\|^2) = -2\mathbf{r}'.$$

Let us perform a regression of $A^t_{i\alpha}$ on $A^s_{i\alpha}$, that is, write the orthogonal decompositions

$$A^t_{i\alpha} = B^{t,s}_{i\alpha} + C_\alpha A^s_{i\alpha} \quad (i, \alpha = 1, d),$$

where $B^{t,s}_{i\alpha}$ is a centered Gaussian independent of the matrix A^s with $\mathbb{E}\left(B^{t,s}_{i\alpha} B^{t,s}_{i\beta}\right) = 0 \forall \alpha \neq \beta$ and

$$\text{for } \alpha = 1, \;\; C_1 = \frac{U_1}{U_0}, \;\; \mathrm{Var}(B^{t,s}_{i1}) = U_0\big(1 - \frac{U_1^2}{U_0^2}\big);$$

$$\text{for } \alpha > 1, \;\; C_\alpha = \frac{V_1}{V_0}, \;\; \mathrm{Var}(B^{t,s}_{i\alpha}) = V_0\big(1 - \frac{V_1^2}{V_0^2}\big).$$

Conditioning we have:

$$\mathbb{E}\big(|\det(A^s)||\det(A^t)|\big) = \mathbb{E}\big[|\det(A^s)|\mathbb{E}\big(|\det((B^{t,s}_{i\alpha} + C_\alpha A^s_{i\alpha})_{i,\alpha=1,\ldots,d})| \;\big|\; A^s\big)\big].$$

For the inner conditional expectation, we can proceed in the same way as we did in the proof of Lemma 2.18 to compute the determinant, obtaining a product of expectations of Euclidean norms of non-centered Gaussian vectors in $\mathbb{R}^k$ for $k = 1, \ldots, d$. Now, using for example classical results on non-centered χ-distributions, we have

$$\mathbb{E}\big(\|\xi + v\|\big) \geq \mathbb{E}\big(\|\xi\|\big)$$

valid for ξ standard normal in $\mathbb{R}^k$ and v any vector in $\mathbb{R}^k$, and it follows that

$$\mathbb{E}\big(|\det(A^s)||\det(A^t)|\big) \geq \mathbb{E}\big(|\det(A^s)|\big)\mathbb{E}\big(|\det(B^{t,s})|\big).$$

Since the elements of A^s (resp. $B^{t,s}$) are independent, centered Gaussian with known variance, we obtain

$$\mathbb{E}\left|\det(A^s)\det(A^t)\right| \geq U_0 V_0^{d-1}\big(1-\frac{U_1^2}{U_0^2}\big)^{1/2}\big(1-\frac{V_1^2}{V_0^2}\big)^{(d-1)/2} L_{d,d}^2,$$

where $L_{d,d}$ is given by (2.32). Going back to (3.44) and on account of (3.42) and (3.45) we have

$$\frac{\mathbb{E}\big(N_0(\mathbf{X},V)\,(N_0(\mathbf{X},V)-1)\big)}{\mathbb{E}\big(N_0(\mathbf{X},V)\big)^2} \geq |B|^{-2}\iint_{B\times B} \mathrm{d}s\mathrm{d}t\left[\frac{1-V_1^2(\|\tau\|^2)V_0^{-2}}{1-\mathbf{r}^2}\right]^{m/2} H(\|\tau\|^2), \tag{3.46}$$

where

$$H(x)=\left(\frac{U_0^2(x)-U_1^2(x)}{V_0^2-V_1^2(x)}\right)^{1/2}.$$

After some lengthy calculation, following Azaïs and Wschebor (2009, pp. 338–339), we obtain (3.43). □

3.5.3 Distributions in the Spherical Case

In this section we consider isotropic random fields defined on the sphere S^d, $d \geq 1$.

Proposition 3.26 *According to Schoenberg's theorem (Schoenberg 1942), a covariance function on the sphere S^d is a valid covariance in any dimension of an isotropic random field if and only if*

$$c(s,t)=\mathbb{E}(X(s)X(t))=\nu(\langle s,t\rangle)$$

where

$$\nu(\langle s,t\rangle)=\sum_{k=0}^{+\infty} c_k\langle s,t\rangle^k \text{ with } c_k \geq 0, \sum_{k=0}^{+\infty} c_k < +\infty. \tag{3.47}$$

Definition 3.4 A covariance function satisfying Proposition 3.26 is called a Schoenberg covariance on the sphere.

In this section we consider the following assumption:

Assumption (A.3.1) $\{X(t) : t \in S^d \subset \mathbb{R}^{d+1}\}$ is a real-valued centered isotropic Gaussian field defined on the unit sphere $S^d \subset \mathbb{R}^{d+1}$. Its covariance is a Schoenberg covariance as in (3.47) and in addition

$$c_0 = c_1 = 0, \quad \sum_{k=2}^{+\infty} c_k = 1, \quad \sum_{k=2}^{+\infty} k c_k > 2, \quad \sum_{k=2}^{+\infty} k^5 c_k < +\infty. \tag{3.48}$$

Remark 3.8 The sum in (3.48) begins at $k = 2$ to be consistent with, for example, Auffinger et al. (2013). It is possible to add a term c_0 and a term c_1, but this would imply some additional technical hypotheses.

3.5.3.1 Differentiability

First note that (3.48) defines a valid covariance function of the whole $\mathbb{R}^{d+1}$. Note also that $\nu(\cdot)$ is C^5. Consider a neighborhood $\bar{S}^d$ of the sphere S^d that does not contain zero. On this set the scalar product $s, t \mapsto \langle s, t\rangle$ is infinitely differentiable and the covariance

$$c(s,t) = \nu(\langle s, t\rangle) \text{ is of class } C^5. \tag{3.49}$$

Let us consider the random field $\bar{X}(\cdot)$ with such a covariance on $\bar{S}^d$. (3.49) implies that $\bar{X}(\cdot)$ is twice differentiable in quadratic mean on $\bar{S}^d$ and that its Euclidean Hessian has a covariance function given by

$$\frac{\partial^4 c(s,t)}{\partial^2 s \partial^2 t},$$

which is a C^1 function. Then using, for example, criterion Adler and Taylor (2009, (1.4.4)), we have for $1 \le i, j \le d+1$

$$\mathrm{Var}(\bar{X}''_{ij}(s) - \bar{X}''_{ij}(t)) \le (\mathtt{const}) - \log(\|s-t\|)^{-(1+\alpha)}, \quad \alpha > 0,$$

for $\|s-t\|$ sufficiently small.

This implies that $\bar{X}''(\cdot)$ has a continuous version and by integration that $\bar{X}(\cdot)$ has a C^2 version on $\bar{S}^d$. The trace of $\bar{X}(\cdot)$ on S^d coincide with $X(\cdot)$. By consequence, $X(\cdot)$ has C^2 sample paths on S^d.

3.5.3.2 Gradient and Hessian

Let us first compute the Euclidean gradient and Hessian for $\bar{X}(\cdot)$. Since the problem is invariant under isometries, we can use the "Santa Claus method".

We set $\mathbf{n} := (0, \ldots, 0, 1)$, the north pole of the unit sphere S^d, $\nabla X(t)$ the Riemannian gradient of X at t with coordinates $\nabla X_i(t)$, $\nabla^2 X(t)$ the Riemannian Hessian matrix of X at t with elements $\nabla^2 X_{ij}(t)$ for $i, j = 1, \ldots, d$. We have

$$\begin{aligned}
\frac{\partial c(s,t)}{\partial t_i} &= s_i \nu'(\langle s, t\rangle), \\
\frac{\partial^2 c(s,t)}{\partial t_i \partial t_j} &= s_i s_j \nu''(\langle s, t\rangle), \\
\frac{\partial^2 c(s,t)}{\partial t_i \partial s_k} &= \nu'(\langle s, t\rangle)\delta_{ik} + s_i t_k \nu''(\langle s, t\rangle), \\
\frac{\partial^3 c(s,t)}{\partial t_i \partial t_j \partial s_k} &= \nu''(\langle s, t\rangle)\left(s_j \delta_{ik} + s_i \delta_{jk}\right) + \nu^{(3)}(\langle s, t\rangle)\left(s_i s_j t_k\right), \\
\frac{\partial^4 c(s,t)}{\partial t_i \partial t_j \partial s_k \partial s_l} &= \nu''(\langle s, t\rangle)\left(\delta_{ik}\delta_{jl} + \delta_{il}\delta_{jk}\right) + \nu^{(4)}(\langle s, t\rangle)\left(s_i s_j t_l t_k\right) \\
&\quad + \nu^{(3)}(\langle s, t\rangle)\left(t_l s_j \delta_{ik} + t_l s_i \delta_{jk} + s_i t_k \delta_{jl} + s_j t_k \delta_{il}\right).
\end{aligned}$$

The Riemannian gradient consists of the first d coordinates of the Euclidean gradient and it is easy to deduce that for $i, k = 1, \ldots, d$ and for all t

$$\begin{aligned}
\mathbb{E}(X(t)\nabla X_i(t)) &= \left.\frac{\partial c(s,t)}{\partial t_i}\right|_{s=t=\mathbf{n}} = 0, \\
\mathbb{E}(\nabla X_i(t)\nabla X_k(t)) &= \left.\frac{\partial^2 c(s,t)}{\partial t_i \partial s_k}\right|_{s=t=\mathbf{n}} = \nu_2 \delta_{ik},
\end{aligned}$$

where

$$\nu_2 := \mathrm{Var}(\nabla X_i(t)) = \nu'(1).$$

Let us turn now to the computation of the Hessian. Instead of using the method of Auffinger and Ben Arous (2013) that was based on the parameterization by the projection on the tangent space, we will use Formula (2.4), where the Riemannian Hessian matrix is given by the sum of the Hessian matrix on the tangent plane and the identity matrix times the normal derivative of X oriented towards the inside at the north pole for $i = 1, \ldots, d$, $j \neq i = 1, \ldots, d$

$$\begin{aligned}
\nabla^2 X_{ii}(t) &= X''_{ii}(t) - X'_{d+1}(t), \\
\nabla^2 X_{ij}(t) &= X''_{ij}(t).
\end{aligned}$$

We obtain

$$\mathbb{E}(X(t)\nabla^2 X_{ij}(t)) = \left.\frac{\partial^2 c(s,t)}{\partial t_i \partial t_j}\right|_{s=t=\mathbf{n}} = 0,$$

$$\mathbb{E}(X(t)\nabla^2 X_{ii}(t)) = \left.\frac{\partial^2 c(s,t)}{\partial^2 t_i}\right|_{s=t=\mathbf{n}} - \mathbb{E}(X(t)X'_{d+1}(t)) = 0 + \nu_2.$$

Thus

$$\mathbb{E}(X(t)\nabla^2 X(t)) = -\nu_2 \mathrm{Id}_d.$$

Set

$$\nu_4 := \mathrm{Var}(\nabla^2 X_{ii}(t)) = 3\nu''(1) + \nu'(1).$$

ν_2 and ν_4 play the role of the λ_2 and λ_4 in the "flat case", but they are not spectral moments, so we use a different notation.

Note that Assumption **(A.3.1)** implies

$$\nu_4 + 2\nu_2 - 3\nu_2^2 \geq 0.$$

In the sequel we will distinguish two cases: the case where the quantity above vanishes, and the case where it is positive. Using the above type of calculation we obtain the following two lemmas, by way of Auffinger and Ben Arous (2013, Lemma 1), describing the joint distribution of $(X(t), \nabla X(t), \nabla^2 X(t))$.

Lemma 3.27 *Under Assumption* ***(A.3.1)****:*

(a) For all $t \in S^d$ $X(t)$, $\nabla X(t)$ and $\nabla^2 X(t)$ are centered Gaussian random variables whose joint distribution is determined by

$$\mathbb{E}(X(t)^2) = 1 \qquad \mathbb{E}(X(t)\nabla X_i(t)) = \mathbb{E}(\nabla X_i(t)\nabla^2 X_{jk}(t)) = 0$$
$$\mathbb{E}(X(t)\nabla^2 X_{ij}(t)) = -\nu_2\delta_{ij} \quad \mathbb{E}(\nabla X_i(t)\nabla X_j(t)) = \nu_2\delta_{ij}$$

and

$$\mathbb{E}(\nabla^2 X_{ij}(t)\nabla^2 X_{kl}(t)) = \nu_2\delta_{ij}\delta_{kl} + \left(\frac{\nu_4 - \nu_2}{3}\right)(\delta_{ij}\delta_{kl} + \delta_{ik}\delta_{jl} + \delta_{il}\delta_{jk}).$$

(b) The conditional joint distribution of $X_{ij}(t)$ given $X(t) = x$ is determined by

$$\mathbb{E}(\nabla^2 X_{ij}(t) \mid X(t) = x) = -x\nu_2\delta_{ij}, \tag{3.50}$$

and

$$\operatorname{Cov}(\nabla^2 X_{ij}(t), \nabla^2 X_{kl}(t)/X(t) = x)$$
$$= \left(\frac{\nu_4 - \nu_2}{3}\right)\left(\delta_{ij}\delta_{kl} + \delta_{ik}\delta_{jl} + \delta_{il}\delta_{jk}\right) + \left(\nu_2 - \nu_2^2\right)\delta_{ij}\delta_{kl}. \tag{3.51}$$

Lemma 3.28 *The Gaussian distributions given in Lemma 3.27 can be represented by the following expressions.*

The unconditional distribution of $\nabla^2 X(t)$ given in case (a) can be represented by

$$\sqrt{2\left(\frac{\nu_4 - \nu_2}{3}\right)}\left(G - Z\mathrm{Id}_d\right),\ \text{with } Z \sim \mathcal{N}\left(0, \frac{\nu_4 + 2\nu_2}{2(\nu_4 - \nu_2)}\right). \tag{3.52}$$

The conditional distribution of $\nabla^2 X(t) \mid X(t) = x$ given in case (b) can be represented by

$$\sqrt{2\left(\frac{\nu_4 - \nu_2}{3}\right)}\left(G - \frac{\sqrt{3}x\nu_2}{\sqrt{2(\nu_4 - \nu_2)}}\mathrm{Id}_d\right) \text{ when } \nu_4 + 2\nu_2 - 3\nu_2^2 = 0, \tag{3.53}$$

$$\sqrt{2\left(\frac{\nu_4 - \nu_2}{3}\right)}\left(G - V\ \mathrm{Id}_d\right) \text{ when } \nu_4 + 2\nu_2 - 3\nu_2^2 > 0, \tag{3.54}$$

*where V is a Gaussian variable with mean $\dfrac{\sqrt{3}x\nu_2}{\sqrt{2(\nu_4 - \nu_2)}}$ and variance $\dfrac{\nu_4 + 2\nu_2 - 3\nu_2^2}{2(\nu_4 - \nu_2)}$ and G is an independent d-*GOE *matrix as defined in Definition 3.3.*

Remark 3.9 Because of independence properties, the conditional distribution of $\nabla^2 X(t)$ is the same if we add a condition on $\nabla X(t)$.

3.5.4 Mean Number of Local Maxima Above a High Level in the Spherical Case

We denote by $M_u^X(S^d)$ the number of local maxima above the level u on S^d,

$$M_u^X(S^d) := \#\left\{t \in S^d : X(t) \geq u, \nabla X(t) = 0, i(\nabla^2 X(t)) = d\right\}.$$

Let $H_d(z)$ be the dth Hermite polynomial defined in Sect. 4.4.

Theorem 3.29 *Under Assumption (A.3.1), for $\delta > 0$ and $u \to +\infty$,*

$$\mathbb{E}\left(M_u^X(S^d)\right) = \frac{(\nu_2 - 1)^{d/2}}{2^{(d-1)/2}\Gamma\left(\frac{d+1}{2}\right)} \times \int_u^{+\infty} \exp\left(-\frac{x^2}{2}\right) H_d\left(x\sqrt{\frac{\nu_2}{\nu_2 - 1}}\right) \mathrm{d}x + O\left(\phi\left(u(1+\delta)\right)\right). \tag{3.55}$$

Proof We use the same techniques as those involved in the proof of Theorem 3.15; see Delmas (2001) and Azaïs and Delmas (2002). □

Remark 3.10 When ν_2 is large in Formula (3.55) then $\nu_2 - 1 \simeq \nu_2$ and $\mathbb{E}\left(M_u^X(S^d)\right)$ can be approximated by

$$\mathbb{E}\left(M_u^X(S^d)\right) \simeq \frac{\nu_2^{d/2}}{2^{(d-1)/2}\Gamma((d+1)/2)} H_{d-1}(u) \exp\left(-\frac{u^2}{2}\right),$$

which corresponds to the expression (3.26) of $\mathbb{E}\left(M_u^X(T)\right)$ for X a flat real-valued zero-mean stationary Gaussian random field with variance $\mathrm{Var}(X(t)) = \lambda_0 = 1$ and spectral matrix $\Lambda_2 := \nu_2 \mathrm{Id}_d$, defined on a compact subset T of $\mathbb{R}^d$ whose measure is the same as that of the unit sphere S^d, $|T| = |S^d| = \dfrac{2\pi^{(d+1)/2}}{\Gamma((d+1)/2)}$.

Chapter 4
Hermite Expansions of Level Functionals

This chapter provides a highly accessible, elementary approach to the Hermite decomposition of random variables. By intentionally avoiding overly complex stochastic calculus tools like multiple Wiener–Itô integrals or Malliavin calculus, we present a clear pathway using Gaussian Hilbert spaces, Wick products, and Feynman diagram formulas. The chapter uses these combinatoric arguments to present an elementary proof of the celebrated Fourth Moment Theorem within a fixed chaos, leveraging Arcones' inequality to establish Central Limit Theorems (CLTs) for the measure of level sets.

4.1 Introduction

In the present chapter we introduce Hermite expansions of level functionals of a stationary Gaussian process. These expansions, also known as chaotic or Wiener–Itô expansions, are well suited for obtaining limit theorems for such functionals as well as for ascertaining the positivity of asymptotic variance. Our approach is completely elementary, relying on some combinatorial arguments and avoiding the more sophisticated tools of stochastic analysis. Needless to say, deeper results follow from the use of the latter techniques: see the monograph Nourdin and Peccati (2012) for a general view of the field. Here, we closely follow Janson (1997), Kratz and León (1997), Arcones (1994), Nourdin (2011) and push-forward their ideas to get a new proof of the so-called *Fourth Moment Theorem* in Sect. 4.7.

Hermite expansions (Slud 1991b; Kratz and León 1997) were used in the 1990s and early 2000s in addition to Malevich's m-dependent approximation (Malevich 1969) method to get Central Limit Theorems for level crossings, maxima and curve lengths based only on first and second moments computations, see e.g. Kratz and León (2000) and Kratz and León (2001).

D. Armentano et al., *Geometry of Level Sets of Random Fields, Kac–Rice Formulas, Hermite Expansions and Applications*, Probability Theory and Stochastic Modelling 111, https://doi.org/10.1007/978-3-032-29144-8_4

A new systematic approach based on Gaussian analysis was introduced 20 years ago by Nualart, Peccati, Tudor and Nourdin, among others. The so-called *Fourth Moment Theorem* is now a classical result. It states that, on a finite sum of chaos, joint convergence towards a Gaussian distribution is equivalent to marginal convergence of each component to the corresponding one-dimensional Gaussian distribution, and that the latter is equivalent to the convergence of the second and fourth moments to the corresponding moments of the Gaussian limit. This is to be compared with the classical *moments method* which requires the convergence of the moments of every order. The Fourth Moment Theorem was first obtained in Nualart and Peccati (2005) and Peccati and Tudor (2005) by means of stochastic calculus and martingale methods. Later on, new proofs using Malliavin Calculus (Nualart and Ortiz-Latorre 2008) and Stein's method for Normal approximation (Nourdin and Peccati 2009) were given. The generalizations include alternative conditions for the asymptotic normality in terms of the so-called contractions and Malliavin derivatives; bounds for distances from the normal distribution can be also obtained, see Nourdin and Peccati (2012) and references therein; and analogous results for functionals of other underlying processes such as Poisson, see e.g. Döbler and Peccati (2018).

4.2 An Intuitive Presentation

To begin, let us informally describe the main ideas of this chapter in the one-dimensional framework. The rest of the sections make these arguments precise and exploit the expansion to study the asymptotic behavior of level functionals of Gaussian processes.

As in the case of Rice formulas, the starting point of our analysis is Kac's counting formula (Lemma 2.15), which can be formally written as

$$N_0(f, [0, T]) = \int_0^T \delta_0(f(t))|f'(t)|\mathrm{d}t,$$

where f is a nice function and δ_0 is Dirac's delta at 0. At this point, we take a different route: we expand δ_0 and $|\cdot|$ in a convenient L^2-basis and interchange sums and integrals:

$$N_0(f, [0, T]) = \sum_{k,\ell=0}^{\infty} a_k b_\ell \int_0^T H_\ell(f(t))H_k(f'(t))\mathrm{d}t.$$

Here, $H_k : k \geq 0$ stands for the Hermite polynomial of degree k and the $a_k, b_k : k \geq 0$ are $|\cdot|$ and δ_0 Hermite coefficients respectively.

It follows that grouping the terms according to the total degree $q = k + \ell$ of the polynomials above is of great convenience since the resulting terms:

$$I_q(T) := \sum_{k=0}^{q} a_k b_{q-k} \int_0^T H_{q-k}(f(t)) H_k(f'(t)) \mathrm{d}t, \qquad q \geq 0,$$

are orthogonal and well suited for the asymptotic analysis as $T \to \infty$. Actually, when specialized to $I_q(T)$ for a fixed q (or to a finite sum of them), the classical moments method reduces to the analysis of just the second and the fourth moments. Provided that one can control the tail of the expansion, this fact entails a considerable simplification when proving central limit theorems. Furthermore, the orthogonality implies that in order to check the positivity of the limit variance, we can restrict the analysis to one convenient term of the expansion. In practice, there are essentially no tools for proving the positivity of the limit variance other than this one.

In this chapter we give a completely elementary proof of the Fourth Moment Theorem within a fixed chaos, based only on combinatoric arguments, thus avoiding stochastic integrals and more sophisticated tools. We give sufficient conditions based on integrals of the covariances of the underlying process and its derivatives for the asymptotic normality, after standardization, of each $I_q(T)$ as $T \to \infty$. We give references but we do not prove the equivalence between joint and marginal convergence of each component of a vector of chaotic random variables. Finally, we present the classical Arcones inequality (Arcones 1994) to deal with the tail of the expansion in order to get central limit theorems for level functionals of stationary Gaussian processes.

We use the following notation: $\mathbb{N}^* = \mathbb{N} \setminus \{0\} = \{1, 2, \dots\}$, for $\boldsymbol{\alpha} = (\alpha_1, \dots, \alpha_m) \in \mathbb{N}^m$ we put $|\boldsymbol{\alpha}| = \sum_{i=1}^m \alpha_i$ and $\boldsymbol{\alpha}! = \prod_{i=1}^m \alpha_i!$. For $\boldsymbol{x} = (x_1, \dots, x_m) \in \mathbb{R}^m$ and $\boldsymbol{\alpha} \in \mathbb{N}^m$,

$$\boldsymbol{x}^{\boldsymbol{\alpha}} = \prod_{i=1}^{m} x_i^{\alpha_i}. \tag{4.1}$$

For $\boldsymbol{x} = (x_{ij}) \in \mathbb{R}^{m \times n}$ and $\boldsymbol{a} = (a_{ij}) \in \mathbb{N}^{m \times n}$, we put

$$\boldsymbol{x}^{\boldsymbol{a}} = \prod_{i=1}^{m} \prod_{j=1}^{n} x_{ij}^{a_{ij}} \tag{4.2}$$

and $\boldsymbol{a}! = \prod_{i=1}^m \prod_{j=1}^n a_{ij}!$. Finally, we use the conventions $\sum_{\varnothing} = 0$ and $\prod_{\varnothing} = 1$.

4.3 Gaussian Moments and Feynman Diagrams

We begin with a basic result that captures much of the spirit of the chapter: the computation of the mixed moments of jointly Gaussian random variables.

Let $m \in \mathbb{N}^*$ and $\mathcal{PP}_m$ be the set of partitions into pairs of $\{1, \dots, m\}$.

Theorem 4.1 (Wick's Formula) *Let $\xi_1, \dots, \xi_m$ be centered jointly Gaussian random variables. Then,*

$$\mathbb{E}(\xi_1 \dots \xi_m) = \sum_{P \in \mathcal{PP}_m} \prod_{\{i,j\} \in P} \mathbb{E}(\xi_i \xi_j).$$

Example 4.1 For $m = 4$, the partitions into pairs of $\{1, 2, 3, 4\}$ are

$$\{\{1, 2\}, \{3, 4\}\}; \{\{1, 3\}, \{2, 4\}\}; \{\{1, 4\}, \{2, 3\}\}.$$

Hence,

$$\mathbb{E}(\xi_1\xi_2\xi_3\xi_4) = \mathbb{E}(\xi_1\xi_2)\mathbb{E}(\xi_3\xi_4) + \mathbb{E}(\xi_1\xi_3)\mathbb{E}(\xi_2\xi_4) + \mathbb{E}(\xi_1\xi_4)\mathbb{E}(\xi_2\xi_3).$$

For $m = 6$, there are 15 partitions into pairs of $\{1, 2, 3, 4, 5, 6\}$ and $\mathbb{E}(\xi_1\xi_2\xi_3\xi_4\xi_5\xi_6)$ is the sum of the 15 corresponding terms:

$$\begin{aligned}
&\mathbb{E}(\xi_1\xi_2\xi_3\xi_4\xi_5\xi_6) \\
&= \mathbb{E}(\xi_1\xi_2)\mathbb{E}(\xi_3\xi_4)\mathbb{E}(\xi_5\xi_6) + \mathbb{E}(\xi_1\xi_2)\mathbb{E}(\xi_3\xi_5)\mathbb{E}(\xi_4\xi_6) + \mathbb{E}(\xi_1\xi_2)\mathbb{E}(\xi_3\xi_6)\mathbb{E}(\xi_4\xi_5) \\
&\quad + \mathbb{E}(\xi_1\xi_3)\mathbb{E}(\xi_2\xi_4)\mathbb{E}(\xi_5\xi_6) + \mathbb{E}(\xi_1\xi_3)\mathbb{E}(\xi_2\xi_5)\mathbb{E}(\xi_4\xi_6) + \mathbb{E}(\xi_1\xi_3)\mathbb{E}(\xi_2\xi_6)\mathbb{E}(\xi_4\xi_5) \\
&\quad + \mathbb{E}(\xi_1\xi_4)\mathbb{E}(\xi_2\xi_3)\mathbb{E}(\xi_5\xi_6) + \mathbb{E}(\xi_1\xi_4)\mathbb{E}(\xi_2\xi_5)\mathbb{E}(\xi_3\xi_6) + \mathbb{E}(\xi_1\xi_4)\mathbb{E}(\xi_2\xi_6)\mathbb{E}(\xi_3\xi_5) \\
&\quad + \mathbb{E}(\xi_1\xi_5)\mathbb{E}(\xi_2\xi_3)\mathbb{E}(\xi_4\xi_6) + \mathbb{E}(\xi_1\xi_5)\mathbb{E}(\xi_2\xi_4)\mathbb{E}(\xi_3\xi_6) + \mathbb{E}(\xi_1\xi_5)\mathbb{E}(\xi_2\xi_6)\mathbb{E}(\xi_3\xi_4) \\
&\quad + \mathbb{E}(\xi_1\xi_6)\mathbb{E}(\xi_2\xi_3)\mathbb{E}(\xi_4\xi_5) + \mathbb{E}(\xi_1\xi_6)\mathbb{E}(\xi_2\xi_4)\mathbb{E}(\xi_3\xi_5) + \mathbb{E}(\xi_1\xi_6)\mathbb{E}(\xi_2\xi_5)\mathbb{E}(\xi_3\xi_4).
\end{aligned}$$

Remark 4.1 Note that if m is odd, $\mathcal{PP}_m = \varnothing$. Thus,

$$\mathbb{E}(\xi_1 \dots \xi_{2n+1}) = 0, \quad \text{for all } n \in \mathbb{N}.$$

Proof of Theorem 4.1 Let $\boldsymbol{t} = (t_1, \dots, t_m) \in \mathbb{R}^m$ and $\langle \cdot, \cdot \rangle$ denote the Euclidean inner product in $\mathbb{R}^m$. Set also $\boldsymbol{\xi} = (\xi_1, \dots, \xi_m)$ and write Σ for the variance-covariance matrix of $\boldsymbol{\xi}$, i.e. $\Sigma = \mathbb{E}(\boldsymbol{\xi}\boldsymbol{\xi}^\top)$. Then, the generating function of $\boldsymbol{\xi}$ is

$$\mathbb{E}(\mathrm{e}^{\langle \boldsymbol{t}, \boldsymbol{\xi} \rangle}) = \mathrm{e}^{\frac{1}{2}\boldsymbol{t}^\top \Sigma \boldsymbol{t}} = \mathrm{e}^{\frac{1}{2}\sum_{i,j=1}^m \mathbb{E}(\xi_i\xi_j)t_i t_j}.$$

Now, we expand $\mathrm{e}^{t_i\xi_i} = \sum_{k=0}^\infty \frac{(t_i\xi_i)^k}{k!}$ and, after a little bit of algebra and the interchange of the expectation with the series, we see that in the l.h.s. $\mathbb{E}(\xi_1 \dots \xi_m)$ is the coefficient of $t_1 \dots t_m$. Moving to the r.h.s., we expand $\mathrm{e}^{\mathbb{E}(\xi_i\xi_j)t_i t_j} = \sum_{k=0}^\infty \frac{(\mathbb{E}(\xi_i\xi_j)t_i t_j)^k}{k!}$. Thus, to get $t_1 \dots t_m$ in the r.h.s. we have to multiply non-overlapping pairs $t_i t_j$. Hence, since the cofactor of $t_i t_j$ is $\mathbb{E}(\xi_i\xi_j)$, each partition into pairs $P \in \mathcal{PP}_m$ contributes to the term in $t_1 \dots t_m$ by $\prod_{\{i,j\} \in P} \mathbb{E}(\xi_i\xi_j)$. The result follows. □

Corollary 4.2 *Let ξ be a standard Gaussian random variable. Then,*

$$\mathbb{E}(\xi^m) = \#\mathcal{PP}_m = \begin{cases} (m-1)(m-3)\dots 1 =: (m-1)!! & \text{if } m \text{ is even} \\ 0 & \text{otherwise} \end{cases}.$$

We now present a graphic device which helps to visualize this kind of formula.

Definition 4.1 A *Feynman diagram* (a diagram for short) is a non-oriented graph such that each vertex has degree at most 1. The *order* $m \in \mathbb{N}$ and the *rank* $\mathrm{rk} \in \mathbb{N}$ of the diagram $\gamma = (V_\gamma, E_\gamma)$ are defined as the cardinalities of the set of vertices V_γ and of the set of edges E_γ of γ respectively. A diagram γ is said to be *complete* if there are no isolated vertices.

Notation 4.1 We write $\Gamma(m)$ (resp. $\Gamma_r(m)$, $\Gamma^o(m)$) for the set of complete (resp. rank r, arbitrary) diagrams of order m.

Usually the vertices of a diagram are indexed by jointly (centered) Gaussian random variables $\xi_1, \dots, \xi_m$ in the sense that each vertex is associated to one of these random variables.

Definition 4.2 The value of a diagram γ indexed by $\xi_1, \dots, \xi_m$ is

$$\nu_\gamma = \nu_\gamma(\xi_1, \dots, \xi_m) = \prod_{\{i,j\} \in E_\gamma} \mathbb{E}(\xi_i \xi_j) \cdot \prod_{k \in I_\gamma} \xi_k,$$

where I_γ stands for the set of isolated vertices in γ.

Note that if γ is a complete diagram, its value is not random.

Notation 4.2 We denote $v_\gamma \cdot \tilde{v}_\gamma$ by ν_γ, where v_γ and $\tilde{v}_\gamma$ are the products over E_γ and I_γ, respectively, in the definition of ν_γ.

Example 4.2 The value of the diagram γ in Fig. 4.1 hereunder equals $v_\gamma = \mathbb{E}(\xi_1\xi_2)\mathbb{E}(\xi_2\xi_4) \cdot \xi_3\xi_4$.

Now, we can write the formula in Theorem 4.1 as

$$\mathbb{E}\big(\xi_1 \dots \xi_m\big) = \sum_{\gamma \in \Gamma(m)} \nu_\gamma(\xi_1, \dots, \xi_m).$$

Indeed, complete diagrams indexed by $\xi_1, \dots, \xi_m$ are just pairings of $\xi_1, \dots, \xi_m$.

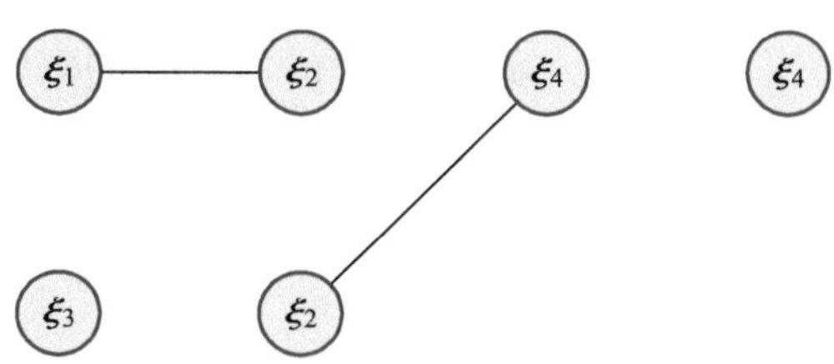

Fig. 4.1 Diagram of Example 4.2

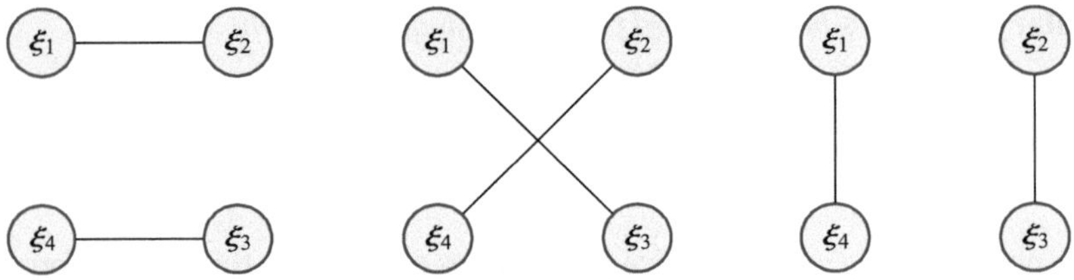

Fig. 4.2 Diagrams for the partitions into pairs of $\{1, 2, 3, 4\}$

Example 4.3 (Continuation of Example 4.1) The complete diagrams corresponding to the three partitions into pairs of the set $\{1, 2, 3, 4\}$ are $\{\{1, 2\}, \{3, 4\}\}$; $\{\{1, 3\}, \{2, 4\}\}$; $\{\{1, 4\}, \{2, 3\}\}$, see Fig. 4.2.

The value of each one of these diagrams is obtained by multiplying the covariances of the random variables associated to incident vertices. For instance, the value of the leftmost diagram in Fig. 4.2 is $\mathbb{E}(\xi_1\xi_2) \cdot \mathbb{E}(\xi_3\xi_4)$. Summing up with the value of the remaining two diagrams we recover the expression for $\mathbb{E}(\xi_1\xi_2\xi_3\xi_4)$ given in Example 4.1.

4.4 Hermite Polynomials

As mentioned in the introduction, a key role is played by polynomials in Gaussian random variables.

Hermite polynomials are the orthogonal polynomials in $\mathbb{R}$ with respect to the standard Gaussian measure $\phi(\mathrm{d}x)$. They are defined in terms of its exponential generating function.

Definition 4.3 For $m \in \mathbb{N}$, the Hermite polynomials H_m are defined by the equation

$$\mathrm{e}^{tx-t^2/2} = \sum_{m=0}^{\infty} \frac{H_m(x)}{m!} t^m. \tag{4.3}$$

Remark 4.2 There are several alternative definitions of the Hermite polynomials. For instance, they can be defined by repeated differentiation of the standard Gaussian density function ϕ:

$$\phi^{(m)}(x) = (-1)^m H_m(x)\phi(x), \quad x \in \mathbb{R}, m \in \mathbb{N}.$$

Example 4.4 The first few Hermite polynomials are:

$$\begin{aligned}
&H_0(x) = 1, && H_5(x) = x^5 - 10x^3 + 15x,\\
&H_1(x) = x, && H_6(x) = x^6 - 15x^4 + 45x^2 - 15,\\
&H_2(x) = x^2 - 1, && H_7(x) = x^7 - 21x^5 + 105x^3 - 105x,\\
&H_3(x) = x^3 - 3x, && H_8(x) = x^8 - 28x^6 + 210x^4 - 420x^2 + 105.\\
&H_4(x) = x^4 - 6x^2 + 3, &&
\end{aligned}$$

With the same kind of arguments as in the proof of Theorem 4.1, see Janson (1997) one can deduce from (4.3) that:

1. For ξ standard normal, $\mathbb{E}(H_m(\xi)) = 0$ for all $m \in \mathbb{N}^*$.
2. **Mehler's formula I**: for ξ_1, ξ_2 standard jointly Gaussian with $\mathbb{E}(\xi_1\xi_2) = \rho$, we have

$$\mathbb{E}\big(H_m(\xi_1)H_n(\xi_2)\big) = \delta_{mn}\, m!\, \rho^m. \tag{4.4}$$

3. **Mehler's formula II**: let $\xi_1, \xi_2, \xi_3, \xi_4$ be jointly Gaussian random variables such that (ξ_1, ξ_2) and (ξ_3, ξ_4) are standard Gaussian with

$$\operatorname{Cov}\big((\xi_1, \xi_2), (\xi_3, \xi_4)\big) = \boldsymbol{\rho} = \begin{pmatrix} \rho_{13} & \rho_{14} \\ \rho_{23} & \rho_{24} \end{pmatrix}.$$

Then, for $\boldsymbol{m} = (m_1, m_2, m_3, m_4) \in \mathbb{N}^4$, we have

$$\mathbb{E}\left[\prod_{i=1}^{4} H_{m_i}(\xi_i)\right] = \sum_{\alpha \in \Lambda} \frac{\boldsymbol{m}!}{\boldsymbol{\alpha}!}\, \boldsymbol{\rho}^{\boldsymbol{a}}, \tag{4.5}$$

where $\Lambda = \left\{\boldsymbol{a} = \left(\begin{smallmatrix} a_1 & a_2 \\ a_3 & a_4 \end{smallmatrix}\right) : \boldsymbol{a}\left(\begin{smallmatrix} 1 \\ 1 \end{smallmatrix}\right) = \left(\begin{smallmatrix} m_1 \\ m_2 \end{smallmatrix}\right) \text{ and } \boldsymbol{a}^\top\left(\begin{smallmatrix} 1 \\ 1 \end{smallmatrix}\right) = \left(\begin{smallmatrix} m_3 \\ m_4 \end{smallmatrix}\right)\right\}$. Note that if $m_1 + m_2 \neq m_3 + m_4$, we get $\Lambda = \emptyset$. In this case, the expectation vanishes.

We will return to these formulas from a more general point of view further on, see Theorem 4.4.

Example 4.5 Let us compute $\mathbb{E}\big(H_2(\xi_1)H_5(\xi_2)H_3(\xi_3)H_4(\xi_4)\big)$ for $(\xi_1, \xi_2, \xi_3, \xi_4)$ as above.

Note that $m_1 + m_2 = 7 = m_3 + m_4$ and that the conditions for the exponents $\boldsymbol{a} = (a_1, a_2, a_3, a_4)$ boil down to:

$$\left(\begin{smallmatrix} a_1 & a_2 \\ a_3 & a_4 \end{smallmatrix}\right)\left(\begin{smallmatrix} 1 \\ 1 \end{smallmatrix}\right) = \left(\begin{smallmatrix} 2 \\ 5 \end{smallmatrix}\right), \text{ and } \left(\begin{smallmatrix} a_1 & a_3 \\ a_2 & a_4 \end{smallmatrix}\right)\left(\begin{smallmatrix} 1 \\ 1 \end{smallmatrix}\right) = \left(\begin{smallmatrix} 3 \\ 4 \end{smallmatrix}\right).$$

The solutions to this system are $\boldsymbol{a} = (0, 2, 3, 2), (1, 1, 2, 3)$ and $(2, 0, 1, 4)$. Hence,

$$\begin{aligned}
&\mathbb{E}\big(H_2(\xi_1)H_5(\xi_2)H_3(\xi_3)H_4(\xi_4)\big)\\
&= \frac{34560}{24}\rho_{13}^0\rho_{14}^2\rho_{23}^3\rho_{24}^2 + \frac{34560}{12}\rho_{13}^1\rho_{14}^1\rho_{23}^2\rho_{24}^3 + \frac{34560}{48}\rho_{13}^2\rho_{14}^0\rho_{23}^1\rho_{24}^4\\
&= 1440\rho_{14}^2\rho_{23}^3\rho_{24}^2 + 2880\rho_{13}\rho_{14}\rho_{23}^2\rho_{24}^3 + 720\rho_{13}^2\rho_{23}\rho_{24}^4.
\end{aligned}$$

Example 4.6 Some usual cases are, for $\mathbb{E}(\xi_1\xi_2) = \mathbb{E}(\xi_3\xi_4) = 0$:

$$\begin{aligned}
\mathbb{E}\big(H_2(\xi_1)H_2(\xi_2)H_2(\xi_3)H_2(\xi_4)\big) &= 4\rho_{13}^2\rho_{24}^2 + 4\rho_{14}^2\rho_{23}^2 + 16\rho_{13}\rho_{14}\rho_{23}\rho_{24};\\
\mathbb{E}\big(H_2(\xi_1)H_2(\xi_2)H_4(\xi_3)\big) &= 24\rho_{13}^2\rho_{23}^2.
\end{aligned}$$

As a consequence of Mehler's formula I, the set $\{H_m : m \in \mathbb{N}\}$ is an orthogonal set in $L^2(\mathbb{R}, \phi(\mathrm{d}x))$. It can be shown in the same way that it is complete. Hence, if $f \in L^2(\mathbb{R}, \phi(\mathrm{d}x))$ we have in the L^2-sense:

$$f = \sum_{m=0}^{\infty} c_m H_m, \qquad \text{with} \qquad c_m = \frac{1}{m!}\mathbb{E}\big(f(\xi)H_m(\xi)\big), \tag{4.6}$$

ξ being standard Gaussian. This is the *Hermite expansion* of f. Note that $c_0 = \mathbb{E}(f(\xi))$ and

$$\sum_{m=0}^{\infty} c_m^2\, m! = \|f\|_2^2 < \infty. \tag{4.7}$$

Analogously, for $q \in \mathbb{N}$, the sets

$$\Big\{\overline{H}_{\boldsymbol{\alpha}}(\boldsymbol{x}) := \prod_{i=1}^{d} H_{\alpha_i}(x_i) : |\boldsymbol{\alpha}| = q\Big\} \tag{4.8}$$

are orthogonal in $L^2(\mathbb{R}^d, \phi_d(\mathrm{d}x))$ and their union over q is complete.

4.5 Gaussian Hilbert Spaces

We move to a more general setting.

Definition 4.4 A Gaussian Hilbert space G is a sub-Hilbert space of $L^2(\Omega, \mathcal{U}, \mathbb{P})$ whose elements are centered Gaussian random variables.

Remark 4.3 Some direct properties of Gaussian Hilbert spaces are:

1. Any set of elements of G has joint Gaussian distribution.
2. $G \subset L^p(\Omega, \mathcal{U}, \mathbb{P})$ for $0 < p < \infty$. By Hölder's inequality, any product of finitely many random variables in G has moments of any order.
3. All L^p-norms, $0 < p < \infty$ are equivalent on G. Moreover, the topology induced by these norms coincides with the topology induced by the convergence in probability.

Though we are mainly concerned with the L^2-structure of Gaussian Hilbert spaces, we need to deal with higher order moments of random variables in G.

Now, we introduce the polynomial spaces which are the basic blocks of our expansions.

Definition 4.5 Let G be a Gaussian Hilbert space.

1. For $m \in \mathbb{N}^*$, the m-th *polynomial chaos* in G, $\overline{\mathcal{P}_m}(G)$, is the closure in $L^2(\Omega, \mathcal{U}, \mathbb{P})$ of

$$\mathcal{P}_m(G) = \{p(\xi_1, \ldots, \xi_k) : k < \infty; \xi_1, \ldots, \xi_k \in G; p \text{ is a polynomial of degree } \leq m\}.$$

2. For $m \in \mathbb{N}^*$, the *homogeneous chaos*, C_m, is defined by

$$\overline{\mathcal{P}_m}(G) = \overline{\mathcal{P}_{m-1}}(G) \oplus C_m,$$

where $\oplus$ stands for orthogonal direct sum.
3. For $m = 0$, we set $\mathcal{P}_0(G) = C_0 = \mathbb{R}$.

Remark 4.4 Without loss of generality one may assume that the random variables in the definition of $\mathcal{P}_m(G)$ are orthogonal.

The next theorem states that the sum of the homogeneous chaos coincides with the subspace of $L^2(\Omega, \mathcal{U}, \mathbb{P})$ consisting of those random variables which are measurable with respect to G.

Theorem 4.3 *With the above notation, we have*

$$\bigoplus_{m=0}^{\infty} C_m = L^2(\Omega, \mathcal{U}_G, \mathbb{P}),$$

where $\mathcal{U}_G = \sigma(G) \subset \mathcal{U}$ *is the* σ*-field generated by all random variables in* G.

Proof Clearly $\overline{\cup_{m=0}^{\infty} \mathcal{P}_m(G)} = \bigoplus_{m=0}^{\infty} C_m \subset L^2(\Omega, \mathcal{U}_G, \mathbb{P})$. Thus, it suffices to check that if $X \in L^2(\Omega, \mathcal{U}_G, \mathbb{P})$ is orthogonal to $\mathcal{P}_m(G)$ for every $m \in \mathbb{N}$, then $X = 0$ a.s. Moreover, without loss of generality, we can consider Hermite polynomials instead of generic ones. Thus, let $X \in L^2(\Omega, \mathcal{U}_G, \mathbb{P})$ satisfy

$$\mathbb{E}(H_{k_1}(\xi_1) \dots H_{k_n}(\xi_n) \cdot X) = 0,$$

for any $n, k_1, \dots, k_n \in \mathbb{N}$ and independent standard normals $\xi_i, \dots, \xi_n \in G$.

For $n \in \mathbb{N}$, let $\mathcal{G}_n = \sigma\big(\xi_1, \dots, \xi_n\big)$. Thus, for all $n, k_1, \dots, k_n \in \mathbb{N}$ we have

$$\mathbb{E}\big(H_{k_1}(\xi_1) \dots H_{k_n}(\xi_n) \cdot \mathbb{E}(X \mid \mathcal{G}_n)\big) = 0.$$

But, as $\mathbb{E}(X \mid \mathcal{G}_n)$ is $\mathcal{G}_n$-measurable, there exists a measurable function Θ_n defined on $\mathbb{R}^n$ such that $\mathbb{E}(X \mid \mathcal{G}_n) = \Theta_n(\xi_1, \dots, \xi_n)$. Therefore,

$$\begin{aligned} 0 &= \mathbb{E}\big(H_{k_1}(\xi_1) \dots H_{k_n}(\xi_n) \cdot \mathbb{E}(X \mid \mathcal{G}_n)\big) \\ &= \int_{\mathbb{R}^n} H_{k_1}(x_1) \dots H_{k_n}(x_n)\, \Theta_n(x_1, \dots, x_n)\, \phi_n(\mathrm{d}x_1, \dots, \mathrm{d}x_n), \end{aligned}$$

for any $k_1, \dots, k_n \in \mathbb{N}$. Since Hermite polynomials form an orthogonal basis of $L^2(\mathbb{R}^n, \phi_n(\mathrm{d}x))$, the above equality implies that $\Theta_n = 0$ ϕ_n-a.s. It follows that $\mathbb{E}(X \mid \mathcal{G}_n) = 0$ $\mathbb{P}$-a.s. for all n and passing to the limit as $n \to \infty$ we get $X = \lim_n \mathbb{E}(X \mid \mathcal{G}_n) = 0$ a.s. This concludes the proof. □

Notation 4.3 For $m \geq 0$, we denote by π_m and $\pi_{\leq m}$ the (orthogonal) *projections* onto C_m and $\oplus_{k=0}^{m} C_k = \overline{\mathcal{P}}_m(G)$ respectively.

Example 4.7 Projecting powers ξ^m gives Hermite polynomials:

$$\pi_m(\xi^m) = H_m(\xi).$$

This fact is justified in Remark 4.6 below.

As a consequence of Theorem 4.3, if $\xi \in L^2(\Omega, \mathcal{U}_G, \mathbb{P})$, for $m \geq 0$ there exist $\xi_m \in C_m$ such that

$$\xi = \sum_{m=0}^{\infty} \xi_m; \quad (\xi_m = \pi_m(\xi)).$$

This expansion is called the *Wiener chaos expansion* of ξ. In Sect. 4.9 we show how to find the ξ_m explicitly in the case when ξ is a level functional of a stationary Gaussian process.

4.6 Wick Products and Diagram Formulas

The above expansions express a random variable $\xi \in L^2(\Omega, \mathcal{U}_G, \mathbb{P})$ in terms of "polynomial" random variables $\pi_m(\xi) : m \in \mathbb{N}$. Recalling Example 4.7, it is natural to consider the multiplication and mixed moment formula for Hermite polynomials in Gaussian random variables.

We need to add some structure to diagrams.

Notation 4.4 Denote by $\boldsymbol{\alpha} \diamond \boldsymbol{\beta}$ the concatenation of multi indexes $\boldsymbol{\alpha} \in \mathbb{N}^m$ and $\boldsymbol{\beta} \in \mathbb{N}^n$, i.e.:

$$\boldsymbol{\alpha} \diamond \boldsymbol{\beta} = (\alpha_1, \ldots, \alpha_m; \beta_1, \ldots, \beta_n) \in \mathbb{N}^{m+n}.$$

We use the same notation to concatenate real vectors. Besides, for $\boldsymbol{\xi} = (\xi_1, \ldots, \xi_m) \in \mathbb{R}^m$ and $\boldsymbol{\alpha} \in \mathbb{N}^m$ with $|\boldsymbol{\alpha}| = q$, we define the vector $\boldsymbol{\xi}^{(\boldsymbol{\alpha})}$ in $\mathbb{R}^q$ by

$$\boldsymbol{\xi}^{(\boldsymbol{\alpha})} := \diamond_{i=1}^m \xi_i^{\diamond \alpha_i}, \qquad \text{with} \qquad \xi_i^{\diamond \alpha_i} = \diamond_{j=1}^{\alpha_i} (\xi_i).$$

The j-th coordinate of $\boldsymbol{\xi}^{(\boldsymbol{\alpha})}$ is denoted by $\boldsymbol{\xi}^{(\boldsymbol{\alpha}),j}$, $j = 1, \ldots, q$.

Observe that $\boldsymbol{\xi}^{(\boldsymbol{\alpha})}$ is a vector while $\boldsymbol{\xi}^{\boldsymbol{\alpha}}$ defined by (4.1) is a real number. For instance, for $\boldsymbol{\xi} = (\xi_1, \xi_2, \xi_3)$ and $\boldsymbol{\alpha} = (1, 3, 2)$ we have

$$\boldsymbol{\xi}^{(\boldsymbol{\alpha})} = (\xi_1, \xi_2, \xi_2, \xi_2, \xi_3, \xi_3) \text{ and } \boldsymbol{\xi}^{\boldsymbol{\alpha}} = \xi_1 \cdot \xi_2^3 \cdot \xi_3^2.$$

Notation 4.5 Let $\Gamma(p, q)$ (resp. $\Gamma_r(p, q)$; $\Gamma^o(p, q)$) stand for the set of complete (resp. of rank r; arbitrary) diagrams of order pq with vertices of the form $\{(i, j) : 1 \le i \le p; 1 \le j \le q\}$ and without "horizontal edges", i.e. without edges joining vertices with equal first coordinates. The vertex sets

$$L_i = \{(i, j) : 1 \le j \le q\}, \quad i = 1, \ldots, p,$$

are the levels of the diagram.

Consider $\gamma \in \Gamma^o(p, q)$ whose levels are indexed respectively by $\boldsymbol{\xi}_i^{(\boldsymbol{\alpha}_i)} : i = 1, \ldots, p$. For an edge $e = \{(i, j), (i', j')\} \in E_\gamma$ set $d_1(e) = \boldsymbol{\xi}_i^{(\boldsymbol{\alpha}_i),j}$ and $d_2(e) = \boldsymbol{\xi}_{i'}^{(\boldsymbol{\alpha}'_i),j'}$, then the value of γ is:

$$\nu_\gamma = \nu_\gamma(\boldsymbol{\xi}_1^{(\boldsymbol{\alpha}_1)}, \ldots, \boldsymbol{\xi}_p^{(\boldsymbol{\alpha}_p)}) = \prod_{e \in E_\gamma} \mathbb{E}(d_1(e) d_2(e)) \cdot \prod_{(i,j) \in I_\gamma} \boldsymbol{\xi}_i^{(\boldsymbol{\alpha}_i),j}, \tag{4.9}$$

where I_γ stands, as above, for the isolated vertices in γ. We also write $\nu_\gamma = \upsilon_\gamma \cdot \tilde{\nu}_\gamma$, where υ_γ and $\tilde{\nu}_\gamma$ are the products over E_γ and I_γ, respectively, in the r.h.s of (4.9).

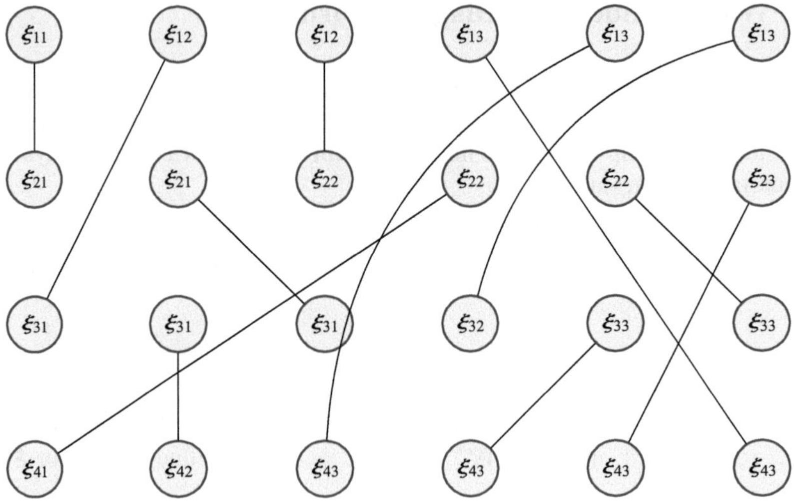

Fig. 4.3 Diagram of Example 4.8

Example 4.8 Let $m = 3$, $p = 4$ and $q = 6$, that is, we consider four Hermite polynomials with total degree 6 in three random variables. Consider $\boldsymbol{\alpha}_1 = (1, 2, 3)$, $\boldsymbol{\alpha}_2 = (2, 3, 1)$, $\boldsymbol{\alpha}_3 = (3, 1, 2)$ and $\boldsymbol{\alpha}_4 = (1, 1, 4)$. Let $\boldsymbol{\xi}_i = (\xi_{i1}, \xi_{i2}, \xi_{i3})$ for $i = 1, 2, 3, 4 = p$. Thus,

$$\boldsymbol{\xi}_1^{(\boldsymbol{\alpha}_1)} = (\xi_{11}, \xi_{12}, \xi_{12}, \xi_{13}, \xi_{13}, \xi_{13});$$

$$\boldsymbol{\xi}_2^{(\boldsymbol{\alpha}_2)} = (\xi_{21}, \xi_{21}, \xi_{22}, \xi_{22}, \xi_{22}, \xi_{23});$$

$$\boldsymbol{\xi}_3^{(\boldsymbol{\alpha}_3)} = (\xi_{31}, \xi_{31}, \xi_{31}, \xi_{32}, \xi_{33}, \xi_{33});$$

$$\boldsymbol{\xi}_4^{(\boldsymbol{\alpha}_4)} = (\xi_{41}, \xi_{42}, \xi_{43}, \xi_{43}, \xi_{43}, \xi_{43}).$$

Consider the (complete) diagram γ described in Fig. 4.3. The value of this diagram is

$$\nu_\gamma\left(\boldsymbol{\xi}_1^{(\boldsymbol{\alpha}_1)}, \boldsymbol{\xi}_2^{(\boldsymbol{\alpha}_2)}, \boldsymbol{\xi}_3^{(\boldsymbol{\alpha}_3)}, \boldsymbol{\xi}_4^{(\boldsymbol{\alpha}_4)}\right) = \mathbb{E}(\xi_{11}\xi_{21})\mathbb{E}(\xi_{21}\xi_{31})\mathbb{E}(\xi_{31}\xi_{12})\mathbb{E}(\xi_{31}\xi_{42})\mathbb{E}(\xi_{41}\xi_{22})$$
$$\cdot\mathbb{E}(\xi_{12}\xi_{22})\mathbb{E}(\xi_{22}\xi_{33})\mathbb{E}(\xi_{13}\xi_{32})\mathbb{E}(\xi_{13}\xi_{43})^2\mathbb{E}(\xi_{23}\xi_{43})\mathbb{E}(\xi_{33}\xi_{43}).$$

The next theorem gives the formula for the expectation of the product of Hermite polynomials in Gaussian random variables. Recall that $\overline{H}$ is defined in (4.8) and that $\Gamma(p, q)$ is the set of complete diagrams.

Theorem 4.4 (Diagram Formula) *Let $\boldsymbol{\xi}_1, \ldots, \boldsymbol{\xi}_p$ be standard (jointly) Gaussian vectors in G^m and $\boldsymbol{\alpha}_1, \ldots, \boldsymbol{\alpha}_p \in \mathbb{N}^m$ such that $|\boldsymbol{\alpha}_i| = q$. Then,*

$$\mathbb{E}(\overline{H}_{\boldsymbol{\alpha}_1}(\boldsymbol{\xi}_1) \ldots \overline{H}_{\boldsymbol{\alpha}_p}(\boldsymbol{\xi}_p)) = \sum_{\gamma \in \Gamma(p,q)} \nu_\gamma(\boldsymbol{\xi}_1^{(\boldsymbol{\alpha}_1)}, \ldots, \boldsymbol{\xi}_p^{(\boldsymbol{\alpha}_p)}).$$

The proof is presented below, after some preliminary facts are established. Roughly speaking, the expectation of a product of Hermite polynomials in Gaussian random variables is computed by summing the values of all the complete diagrams which associate one level with each one of the Hermite polynomials, each level being indexed by the vector $\xi_i^{(\alpha_i)}$.

Remark 4.5 We take $|\boldsymbol{\alpha}_i| = q$ for all i, for simplicity, and since it suffices for our applications in the forthcoming sections. However, the theorem still holds if the $\boldsymbol{\alpha}_i$'s have different norms. In that case, the sum on the r.h.s. of the diagram formula runs over the set of complete diagrams, with no horizontal edges, whose levels are indexed by the $\xi_i^{(\alpha_i)}$, see Janson (1997, Th. 3.12).

Note that Mehler's formulas (4.4) and (4.5) are particular cases of this extension to Theorem 4.4. For instance, in the case of Mehler's formula I (4.4), one has $p = 2$ and $m = 1$. Besides, the complete diagrams with no horizontal edges joining two levels are just bijections between those levels. In particular, if $|\boldsymbol{\alpha}_1| \neq |\boldsymbol{\alpha}_2|$ there are no such diagrams and the expectation vanishes.

Before giving other examples and proving Theorem 4.4 we need to complete the relation between Hermite polynomials and projections onto the chaos.

Definition 4.6 The Wick product of $\xi_1, \ldots, \xi_m \in G$ is defined to be

$$\langle \xi_1, \ldots, \xi_m \rangle = \pi_m(\xi_1, \ldots, \xi_m).$$

Remark 4.6 In the case that $\xi_i = \xi$ for $i = 1, \ldots, m$, we recover the Hermite polynomials

$$\langle \xi^m \rangle = H_m(\xi).$$

Indeed, in $G = \mathbb{R}\xi$ the chaos C_m, $m \geq 0$, are uni-dimensional and thus generated by $H_m(\xi)$. Remark 4.7 below extends this relation to the tensorial case.

The next lemma gives an explicit formula for the Wick product of $\xi_1, \ldots, \xi_m \in G$ in terms of diagrams.

Lemma 4.5 *Let $\xi_1, \ldots, \xi_m \in G$, then*

$$\langle \xi_1, \ldots, \xi_m \rangle = \sum_{\gamma \in \Gamma(m)} (-1)^{\mathrm{rk}(\gamma)} \nu_\gamma,$$

where $\mathrm{rk}(\gamma)$ *is the rank of* γ.

Proof Let $\psi = \sum_\gamma (-1)^{\mathrm{rk}(\gamma)} v_\gamma$. It is clear that $\psi \in \mathcal{P}_m(G)$ and that the term corresponding to the diagram with rank 0 equals $\xi_1 \dots \xi_m$. Thus, $\xi_1 \dots \xi_m - \psi \in \mathcal{P}_{m-2}(G) \subset \mathcal{P}_{m-1}(G)$.

Since $\overline{\mathcal{P}_m(G)} = \overline{\mathcal{P}_{m-1}(G)} \oplus C_m$, it suffices to show that $\xi_1 \dots \xi_m - \psi = \pi_{\leq m-1}(\xi_1 \dots \xi_m)$, or equivalently, that ψ is orthogonal to $\mathcal{P}_{m-1}(G)$. Since $\mathcal{P}_{m-1}(G)$ is generated by monomials, it suffices to consider arbitrary $\eta_1, \dots, \eta_n \in G$, $n < m$ and to show that

$$\mathbb{E}(\psi \cdot \eta_1 \dots \eta_n) = 0.$$

For the moment, let $n \in \mathbb{N}^*$. We have,

$$\mathbb{E}(\psi \cdot \eta_1 \dots \eta_n) = \sum_\gamma (-1)^{r(\gamma)} v_\gamma \mathbb{E}(\tilde{v}_\gamma \eta_1 \dots \eta_n) = \sum_\gamma (-1)^{r(\gamma)} v_\gamma \sum_{\gamma'} \nu_{\gamma'},$$

where the second sum ranges over all complete diagrams indexed by the isolated vertices in γ and $\eta_1, \dots, \eta_n$. Defining $\gamma'' = \gamma \cup \gamma'$ yields

$$\mathbb{E}(\psi \cdot \eta_1 \dots \eta_n) = \sum_\gamma (-1)^{r(\gamma)} \sum_{\gamma'' \supset \gamma} \nu_{\gamma''},$$

where the inner sum runs over all complete diagrams, indexed by $\xi_1, \dots, \xi_m$; $\eta_1, \dots, \eta_n$, which extend γ. Now, we reverse the order of summation and group diagrams γ'' according to how many edges lie among vertices indexed by $\xi_1, \dots, \xi_m$. Thus,

$$\mathbb{E}(\psi \cdot \eta_1 \dots \eta_n) = \sum_{\gamma''} \left(\sum_{\gamma \subset \gamma''} (-1)^{r(\gamma)} \right) \nu_{\gamma''},$$

where γ'' ranges over all complete diagrams indexed by $\xi_1, \dots, \xi_m; \eta_1, \dots, \eta_n$. Now if γ'' has exactly $\ell \geq 0$ edges joining vertices indexed by $\xi_1, \dots, \xi_m$, for each $k \leq \ell$ there are $\binom{\ell}{k}$ choices for a subdiagram γ with rank k. Hence,

$$\sum_{\gamma \subset \gamma''} (-1)^{r(\gamma)} = \sum_{k=0}^{\ell} \binom{\ell}{k} (-1)^k = (1 + (-1))^\ell = \begin{cases} 1 & \text{if } \ell = 0 \\ 0 & \text{otherwise} \end{cases}.$$

Hence, to get a non-vanishing term, the rank of γ must be 0. Thus,

$$\mathbb{E}(\psi \cdot \eta_1 \dots \eta_n) = \sum_{\gamma''} \nu_{\gamma''}, \tag{4.10}$$

where the sum ranges over all complete diagrams γ'' indexed by $\xi_1, \dots, \xi_m$, $\eta_1, \dots, \eta_n$ with no edges joining two vertices among $\xi_1, \dots, \xi_m$. If $n < m$, there are no such diagrams γ''. Hence,

$$\mathbb{E}(\psi \cdot \eta_1 \dots \eta_n) = 0.$$

The result follows. □

Now, we can write the projections $\pi_m : m \geq 0$ or, equivalently, the Wick products in terms of (tensorial) Hermite polynomials.

Remark 4.7 As said above, by definition of C_m, if $n \neq m$ $\langle \xi_1 \dots \xi_n \rangle$ and $\langle \eta_1 \dots \eta_m \rangle$ are orthogonal for any $\xi_1, \dots, \xi_n, \eta_1, \dots, \eta_m \in G$. Besides, if $\mathbb{E}(\xi_i \eta_j) = 0$ for all i, j, it follows from Lemma 4.5 that

$$\langle \xi_1 \dots \xi_n \rangle \cdot \langle \eta_1 \dots \eta_m \rangle = \langle \xi_1 \dots \xi_n, \eta_1 \dots \eta_m \rangle.$$

Together with Remark 4.6 this fact implies that for orthogonal and unit variance $\xi_1, \dots, \xi_m$, we have (with the usual notation $\boldsymbol{\alpha} = (\alpha_1, \dots, \alpha_m)$, $\boldsymbol{\xi} = (\xi_1, \dots, \xi_m)$):

$$\overline{H}_{\boldsymbol{\alpha}}(\boldsymbol{\xi}) = \prod_{i=1}^{m} H_{\alpha_i}(\xi_i) = \Big\langle \prod_{i=1}^{m} \xi_i^{\alpha_i} \Big\rangle = \langle \boldsymbol{\xi}^{\boldsymbol{\alpha}} \rangle.$$

The next example specializes Theorem 4.4 to the case of two Hermite polynomials, extending Mehler's formula I (4.4), which is recovered by setting $\xi_1 = \dots = \xi_m$ and $\xi_{m+1} = \dots = \xi_{2m}$.

Example 4.9 Let $p = 2$, the complete diagrams in $\Gamma(2, m)$ are the bijections between $\{\xi_1, \dots, \xi_m\}$ and $\{\xi_{m+1}, \dots, \xi_{2m}\}$. Thus, we can write

$$\mathbb{E}\big(\langle \xi_1 \dots \xi_m \rangle \langle \xi_{m+1} \dots \xi_{2m} \rangle\big) = \sum_{\pi \in \mathcal{S}_m} \prod_{i=1}^{m} \mathbb{E}(\xi_i \xi_{m+\pi(i)}),$$

where $\mathcal{S}_m$ stands for the symmetric group of order m.

Proof of Theorem 4.4 We begin with the case $p = 2$. By definition of C_q, $q \geq 0$, these spaces are orthogonal and, thus, so are the Wick products $\langle \xi_1 \dots \xi_q \rangle$ and $\langle \eta_1 \dots \eta_{q'} \rangle$ whenever $q \neq q'$. Moreover, $\xi_{q+1} \dots \xi_{2q} - \langle \xi_{q+1} \dots \xi_{2q} \rangle$ is orthogonal to C_q. Thus,

$$\mathbb{E}\big(\langle \xi_1 \dots \xi_q \rangle \langle \xi_{q+1} \dots \xi_{2q} \rangle\big) = \mathbb{E}\big(\langle \xi_1 \dots \xi_q \rangle \xi_{q+1} \dots \xi_{2q}\big) = \sum_{\gamma} \nu_\gamma,$$

where the sum ranges over all complete diagrams with no edges joining two vertices ξ_i, ξ_j with $i, j \leq q$. The last equality is a consequence of the proof of Lemma 4.5, see (4.10).

Now, we move to the case $p > 2$, which is in the same spirit as that of the case $p = 2$. By Remark 4.7 and Lemma 4.5 we have

$$\mathbb{E}\Big(\overline{H}_{\alpha_1}(\boldsymbol{\xi}_1)\dots\overline{H}_{\alpha_p}(\boldsymbol{\xi}_p)\Big) = \mathbb{E}\Big(\langle\boldsymbol{\xi}_1{}^{\alpha_1}\rangle\dots\langle\boldsymbol{\xi}_p{}^{\alpha_p}\rangle\Big)$$

$$= \mathbb{E}\Big(\prod_{i=1}^{p}\sum_{\bar{\gamma}_i\in\Gamma^o(q)}(-1)^{r(\bar{\gamma}_i)}\nu_{\bar{\gamma}_i}(\xi_i^{(\alpha_i)})\Big).$$

Now, the value of any non-complete diagram is centered. Thus, it does not contribute to the last expectation. Hence,

$$\mathbb{E}\Big(\overline{H}_{\alpha_1}(\boldsymbol{\xi}_1)\dots\overline{H}_{\alpha_p}(\boldsymbol{\xi}_p)\Big)$$

$$= \sum_{\gamma_1,\dots,\gamma_p\in\Gamma^o(q)}(-1)^{r(\gamma_1)+\dots+r(\gamma_p)}\sum_{\gamma'\in\Gamma(pq):\gamma_i\subset\gamma',\forall i}\nu_{\gamma'}(\xi_1^{(\alpha_1)},\dots,\xi_p^{(\alpha_p)}).$$

Here, γ' extends each γ_i. More precisely, γ' is a diagram defined on a vertex set with p levels indexed respectively by $\boldsymbol{\xi}_i^{(\alpha_i)} : i = 1,\dots,p$ and each sub-diagram γ_i is defined within level i.

Again, we change the order of the sums and consider

$$\sum_{\gamma'\in\Gamma(pq)}\sum_{\gamma_i\subset\gamma',\forall i}(-1)^{r(\gamma_1)+\dots+r(\gamma_p)} = \sum_{\gamma\in\Gamma(pq)}\prod_{i=1}^{p}\sum_{\gamma_i\subset\gamma}(-1)^{r(\gamma_i)}.$$

Now, proceeding as in the case $p = 2$ we get that if γ' has at least one edge joining vertices corresponding to $\boldsymbol{\xi}_i$ then its value vanishes. The result follows. □

Corollary 4.6 *Let $\xi_1,\dots,\xi_m,\eta_1,\dots,\eta_n\in G$. Then,*

$$\langle\xi_1\dots\xi_m\rangle\langle\eta_1\dots\eta_n\rangle = \sum_{\gamma}\langle\nu_\gamma\rangle,$$

where the sum ranges over all diagrams with one level indexed by $\xi_1,\dots,\xi_m$ and another one indexed by $\eta_1,\dots,\eta_n$.

Proof Consider $k \geq 0$ and $\varepsilon_1,\dots,\varepsilon_k \in G$, multiply both sides of the desired equality by $\langle\varepsilon_1,\dots,\varepsilon_k\rangle$ and take expectations. Both sides yield $\sum_{\gamma'}\nu_{\gamma'}$, where the sum ranges over all complete diagrams indexed by all of $\xi_i,\eta_j,\varepsilon_k$ without edges among vertices indexed by the same letter. Since the products of the form $\varepsilon_1\dots\varepsilon_k$ are dense in G, the result follows. □

Corollary 4.7 *The m-th homogeneous chaos C_m is the closure of the linear span of Hermite polynomials on elements of G. That is,*

$$C_m = \overline{\operatorname{span}\{\overline{H}_\alpha(\xi_1, \dots, \xi_k) : k < \infty; \xi_1, \dots, \xi_k \in G; |\alpha| = m\}}.$$

Proof It is easy to see that the Hermite polynomials span the same spaces as general polynomials of the same degree. Moreover, Remark 4.7 shows that they are orthogonal. The result follows. □

We end this section with a key property of chaotic random variables: *hypercontractivity*. It states that in a fixed chaos $(\overline{\mathcal{P}_m}, C_m)$ all the L^p-norms are equivalent. We consider only $p \geq 2$, see Janson (1997) for the remaining cases and for a discussion about the values of the constants.

Theorem 4.8 *Let $X \in \overline{\mathcal{P}_m}(G)$. Then, for $2 \leq p, q$,*

$$\|X\|_q \leq c\, \|X\|_p,$$

for some constant $c = c(m, p, q)$ depending only on m, p and q.

Proof It suffices to consider $X \in \mathcal{P}_m(G)$. Write

$$X = \sum_{|\alpha| \leq m} c_\alpha \langle \boldsymbol{\xi}^\alpha \rangle,$$

where $\boldsymbol{\xi} = (\xi_1, \dots, \xi_m)$ is standard normal.

We start with the case $p = 2$, $q = 4$. We compute X^2 using Corollary 4.6 and some standard manipulations,

$$X^2 = \sum_{|\alpha|,|\beta| \leq m} c_\alpha c_\beta \langle \boldsymbol{\xi}^\alpha \rangle \langle \boldsymbol{\xi}^\beta \rangle = \sum_{|\alpha|,|\beta| \leq m} c_\alpha c_\beta \sum_{\gamma \leq \alpha \wedge \beta} d_\gamma \langle \boldsymbol{\xi}^{\alpha - \gamma + \beta - \gamma} \rangle.$$

Here $\wedge$ must be read as the coordinate-wise minimum. The combinatorial coefficients d_γ can be computed explicitly and are bounded by a constant depending only on m. Write $\boldsymbol{\delta} = \boldsymbol{\alpha} + \boldsymbol{\beta} - 2\boldsymbol{\gamma}$ so that

$$X^2 = \sum_{|\delta| \leq 2m} b_\delta \langle \boldsymbol{\xi}^\delta \rangle, \quad \text{with} \quad b_\delta = \sum_{(\alpha,\beta,\gamma) \in \Lambda_\delta} c_\alpha c_\beta d_\gamma,$$

where $\Lambda_\delta = \{(\boldsymbol{\alpha}, \boldsymbol{\beta}, \boldsymbol{\gamma}) \in \mathbb{N}^{3m} : |\boldsymbol{\alpha}|, |\boldsymbol{\beta}| \leq m, \boldsymbol{\gamma} \leq \boldsymbol{\alpha} \wedge \boldsymbol{\beta}, \boldsymbol{\alpha} + \boldsymbol{\beta} - 2\boldsymbol{\gamma} = \boldsymbol{\delta}\}$.

Now, let C be a meaningless constant whose actual value may change at each step. Using $d_\gamma \leq C$ and the Cauchy–Schwarz inequality we get

$$|b_\delta|^2 = \Big| \sum_{(\alpha,\beta,\gamma) \in \Lambda_\delta} c_\alpha c_\beta d_{\frac{1}{2}(\delta - \alpha - \beta)} \Big|^2 \leq C \Big(\sum_{(\alpha,\beta,\gamma) \in \Lambda_\delta} |c_\alpha c_\beta| \Big)^2$$

$$\leq C\Big(\sum_{|\alpha|,|\beta|\leq m}|c_\alpha c_\beta|\Big)^2 \leq C\Big(\sum_{|\alpha|\leq m}|c_\alpha|\Big)^4$$

$$\leq C\Big(\sum_{|\alpha|\leq m}c_\alpha^2\Big)^2 \leq C\|X\|_2^4.$$

Thus, plugging this into the bound for $|b_\delta|^2$ we get

$$\|X\|_4^4 = \sum_{|\delta|\leq 2m}|b_\delta|^2\delta! \leq C\|X\|_2^4\sum_{|\delta|\leq 2m}\delta! \leq C\|X\|_2^4.$$

The result follows for $p = 2$ and $q = 4$.

The next step is to consider $p = 2$ and $q = 2k$, $k \in \mathbb{N}$. Hölder's inequality shows that for $X, Y \in \mathcal{P}_m$:

$$\|XY\|_2 \leq C\|X\|_2\|Y\|_2.$$

By induction we get a constant depending on m and $q = 2k$ s.t.

$$\|X\|_{2k} \leq C\|X\|_2.$$

Now, since the L^p norms are increasing with p, we conclude that for $2 \leq p \leq q$ we have

$$\|X\|_2 \leq \|X\|_p \leq \|X\|_q \leq \|X\|_{2k} \leq C\|X\|_2,$$

with k large enough. The result follows. □

4.7 The Fourth Moment Theorem

The celebrated Fourth Moment Theorem is a classical result (Nualart and Peccati 2005) which states that within a fixed chaos, convergence towards the standard Gaussian distribution is equivalent to the convergence of the variances to 1 and of the fourth moments to 3. Thus, within a fixed chaos, the Method of Moments boils down to the verification of the convergence of two numerical sequences.

A similar result holds within a finite sum of chaos, see Peccati and Tudor (2005). As mentioned in the introduction, we do not prove this multivariate version of the Fourth Moment Theorem.

Let $X(\cdot) = \{X(t) : t \in \mathbb{R}^D\}$ be an $\mathbb{R}^{\mathcal{D}}$-valued mean zero unit variance stationary Gaussian field. Since our motivation comes from level functionals, see Sect. 4.9, we write our functionals as integrals of Hermite polynomials rather than just Hermite polynomials. Let $T = [0, 1]^D$ and consider

$$f_{n,q} = \sum_{\alpha \in \mathcal{J}_q} c_{\alpha}^{(n)} \int_{nT} \overline{H}_{\alpha}(X(t))\mathrm{d}t, \tag{4.11}$$

where $\mathcal{J}_q = \{\alpha \in \mathbb{N}^{\mathcal{D}} : |\alpha| = q\}$ and $nT = \{nt : t \in T\}$.

Theorem 4.9 *Fix $q \in \mathbb{N}^*$ and consider the sequence of functionals $(f_{n,q})_{n \in \mathbb{N}}$ as defined in* (4.11). *The following conditions are equivalent:*

1. *$f_{n,q}$ converges in distribution as $n \to \infty$ towards a standard Gaussian random variable, i.e. $f_{n,q} \underset{n\to\infty}{\Rightarrow} N(0, 1)$, where $\Rightarrow$ denotes the convergence in distribution,*
2. *the second and the fourth moments of $f_{n,q}$ converge as $n \to \infty$ to the corresponding moments of a standard Gaussian random variable, i.e. $\mathbb{E}(f_{n,q}^2) \underset{n\to\infty}{\to} 1$ and $\mathbb{E}(f_{n,q}^4) \underset{n\to\infty}{\to} 3$.*

Remark 4.8 The Fourth Moment Theorem is usually proved using stochastic calculus, see Nualart and Peccati (2005). It has several generalizations, including ones that are multidimensional (Peccati and Tudor 2005), quantitative (Peccati 2014), or that involve some other pair of moments (Azmoodeh et al. 2016). Our elementary proof is mainly inspired by Nourdin (2011) and avoids stochastic calculus. In this sense, it is similar in spirit to the approach used in Kratz and León (1997) to get the expansions, in contrast to that of Slud (1991b).

Let us pay attention to one of the results mentioned above. Peccati and Tudor (2005) proved strong results about joint convergence of finitely many chaotic random variables, see also Nourdin and Peccati (2012, Th. 6.2.3.). In particular, joint normal convergence of chaotic random variables of distinct orders is equivalent to the marginal normal convergence of each one of them.

Remark 4.9 As a byproduct of the proof of Theorem 4.9, we see that the asymptotic normality of $f_{n,q}$ follows from the convergence towards zero of some combinations of integrals involving products of covariances of $X(\cdot)$ and its derivatives, see (4.14) below.

Proof of Theorem 4.9 The proof is divided into three steps which correspond to the computation of the second moment, the fourth moment and higher order moments of $f_{n,q}$. For the sake of brevity, we omit the index n in the coefficients.
<u>Second moment:</u> Using the diagram formula, Theorem 4.4, we get

$$\mathbb{E}(f_{n,q}^2) = \sum_{\alpha,\beta \in \mathcal{J}_q} c_{\alpha} c_{\beta} \int_{(nT)^2} \mathbb{E}\Big(\overline{H}_{\alpha}(X(s))\overline{H}_{\beta}(X(t))\Big)\, \mathrm{d}s\mathrm{d}t$$

$$= \sum_{\alpha,\beta\in\mathcal{J}_q} c_\alpha c_\beta \int_{(nT)^2} \sum_{\gamma\in\Gamma(2,q)} \nu_\gamma(X^{(\alpha)}(s), X^{(\beta)}(t))\,\mathrm{d}s\mathrm{d}t$$

$$= \sum_{\gamma\in\Gamma(2,q)} \sum_{\alpha,\beta\in\mathcal{J}_q} c_\alpha c_\beta \int_{(nT)^2} \nu_\gamma(X^{(\alpha)}(s), X^{(\beta)}(t))\,\mathrm{d}s\mathrm{d}t.$$

Keep in mind this expression since it will appear in the next two steps.
Fourth moment: Using the multiplication formula, Corollary 4.6, we get:

$$\overline{H}_\alpha(X(s))\overline{H}_\beta(X(t)) = \langle X^\alpha(s)\rangle\langle X^\beta(t)\rangle = \sum_{\gamma\in\Gamma^o(2,q)} \langle \nu_\gamma(X^{(\alpha)}(s), X^{(\beta)}(t))\rangle.$$

Here,

$$\langle \nu_\gamma(X^{(\alpha)}(s), X^{(\beta)}(t))\rangle = \upsilon_\gamma(X^{(\alpha)}(s), X^{(\beta)}(t))\langle \tilde{\upsilon}_\gamma(X^{(\alpha)}(s), X^{(\beta)}(t))\rangle,$$

where υ_γ and $\tilde{\upsilon}_\gamma$ are the non-random and random parts of the value of γ: i.e.

$$\upsilon_\gamma(X^{(\alpha)}(s), X^{(\beta)}(t)) = \prod_{\{(1,i),(2,j)\}\in E_\gamma} \mathbb{E}(X_i^{(\alpha)}(s)X_j^{(\beta)}(t));$$

$$\tilde{\upsilon}_\gamma(X^{(\alpha)}(s), X^{(\beta)}(t)) = \prod_{j\in I_1(\gamma)} X_j^{(\alpha)}(s)\cdot \prod_{j\in I_2(\gamma)} X_j^{(\beta)}(t).$$

Here $I_1(\gamma)$ (resp. $I_2(\gamma)$) stands for the isolated vertices of level 1 (resp. 2) of γ.
Grouping and taking expectation we have

$$\mathbb{E}\Big(\overline{H}_\alpha(X(s))\overline{H}_\beta(X(t))\cdot \overline{H}_{\alpha'}(X(s'))\overline{H}_{\beta'}(X(t'))\Big)$$

$$= \sum_{\gamma,\gamma'\in\Gamma^o(2,q)} \upsilon_\gamma(X^{(\alpha)}(s)X^{(\beta)}(t))\upsilon_{\gamma'}(X^{(\alpha')}(s')X^{(\beta')}(t'))$$

$$\cdot\,\mathbb{E}\Big(\langle\tilde{\upsilon}_\gamma(X^{(\alpha)}(s)X^{(\beta)}(t))\rangle\,\langle\tilde{\upsilon}_{\gamma'}(X^{(\alpha')}(s')X^{(\beta')}(t'))\rangle\Big).$$

If $\#E_\gamma \neq \#E_{\gamma'}$, then $\langle\tilde{\upsilon}_\gamma\rangle$ y $\langle\tilde{\upsilon}_{\gamma'}\rangle$ are orthogonal and the corresponding terms vanish. Thus, we split the sum according to $m = \#E_\gamma = \#E_\gamma$. Hence, $\mathbb{E}(f_{n,q}^4)$ equals

$$\sum_{m=0}^{q} \sum_{\gamma,\gamma'\in\Gamma_m(2,q)} \sum_{\alpha,\beta,\alpha',\beta'\in\mathcal{J}_q} c_\alpha c_\beta c_{\alpha'} c_{\beta'} \int_{(nT)^4} \upsilon_\gamma(X_s^{(\alpha)}, X_t^{(\beta)})\upsilon_{\gamma'}(X_{s'}^{(\alpha')}, X_{t'}^{(\beta')})$$

$$\cdot\,\mathbb{E}\Big(\langle\tilde{\upsilon}_\gamma(X_s^{(\alpha)}, X_t^{(\beta)})\rangle\langle\tilde{\upsilon}_{\gamma'}(X_{s'}^{(\alpha')}, X_{t'}^{(\beta')})\rangle\Big)\mathrm{d}s\mathrm{d}t\mathrm{d}s'\mathrm{d}t'$$

$$= \sum_{m=0}^{q} \mathbb{E}\left(\sum_{\gamma \in \Gamma_m(2,q)} \sum_{\alpha,\beta \in \mathcal{J}_q} c_\alpha c_\beta \int_{(nT)^2} v_\gamma(X_s^{(\alpha)}, X_t^{(\beta)}) \langle \tilde{v}_\gamma(X_s^{(\alpha)}, X_t^{(\beta)}) \rangle \mathrm{d}s\mathrm{d}t \right)^2 .$$

We will look at the terms $m = 0$ and $m = q$ separately. If $m = q$, the diagram γ is complete and $\nu_\gamma = v_\gamma$ is not random. Thus, the $m = q$-term is

$$\left(\sum_{\gamma \in \Gamma(2,q)} \sum_{\alpha,\beta \in \mathcal{J}_q} c_\alpha c_\beta \int_{(nT)^2} \nu_\gamma(X^{(\alpha)}(s), X^{(\beta)}(t)) \mathrm{d}s\mathrm{d}t \right)^2 \underset{n\to\infty}{\to} 1,$$

since this is exactly the square of $\mathbb{E}(f_{n,q}^2)$.

Now we move to $m = 0$. In $\Gamma_0(2, q)$ there is only one diagram, thus the sum over γ has only one term. Moreover, the value of this diagram is the product of all the coordinates of $X^{(\alpha)}(s)$ and $X^{(\beta)}(t)$: i.e. $X^\alpha(s)X^\beta(t)$. Hence, the term $m = 0$ equals

$$\mathbb{E}\left(\sum_{\alpha,\beta \in \mathcal{J}_q} c_\alpha c_\beta \int_{(nT)^2} \langle X_s^\alpha X_t^\beta \rangle \mathrm{d}s\mathrm{d}t \right)^2$$

$$= \sum_{\alpha,\beta,\alpha',\beta' \in \mathcal{J}_q} c_\alpha c_\beta c_{\alpha'} c_{\beta'} \int_{(nT)^4} \mathbb{E}\left(\langle X_s^\alpha X_t^\beta \rangle \langle X_{s'}^{\alpha'} X_{t'}^{\beta'} \rangle \right) \mathrm{d}s\mathrm{d}t\mathrm{d}s'\mathrm{d}t'$$

$$= \sum_{\alpha,\beta,\alpha',\beta' \in \mathcal{J}_q} c_\alpha c_\beta c_{\alpha'} c_{\beta'} \int_{(nT)^4} \sum_{\gamma \in \Gamma(2,2q)} \nu_\gamma(X_s^{(\alpha)} \diamond X_t^{(\beta)}; X_{s'}^{(\alpha')} \diamond X_{t'}^{(\beta')}) \mathrm{d}s\mathrm{d}t\mathrm{d}s'\mathrm{d}t'. \tag{4.12}$$

We discuss the inner sum according to how many edges join the first q vertices of level 1 with the first q vertices of level 2. More precisely, we split the complete diagram γ into a diagram γ_1 joining the first q vertices in each level, a diagram γ_2 joining the latter q vertices in each level, and the complement of $\gamma_1 \cup \gamma_2$ with respect to γ. Hence, we can write

$$\sum_{\gamma \in \Gamma(2,2q)} \nu_\gamma = \sum_{h=0}^{q} \sum_{\gamma_1,\gamma_2 \in \Gamma_h(2,q)} v_{\gamma_1}(X^{(\alpha)}(s), X^{(\alpha')}(s')) v_{\gamma_2}(X^{(\beta)}(t), X^{(\beta')}(t'))$$

$$\cdot \mathbb{E}\left(\langle \tilde{v}_{\gamma_1}(X^{(\alpha)}(s), \tilde{X}^{(\alpha')}(s')) \rangle \langle \tilde{v}_{\gamma_1}(\tilde{X}^{(\beta)}(t), X^{(\beta')}(t')) \rangle \right). \tag{4.13}$$

Here, $\tilde{X}$ stands for an independent copy of X. Note that the introduction of $\tilde{X}$ avoids the terms involving covariances between random variables evaluated at s (resp. s') with those evaluated at t (resp. t') in the computation of the latter expectation in the expression above, thus, avoiding horizontal edges. Now, replacing (4.13) in (4.12), we can write the term $m = 0$ as a sum of squares.

$$\sum_{h=0}^{q} \mathbb{E}\left(\sum_{\gamma_1 \in \Gamma_h(2,q)} \sum_{\boldsymbol{\alpha},\boldsymbol{\beta} \in \mathcal{J}_q} c_{\boldsymbol{\alpha}} c_{\boldsymbol{\beta}} \int_{(nT)^2} v_{\gamma_1}(X_s^{(\boldsymbol{\alpha})}, X_{s'}^{(\boldsymbol{\alpha}')}) \langle \tilde{v}_{\gamma_1}(X_s^{(\boldsymbol{\alpha})}, \tilde{X}_{s'}^{(\boldsymbol{\alpha}')}) \rangle \mathrm{d}s\mathrm{d}s' \right)^2 .$$

(While squaring the $\mathrm{d}s\mathrm{d}s'$ integral we put $\mathrm{d}t'\mathrm{d}t$ for the second integral).

It follows that the terms $h = 0$ and $h = q$ are analogous to the term $m = q$, thus each one of them tends to 1 as $n \to \infty$. Consequently, the sum of the terms $m = q$, $m = h = 0$ and $m = 0, h = q$ tends to 3 as $n \to \infty$.

Hence, as all the terms are squares, $\mathbb{E}(f_{n,q}^4) \to_{n\to\infty} 3$ if and only if the terms $m = 1, \ldots, q-1$ and $m = 0, h = 1, \ldots, q-1$ tend to 0 as $n \to \infty$.

Terms corresponding to $m = 0, h = 1, \ldots, q-1$ are obtained replacing (4.13) in (4.12):

$$\sum_{\gamma_1,\gamma_2 \in \Gamma_h(2,q)} \sum_{\boldsymbol{\alpha},\boldsymbol{\beta},\boldsymbol{\alpha}',\boldsymbol{\beta}' \in \mathcal{J}_q} c_{\boldsymbol{\alpha}} c_{\boldsymbol{\beta}} c_{\boldsymbol{\alpha}'} c_{\boldsymbol{\beta}'} \int_{(nT)^4} v_{\gamma_1}(X_s^{(\boldsymbol{\alpha})}, X_{s'}^{(\boldsymbol{\alpha}')}) v_{\gamma_2}(X_{t'}^{(\boldsymbol{\beta}')}, X_t^{(\boldsymbol{\beta})})$$

$$\cdot \mathbb{E}\big(\langle \tilde{v}_{\gamma_1}(X_s^{(\boldsymbol{\alpha})}, \tilde{X}_{s'}^{(\boldsymbol{\alpha}')}) \rangle \langle \tilde{v}_{\gamma_2}(X_{t'}^{(\boldsymbol{\beta}')}, \tilde{X}_t^{(\boldsymbol{\beta})}) \rangle\big) \mathrm{d}s\mathrm{d}s'\mathrm{d}t'\mathrm{d}t$$

$$= \sum_{\gamma_1,\gamma_2 \in \Gamma_h(2,q)} \sum_{\tilde{\gamma} \in \Gamma(2,q-h)} \sum_{\boldsymbol{\alpha},\boldsymbol{\beta},\boldsymbol{\alpha}',\boldsymbol{\beta}' \in \mathcal{J}_q} c_{\boldsymbol{\alpha}} c_{\boldsymbol{\beta}} c_{\boldsymbol{\alpha}'} c_{\boldsymbol{\beta}'}$$

$$\int_{(nT)^4} \prod_{i=1}^{h} r_{e_i}(s-s') r_{e_i'}(t'-t) \prod_{i=1}^{q-h} r_{\tilde{e}_i}(s-t') r_{\tilde{e}_{q-h+i}}(t-s') \mathrm{d}s\mathrm{d}s'\mathrm{d}t'\mathrm{d}t. \tag{4.14}$$

Here, r_e is a shorthand for $r_{d_1(e)d_2(e)}$. Besides, edges e_i, e_i' y $\tilde{e}_i$ correspond to γ_1, γ_2 and $\tilde{\gamma}$ respectively. Terms corresponding to $m = 1, \ldots, q-1$ are similar but they involve the product of six groups of covariances instead of four.

Higher order moments: We move to the $2k$-th moment. We have

$$\mathbb{E}(f_{n,q}^{2k}) = \mathbb{E}\left(\sum_{\boldsymbol{\alpha} \in \mathcal{J}} c_{\boldsymbol{\alpha}} \int_{nT} \overline{H}_{\boldsymbol{\alpha}}(X(s)) \mathrm{d}s \right)^{2k}$$

$$= \sum_{\gamma \in \Gamma(2k,q)} \sum_{\boldsymbol{\alpha}_i \in J_q, i \le 2k} \left[\prod_{i=1}^{2k} c_{\boldsymbol{\alpha}_i} \right] \int_{(nT)^{2k}} v_\gamma(X^{(\boldsymbol{\alpha}_1)}(s_1), \ldots, X^{(\boldsymbol{\alpha}_{2k})}(s_{2k})) \mathrm{d}s_1 \ldots \mathrm{d}s_{2k}.$$

We say that the diagram $\gamma \in \Gamma(p, q)$ is regular if γ is complete and there exists a partition into pairs $P \in \mathcal{PP}_p$ such that all edges in γ occur within levels in the same pair of P.

We first consider regular diagrams. It is easy to see that each partition P in $\mathcal{PP}_{2k}$ (i.e. all the diagrams whose edges lie within the pairs in P) contributes in the same proportion to the $2k$-th moment:

$$\sum_{\gamma_j\in\Gamma(2,q),j\leq k}\sum_{\boldsymbol{\alpha}_i\in\mathcal{J}_q,i\leq 2k}\left[\prod_{i=1}^{2k}c_{\boldsymbol{\alpha}_i}\right]\int_{(nT)^{2k}}\prod_{j=1}^{k}\upsilon_{\gamma_j}(X_{s_{2j-1}}^{(\boldsymbol{\alpha}_{2j-1})},X_{s_{2j}}^{(\boldsymbol{\alpha}_{2j})})\mathrm{d}s_1\ldots\mathrm{d}s_{2k}$$

$$=\left[\sum_{\gamma\in\Gamma(2,q)}\sum_{\boldsymbol{\alpha}_1,\boldsymbol{\alpha}_2\in\mathcal{J}_q}c_{\boldsymbol{\alpha}_1}c_{\boldsymbol{\alpha}_2}\int_{(nT)^2}\upsilon_{\gamma_j}(X_{s_1}^{(\boldsymbol{\alpha}_1)},X_{s_2}^{(\boldsymbol{\alpha}_2)})\mathrm{d}s_1\mathrm{d}s_2\right]^k\underset{n\to\infty}{\to}1,$$

where we have used the fact that the sum within the brackets coincides with $\mathbb{E}(f_{n,q}^2)$. Then, the contribution of regular diagrams is $\#\mathcal{PP}_{2k}=(2k-1)!!$, which coincides with the $2k$-th moment of a standard Gaussian random variables.

Let us turn now to the non-regular diagrams. If a diagram is not regular, there must be two levels joined by exactly $0<m<q$ edges, without loss of generality we can assume that these levels are those indexed by $X^{(\boldsymbol{\alpha}_1)}(s_1)$ and $X^{(\boldsymbol{\alpha}_2)}(s_2)$. To compute the contribution of all diagrams joining these levels by $0<m<q$ edges we write $f_{n,q}^k=f_{n,q}^{k-2}f_{n,q}^2$, we use Remark 4.7 and Corollary 4.6 to write $f_{n,q}^2$ and impose the condition there. We have that the contribution of all diagrams joining these levels by $0<m<q$ edges is:

$$\mathbb{E}\Bigg(f_{n,q}^{k-2}\cdot\Bigg[\sum_{\gamma_1\in\Gamma_m(2,q)}\sum_{\boldsymbol{\alpha}_1,\boldsymbol{\alpha}_2\in\mathcal{J}_q}c_{\boldsymbol{\alpha}_1}c_{\boldsymbol{\alpha}_2}\cdot\int_{(nT)^2}\upsilon_{\gamma_1}(X_{s_1}^{(\boldsymbol{\alpha}_1)},X_{s_2}^{(\boldsymbol{\alpha}_2)})\langle\tilde{\upsilon}_{\gamma_1}(X_{s_1}^{(\boldsymbol{\alpha}_1)},X_{s_2}^{(\boldsymbol{\alpha}_2)})\rangle$$
$$\times\,\mathrm{d}s_1\mathrm{d}s_2\Bigg]\Bigg).$$

Now, by the Cauchy–Schwarz inequality, we can bound this expression by

$$\left[\mathbb{E}(f_{n,q}^{2k-4})\right]^{1/2}\cdot\Bigg[\mathbb{E}\Bigg(\sum_{\gamma_1\in\Gamma_m(2,q)}\sum_{\boldsymbol{\alpha}_1,\boldsymbol{\alpha}_2\in\mathcal{J}_q}c_{\boldsymbol{\alpha}_1}c_{\boldsymbol{\alpha}_2}$$
$$\int_{(nT)^2}\upsilon_{\gamma_1}(X_{s_1}^{(\boldsymbol{\alpha}_1)},X_{s_2}^{(\boldsymbol{\alpha}_2)})\langle\upsilon_{\gamma_1}(X_{s_1}^{(\boldsymbol{\alpha}_1)},X_{s_2}^{(\boldsymbol{\alpha}_2)})\rangle\mathrm{d}s_1\mathrm{d}s_2\Bigg)^2\Bigg]^{1/2}.$$

The first factor is bounded by hypercontractivity, see Theorem 4.8. The second factor is analogous to the terms $0<m<q$ or $m=0,0<h<q$ in the fourth moment.

To conclude the proof assume in the first place that condition 2 holds. Note that $\mathbb{E}(f_{n,q}^4)\to_n 3$ implies that the integrals in (4.14) tend to 0 as $n\to\infty$, thus the contribution of non-regular diagrams vanish and the higher order moments of $f_{n,q}$ converge to the corresponding ones of a standard Gaussian random variable.

Hence, $f_{n,q}$ converges in distribution towards a standard Gaussian random variable. Conversely, if condition 1 holds true, we can assume without loss of generality that $\mathbb{E}(f_{n,q}^2) = 1$ for all n, q. Thus, the hypercontractivity property implies that $\mathbb{E}(f_{n,q}^4) \le C$ for all n, q for some constant C. Hence, we can pass to the limit to get $\mathbb{E}(f_{n,q}^4) \to 3$. This concludes the proof. □

4.8 Arcones' Inequality

A key role is played by Arcones' inequality.

Definition 4.7 Consider $g \in L^2(\mathbb{R}^{\mathcal{D}}, \phi_{\mathcal{D}}(\mathrm{d}x))$ such that $\mathbb{E}g(\boldsymbol{\xi}) = 0$. Its Hermite rank is defined by

$$\mathrm{rk}(g) = \min\left\{k \in \mathbb{N} : \exists \boldsymbol{\alpha} \in \mathbb{N}^{\mathcal{D}} \text{ such that } |\boldsymbol{\alpha}| = k,\ \mathbb{E}\big[g(\boldsymbol{\xi}) \cdot \overline{H}_{\boldsymbol{\alpha}}(\boldsymbol{\xi})\big] \neq 0\right\},$$

$\boldsymbol{\xi}$ being a standard Gaussian vector.

Lemma 4.10 (Arcones) *Let* $\boldsymbol{\xi} = (\xi_1, \dots, \xi_{\mathcal{D}})$ *and* $\boldsymbol{\eta} = (\eta_1, \dots, \eta_{\mathcal{D}})$ *be two standard Gaussian vectors with covariances:* $c_{i,j} = \mathbb{E}(\xi_i \eta_j)$, $i, j = 1, \dots, \mathcal{D}$. *Set*

$$\psi := \max\left\{\max_{1\le j\le \mathcal{D}} \sum_{i=1}^{\mathcal{D}} |c_{i,j}|,\ \max_{1\le i\le \mathcal{D}} \sum_{j=1}^{\mathcal{D}} |c_{i,j}|\right\}.$$

Consider $g \in L^2(\mathbb{R}^{\mathcal{D}}, \phi_{\mathcal{D}}(\mathrm{d}x))$, $\mathbb{E}g(\boldsymbol{\xi}) = 0$, *of Hermite rank* τ. *Then, if* $|\psi| \le 1$,

$$\big|\mathbb{E}\big(g(\boldsymbol{\xi})g(\boldsymbol{\eta})\big)\big| \le \mathbb{E}\Big(g(\boldsymbol{\xi})^2\Big)\psi^{\tau}.$$

Remark 4.10 With the notation in Lemma 4.10, we always have the Cauchy–Schwarz bound

$$\big|\mathbb{E}\big(g(\boldsymbol{\xi})g(\boldsymbol{\eta})\big)\big| \le \mathbb{E}\big(g(\boldsymbol{\xi})^2\big) =: \|g\|_2^2,$$

which correspond to the case of rank 0. Thus, Arcones' inequality is a refinement of Cauchy–Schwarz's inequality.

Proof Since $g \in L^2(\mathbb{R}^{\mathcal{D}}, \phi_{\mathcal{D}}(\mathrm{d}x))$, we can expand it in terms of Hermite polynomials.

$$g(x) = \sum_{q=\tau}^{\infty} \sum_{\boldsymbol{\alpha}\in\mathcal{J}_q} c_{\boldsymbol{\alpha}} \overline{H}_{\boldsymbol{\alpha}}(x); \qquad \text{with} \qquad \sum_{q=\tau}^{\infty} \sum_{\boldsymbol{\alpha}\in\mathcal{J}_q} c_{\boldsymbol{\alpha}}^2 \boldsymbol{\alpha}! < \infty.$$

Note that $\mathrm{rk}(g) = \tau$ implies that the terms $|\boldsymbol{\alpha}| < \tau$ vanish.

By Example 4.9 we have

$$\begin{aligned}\mathbb{E}\big(g(\boldsymbol{\xi})g(\boldsymbol{\eta})\big) &= \sum_{q=\tau}^{\infty}\sum_{\boldsymbol{\alpha},\boldsymbol{\beta}\in\mathcal{J}_q} c_{\boldsymbol{\alpha}}c_{\boldsymbol{\beta}}\mathbb{E}\Big(\overline{H}_{\boldsymbol{\alpha}}(\boldsymbol{\xi})\overline{H}_{\boldsymbol{\beta}}(\boldsymbol{\eta})\Big)\\ &= \sum_{q=\tau}^{\infty}\sum_{\boldsymbol{\alpha},\boldsymbol{\beta}\in\mathcal{J}_q} c_{\boldsymbol{\alpha}}c_{\boldsymbol{\beta}} \sum_{\pi\in\mathcal{S}_q}\prod_{i=1}^{q}\mathbb{E}(\boldsymbol{\xi}_i^{(\boldsymbol{\alpha})}\boldsymbol{\eta}_{\pi(i)}^{(\boldsymbol{\beta})}).\end{aligned}$$

Now, we group the inner sum according to how many factors $c_{i,j} = \mathbb{E}(\xi_i\eta_j)$ $(i, j = 1, \ldots, \mathcal{D})$ appear. For $\pi \in \mathcal{S}_q$, the power m_{11} of $c_{1,1}$ is $\#\{\pi(1), \ldots, \pi(\alpha_1)\} \cap \{1, \ldots, \beta_1\}$, the power m_{ij} of $c_{i,j}$ is obtained analogously. Hence, define $\mathcal{M}$ to be the set of matrices $M = (m_{ij})_{ij}$ s.t. $\sum_i m_{ij} = \beta_j$ for $j = 1, \ldots, \mathcal{D}$ and $\sum_j m_{ij} = \alpha_i$ for $i = 1, \ldots, \mathcal{D}$. Note that for each matrix M there are $\frac{\boldsymbol{\alpha}!\boldsymbol{\beta}!}{M!}$ permutations in $\mathcal{S}_q$ which lead to M. Thus, we can write the last sum as

$$\begin{aligned}\big|\mathbb{E}\big(g(\boldsymbol{\xi})g(\boldsymbol{\eta})\big)\big| &= \Big|\sum_{q=\tau}^{\infty}\sum_{\boldsymbol{\alpha},\boldsymbol{\beta}\in\mathcal{J}_q}^{\infty} c_{\boldsymbol{\alpha}}c_{\boldsymbol{\beta}}\sum_{M\in\mathcal{M}}\frac{\boldsymbol{\alpha}!\boldsymbol{\beta}!}{M!}\prod_{i,j=1}^{\mathcal{D}} c_{i,j}^{m_{ij}}\Big|\\ &\le \sum_{q=\tau}^{\infty}\sum_{\boldsymbol{\alpha},\boldsymbol{\beta}\in\mathcal{J}_q}^{\infty} |c_{\boldsymbol{\alpha}}\boldsymbol{\alpha}!c_{\boldsymbol{\beta}}\boldsymbol{\beta}!| \sum_{M\in\mathcal{M}}\frac{1}{M!}\prod_{i,j=1}^{\mathcal{D}}|c_{i,j}^{m_{ij}}|\\ &\le \sum_{q=\tau}^{\infty}\sum_{\boldsymbol{\alpha},\boldsymbol{\beta}\in\mathcal{J}_q}^{\infty}\frac{1}{2}(c_{\boldsymbol{\alpha}}^2(\boldsymbol{\alpha}!)^2 + c_{\boldsymbol{\beta}}^2(\boldsymbol{\beta}!)^2)\sum_{M\in\mathcal{M}}\frac{1}{M!}\prod_{i,j=1}^{\mathcal{D}}|c_{i,j}^{m_{ij}}|.\end{aligned}$$

Both terms are analogous and bounded by the same expression. We consider the term in $c_{\boldsymbol{\alpha}}$,

$$\begin{aligned}\sum_{q=\tau}^{\infty}\sum_{\boldsymbol{\alpha},\boldsymbol{\beta}\in\mathcal{J}_q}^{\infty} c_{\boldsymbol{\alpha}}^2(\boldsymbol{\alpha}!)^2\sum_{M\in\mathcal{M}}\frac{1}{M!}\prod_{i,j=1}^{\mathcal{D}}|c_{i,j}^{m_{ij}}| &= \sum_{q=\tau}^{\infty}\sum_{\boldsymbol{\alpha}\in\mathcal{J}_q}^{\infty} c_{\boldsymbol{\alpha}}^2\boldsymbol{\alpha}!\sum_{\boldsymbol{\beta}\in\mathcal{J}_q}^{\infty}\sum_{M\in\mathcal{M}}\frac{\boldsymbol{\alpha}!}{M!}\prod_{i,j=1}^{\mathcal{D}}|c_{i,j}^{m_{ij}}|\\ &= \sum_{q=\tau}^{\infty}\sum_{\boldsymbol{\alpha}\in\mathcal{J}_q}^{\infty} c_{\boldsymbol{\alpha}}^2\boldsymbol{\alpha}!\prod_{i=1}^{\mathcal{D}}\sum_{(m_{ij}):\sum_j m_{ij}=\alpha_i}\frac{\alpha_i!}{M_i!}\prod_{j=1}^{D}|c_{i,j}^{m_{ij}}|\\ &= \sum_{q=\tau}^{\infty}\sum_{\boldsymbol{\alpha}\in\mathcal{J}_q}^{\infty} c_{\boldsymbol{\alpha}}^2\boldsymbol{\alpha}!\prod_{i=1}^{\mathcal{D}}\Big(\sum_j|c_{i,j}^{m_{ij}}|\Big)^{\alpha_i} \le \sum_{q=\tau}^{\infty}\sum_{\boldsymbol{\alpha}\in\mathcal{J}_q}^{\infty} c_{\boldsymbol{\alpha}}^2\boldsymbol{\alpha}!\,\psi^{|\boldsymbol{\alpha}|}.\end{aligned}$$

Here, $M_i = (m_{i1}, \ldots, m_{i\mathcal{D}})$.

Now, if $\psi \leq 1$ we have $\psi^{|\alpha|} \leq \psi^{\tau}$. Thus, if $\psi \leq 1$, we can bound last sum by

$$\sum_{q=\tau}^{\infty} \sum_{\alpha \in \mathcal{J}_q} c_\alpha^2 \alpha! \cdot \psi^{\tau} = \|g\|_2^2 \, \psi^{\tau}.$$

Taking into account the term in $c_{\boldsymbol{\beta}}$, the result follows. □

4.9 Central Limit Theorem for Level Functionals

In this section, we show how the tools of the present chapter can be combined in order to study the asymptotic normality of the measure of a level set of a Gaussian random field. In the following and final section we offer a few brief comments on the generalizations of these ideas. Elsewhere, in Chap. 8 we present some concrete examples.

Let us recall the outline of the method presented at the beginning of this chapter: we start with an integral representation for the geometric measure of the level set, then we obtain its Hermite expansion. Arcones' inequality helps us to prove the finiteness of the limit variance and to truncate the expansion so that we can use the Fourth Moment Theorem to establish the asymptotic normality of the measure of the level set.

4.9.1 Integral Representation of the Measure of the Level Set

Let $X(\cdot) = \{X(t) : t \in \mathbb{R}^D\}$ be an $\mathbb{R}^d$-valued ($d \leq D$) standardized ($\mathbb{E}(X(t)) = 0$, $\mathrm{Var}(X(t) = \mathrm{Id}_d)$ stationary Gaussian field with a.s. C^1 paths. For every Borel set S and every $u \in \mathbb{R}$, the σ_{D-d} measure of the level set $\mathcal{L}_u(S)$ is denoted by $m_u(S)$.

Consider the following conditions:

Assumption (A.4.1) (Geman's Condition) For every set S, $\mathbb{E}(m_u(S))^2 < +\infty$; and for a sufficiently small set S, $\mathbb{E}(m_u(S))^2$ is continuous as a function of u.

Assumption (A.4.2) (Continuity of the Measure of the Level Set) For every Borel set S, $m_u(S)$ is a.s. continuous as a function of $u \in \mathbb{R}^d$.

The following lemma gives a key representation for $m_u(S)$ both a.s. and in the L^2-sense.

Lemma 4.11 *Assume that $X(\cdot) = \{X(t) : t \in \mathbb{R}^D\}$ is an $\mathbb{R}^d$-valued stationary Gaussian field satisfying Assumptions* **(A.4.1)** *and* **(A.4.2)**. *Let S be a Borel set of $\mathbb{R}^D$. Then, it holds true a.s. and in L^2 that*

$$m_u(S) = \lim_{\delta \to 0} m_{u,\delta}(S) := \lim_{\delta \to 0} \int_S \frac{1}{W(\delta)} \mathbb{1}_{\|X(t)-u\|<\delta} \Delta(s) \mathrm{d}s, \tag{4.15}$$

where $W(\delta)$ stands for the volume of the ball of radius $\delta \in \mathbb{R}^D$ and $\Delta(s) = [\det(X'(s)(X'(s)^\top))]^{1/2}$.

Proof Note that both sides of (4.15) are additive in S. So it suffices to prove it for sufficiently small sets S.

Almost sure convergence: By the Co-area formula (2.8), the Kac's counter $m_{u,\delta}(S)$ is just the mean over the ball $B(u, \delta)$ of $m_v(S)$, so it converges because of the continuity Assumption **(A.4.2)**.

Lower-bound for the limit second moment of $m_{u,\delta}$: Using Fatou's lemma

$$\mathbb{E}(m_u^2(S)) = \mathbb{E}(\liminf_{\delta\to 0} m_{u,\delta}^2(S)) \le \liminf_{\delta\to 0} \mathbb{E}(m_{u,\delta}^2(S)).$$

Upper-bound for the limit second moment of $m_{u,\delta}$: Consider a sufficiently small set S so that **(A.4.1)** holds true. Using Jensen's inequality we get

$$\mathbb{E}(m_{u,\delta}^2(S)) \ge \frac{1}{W(\delta)} \int_{B(u,\delta)} \mathbb{E}(m_v^2(S))\mathrm{d}v.$$

Letting δ tend to zero it follows that

$$\limsup_{\delta\to 0} \mathbb{E}(m_{u,\delta}^2(S)) \ge \lim_{\delta\to 0} \frac{1}{W(\delta)} \int_{B(u,\delta)} \mathbb{E}(m_v^2(S))\mathrm{d}v = \mathbb{E}(m_u^2(S)).$$

Bringing the above together, we obtain the convergence of the second moment and the almost sure convergence. It is well known that these facts imply the L^2 convergence. □

4.9.1.1 Sufficient Conditions for (A.4.1) and (A.4.2)

The situation is different depending on whether $D = d$ or not.

When $D = d$, the condition which was assumed at the beginning of the section implies that the random field $X(\cdot)$ satisfies the conditions of Proposition 2.11, so that the Bulinskaya condition (2.15) is met. This implies **(A.4.2)** since the number of roots is locally constant. As for **(A.4.1)**, we can use the results of Sect. 2.8:

When $D > d$, for **(A.4.2)** we can apply Lemma 2.7, which demands a "strong Bulinskaya condition" (2.15). Sufficient conditions for that can be found in Azaïs and Wschebor (2009, Prop. 6.12): the paths must be C^2 and for all $t \in T$, $X(t)$ and $X'(t)$ must have a joint bounded density. Since the random field is stationary, $X(t)$ and $X'(t)$ are independent and it suffices to have $X'(0)$ N.D. As for **(A.4.1)**, when $X(\cdot)$ satisfies **(A.2.9)**, Theorem 2.28 gives the result.

4.9.2 Hermite Expansion of the Measure of the Level Set

Theorem 4.12 *Assume that $X(\cdot) = \{X(t) : t \in \mathbb{R}^D\}$ is an $\mathbb{R}^d$-valued stationary Gaussian field satisfying Assumptions* **(A.4.1)** *and* **(A.4.2)***. Then, for $u \in \mathbb{R}^d$, a.s. and in the L^2-sense we have*

$$m_u(nT) = \sum_{q=0}^{\infty} \sum_{\alpha \diamond \beta \in \mathcal{J}_{2q}} c_{\alpha,\beta}(u) \int_{nT} \overline{H}_{\alpha}(X(s))\overline{H}_{\beta}(X'(s))\mathrm{d}s.$$

Here, $\alpha \in \mathbb{N}^d$, $\beta \in \mathbb{N}^{dD}$ and $\mathcal{J}_{2q}$ is defined as in (4.11) *with $\mathcal{D} = d(D+1)$. Moreover, $c_{\alpha,\beta}(u) = b_{\alpha}(u) f_{\beta}$, where, see Kratz and León (1997), $b_{\alpha}(u) = b_{(\alpha_1,\dots,\alpha_d)}(u) = \prod_{i=1}^{d} b_{\alpha_i}(u)$ with*

$$b_j(u) = \frac{1}{j!} H_j(u)\phi(u); \qquad \textit{for } j \in \mathbb{N},$$

and f_{β} is the β-th Hermite coefficient of $s \mapsto \Delta(s)$, which was defined in Lemma 4.11.

Proof We consider the case $d < D$, the other case is similar, replacing the second moment by the second factorial moment. Let $\psi_{\delta}(x) = \frac{1}{W(\delta)} \mathbb{1}_{\|x-u\|<\delta}$ and define

$$b_{\alpha,\delta}(u) = \frac{1}{\alpha!} \int_{\mathbb{R}^d} \psi_{\delta}(x) H_{\alpha}(x) \phi_d(x)\mathrm{d}x.$$

Note that $b_{\alpha,\delta}(u) \to_{\delta \to 0} b_{\alpha}(u)$. Consider also $m_{u,\delta}(nT)$ as in Lemma 4.11, i.e.

$$m_{u,\delta}(nT) = \int_{nT} \psi_{\delta}(X(s))\Delta(X'(s))\mathrm{d}s.$$

The expansion for a function belonging to $L^2(\mathbb{R}^{\mathcal{D}}, \phi_{\mathcal{D}}(\mathrm{d}x))$, ($\mathcal{D} = d(D+1)$), implies that

$$m_{u,\delta}(nT) = \sum_{q=0}^{\infty} \sum_{\alpha \diamond \beta \in \mathcal{J}_q} b_{\alpha,\delta}(u) f_{\beta} \int_{nT} \overline{H}_{\alpha}(X(s))\overline{H}_{\beta}(X'(s))\mathrm{d}s.$$

The L^2 convergence of $m_{u,\delta}(nT)$ towards $m_u(nT)$ and the continuity of the projection into the q-th chaos imply the stated expression for the chaotic components of $m_u(nT)$. The result follows. □

4.9.3 Central Limit Theorem

We are interested in the asymptotic distribution, as $n \to \infty$, of the measure $m_0(nT)$ of the zero set of X restricted to $nT = [0, n]^D$:

$$m_0(nT) = \{t \in nT : X(t) = 0\}.$$

The following is the main result of the chapter. Let us introduce the following assumption.

Assumption (A.4.3) For $t \in \mathbb{R}^D$, set

$$\Psi(t) := \max_{i,j,k,\ell} \{|r_{ij}(t)|, |\partial_k r_{ij}(t)|, |\partial_{k,\ell} r_{ij}(t)|\}.$$

We assume that $\Psi \in L^2(\mathbb{R}^D, \mathrm{d}t)$ and that $\Psi(t) \to_{\|t\|\to\infty} 0$.

Theorem 4.13 *Let $X(\cdot) = \{X(t) = (X_1(t), \dots, X_d(t)) : t \in \mathbb{R}^D\}$ be an $\mathbb{R}^d$-valued ($d \leq D$) standardized stationary Gaussian field satisfying* **(A.4.1)**, **(A.4.2)**, *and* **(A.4.3)**. *Then, as $n \to \infty$,*

$$\tilde{m}_0(nT) := \frac{m_0(nT) - \mathbb{E}[m_0(nT)]}{n^{D/2}}$$

has a finite limit variance V and converges in distribution towards a centered Gaussian random variable.

We need an auxiliary bound, which we will prove later.

Lemma 4.14 *With the notation in Lemma 4.11 and writing*

$$g_q(x, x') = \sum_{\alpha+\beta \in \mathcal{J}_{2q}} c_{\alpha,\beta}(0) \overline{H}_\alpha(x) \overline{H}_\beta(x'),$$

we have, $\|g_q\|_2^2 \leq \|\Delta\|_2^2$.

Proof of Theorem 4.13 The pipeline of the proof is the following:

0. Adapt the expansion of $m_0(nT)$ to $\tilde{m}_0(nT)$.
1. Prove that the variance of each chaotic component $I_q(nT)$ has a finite limit V_q.
2. Prove that the variance of the tail of the expansion $\sum_{q=Q+1}^{\infty} I_q(nT)$ is negligible, i.e.

$$\sup_{n\in\mathbb{N}} \sum_{q=Q+1}^{\infty} \mathrm{Var}\big(I_q(nT)\big) \underset{Q\to\infty}{\longrightarrow} 0.$$

   Note that these two facts imply the finiteness of the limit variance of $\tilde{m}_0(nT)$.
3. Prove the asymptotic normality of $\tilde{m}_0(nT)$.

Step 1: We normalize the Hermite expansion of $m_0(nT)$, see Theorem 4.12, to obtain the Hermite expansion of $\tilde{m}_0(nT) = \sum_{q=1}^{\infty} I_q(nT)$ with

$$I_q(nT) = \int_{nT} g_q(X(s), X'(s))\mathrm{d}s = \sum_{\boldsymbol{\alpha}\diamond\boldsymbol{\beta}\in\mathcal{J}_q} b_{\boldsymbol{\alpha}} f_{\boldsymbol{\beta}} \int_{nT} \overline{H}_{\boldsymbol{\alpha}}(X(s))\overline{H}_{\boldsymbol{\beta}}(X'(s))\mathrm{d}s.$$

Here, g_q is defined as in Lemma 4.14 and, for short, we write $b_{\boldsymbol{\alpha}} = b_{\boldsymbol{\alpha}}(0)$.

Note that the coefficients $b_{\boldsymbol{\alpha}}$ vanish if one of the α_j is odd. Furthermore, the function $\Delta(\mathbf{y})$ is even with respect to each row, denoted by $\boldsymbol{\beta}_j : j = 1, \ldots, d$ hereunder, thus, its Hermite coefficients

$$f_{\boldsymbol{\beta}} = f_{\boldsymbol{\beta}_1,\ldots,\boldsymbol{\beta}_d} = \int_{\mathbb{R}^d} \Delta(\mathbf{y})\mathbf{H}_{\boldsymbol{\beta}_1}(y_1)\ldots\mathbf{H}_{\boldsymbol{\beta}_r}(y_d)\phi_d(\mathbf{y})\mathrm{d}\mathbf{y}$$

are zero if at least one of the rows $\boldsymbol{\beta}_j$ satisfies that $|\boldsymbol{\beta}_j|$ is odd. In this form $|\beta_j| = \sum_{\ell=1}^{m} \beta_{\ell j}$ is necessarily even. Hence, as $\boldsymbol{\alpha}\diamond\boldsymbol{\beta} \in \mathcal{J}_q$, we deduce that $I_q(nT)$ vanishes for odd q.

Step 1: Let us consider the variance of $I_q(nT)$ for fixed $q \geq 1$. We have

$$\begin{aligned}\mathrm{Var}\big(I_q(nT)\big) &= \frac{1}{n^D}\int_{(nT)^2} \mathbb{E}\big[g_q(X(s), X'(s))g_q(X(t), X'(t))\big]\mathrm{d}s\mathrm{d}t \\ &= \int_{nT-nT} \mathbb{E}\big[g_q(X(0), X'(0))g_q(X(t), X'(t))\big]\prod_{k=1}^{d}\big[1-\tfrac{|t_k|}{n}\big]\mathrm{d}t.\end{aligned}$$

We used the stationarity to reduce to a simple integral. Now, Lemma 4.10 applied to g_q, which has Hermite rank q, yields

$$\mathrm{Var}\big(I_q(nT)\big) \leq \int_{nT-nT} \|g_q\|_2^2\, \psi(t)^q \mathrm{d}t,$$

with $\psi(t)$ the Arcones' coefficient as in Lemma 4.10. Under Assumption **(A.4.3)** we have $\psi(t) \leq c(d)\Psi(t)$ and using Lemma 4.14 we get

$$\mathrm{Var}\big(I_q(nT)\big) \leq \int_{nT-nT} c(d)^q \|\Delta\|_2^2 \Psi(t)^q \mathrm{d}t \underset{n\to\infty}{\longrightarrow} c(d)^q\|\Delta\|_2^2 \int_{\mathbb{R}^D} \Psi(t)^q\mathrm{d}t < \infty,$$

since $\Psi \in L^2(\mathbb{R}^D)$, again by Assumption **(A.4.3)**.

Step 2: We now check that the tail of the expansion is negligible. Write $G_Q = \sum_{q>Q} g_q$ and $\pi_{>Q} = \mathrm{Id} - \pi_{\leq Q}$. We have,

$$\mathrm{Var}\big(\pi_{>Q}(\tilde{m}_0(nT))\big) = \frac{1}{n^D}\int_{(nT)^2} \mathbb{E}\big[G_Q(X(s), X'(s))G_Q(X(t), X'(t))\big]\mathrm{d}s\mathrm{d}t.$$

We now split the integral into two regions. Consider $0 < \rho < 1$. As $r_{i,j}(\tau)$ and their derivatives tend to 0 as $|\tau| \to \infty$ for all $i, j \le d$, there exists $a \in \mathbb{N}^*$ such that $\psi(\tau) \le \rho < 1$ for $|\tau| > a/\sqrt{2}$. We partition nT into unit boxes: $nT = \cup_{\delta \in \mathbb{N}^D : |\delta| \le n} (T + \delta)$.

In the first place we bound the integral corresponding to pairs $|\delta|, |\delta'| \le n$ which are far enough apart. More precisely, we consider those pairs $|\delta|, |\delta'| \le n$ such that $|\delta - \delta'| > a$. Thus,

$$\frac{1}{n^D}\left|\sum_{|\delta-\delta'|>a} \int_{(T+\delta)\times(T+\delta')} \mathbb{E}\Big(G_Q(X(s), X'(s))G_Q(X(t), X'(t))\Big)\mathrm{d}s\mathrm{d}t\right|$$
$$\le \frac{1}{n^D}\int_{\{(s,t)\in(nT)^2 : |s-t|>a/\sqrt{2}\}} \left|\mathbb{E}\Big(G_Q(X(s), X'(s))G_Q(X(t), X'(t))\Big)\right|\mathrm{d}s\mathrm{d}t.$$

Using the same arguments as in the previous step, the last integral is bounded by

$$\le \int_{|\tau|>a+1} \sum_{q>Q} \|g_q\|_2^2 \psi(\tau)^q \mathrm{d}\tau \le \|\Delta\|_2^2 \sum_{q>Q} \rho^{q-2} \int_{\mathbb{R}^d} \psi^2(\tau)\mathrm{d}\tau.$$

Thus, this term tends to 0, uniformly in n as $Q \to \infty$.

We move to the integral over the complementary region:

$$\frac{1}{n^D}\left|\sum_{|\delta-\delta'|\le a} \mathbb{E}\Big(\int_{T+\delta} G_Q(X(s), X'(s))\mathrm{d}s \int_{T+\delta'} G_Q(X(t), X'(t))\mathrm{d}t\Big)\right|$$
$$\le \frac{\#\{(\delta, \delta') : |\delta - \delta'| \le a\}}{n^D} \int_{T^2} \mathbb{E}\big[G_Q(X(s), X'(s))G_Q(X(t), X'(t))\big]\mathrm{d}s\mathrm{d}t$$
$$\le \frac{(2a)^d}{n^{D-1}} \mathrm{Var}\Big(\pi_{>Q}\big(\tilde{m}_0(T)\big)\Big),$$

where we used the Cauchy–Schwarz inequality and the stationarity to get the first inequality. Hence, this term converges to 0, uniformly in n, as $Q \to \infty$.

Step 3: At this point, we claim that it suffices to prove that the partial sums converge in distribution towards a Gaussian random variable. Let d be any distance that metricizes the weak convergence of probability measures. By the triangle inequality we have

$$\mathrm{d}(\tilde{m}_0(nT), \mathcal{N}(0, V))$$
$$\le \mathrm{d}(\tilde{m}_0(nT), \pi_{\le Q}(\tilde{m}_0(nT))) + \mathrm{d}(\pi_{\le Q}(\tilde{m}_0(nT)), \mathcal{N}(0, V_Q)) + \mathrm{d}(\mathcal{N}(0, V_Q), \mathcal{N}(0, V)),$$

where $V_Q = \sum_{q \le Q} V_q$ is the limit variance of the projection $\pi_{\le Q}(\tilde{m}_0(nT))$ of $\tilde{m}_0(nT)$ over $\bigoplus_{q \le Q} C_q$, V is the limit variance of $\tilde{m}_0(nT)$ and $\mathcal{N}(0, \gamma)$ denotes the law of a centered Gaussian random variable with variance γ. The first two steps above show that the first term in last inequality tends to 0 as $Q \to \infty$ uniformly in n. Moreover, we can prove, as in Lemma 2 of Kratz and León (1997), that

$$\lim_{Q\to\infty} \mathrm{d}(\mathcal{N}(0, V_Q), \mathcal{N}(0, V)) = 0.$$

Our claim follows easily from this.

Furthermore, we can reduce the study of the asymptotic distribution of the partial sums to that of single chaotic terms, see Theorem 6.2.3 and Remark 6.2.4 in Nourdin and Peccati (2012). Hence, we establish the asymptotic normality of each chaotic component. So we fix q, thus, we can forget about sums and coefficients in (4.14) and restrict ourselves to bound the integrals. This allows us to forget the sub-indexes leaving the arguments untouched.

Performing the isometric change-of-variables $u = s - s'$, $v = t' - s$, $w = t - t'$ and $x = t$ we see that each integral in (4.14) is bounded in absolute value by a constant times an integral of the form

$$J := \frac{1}{n^{2D}} \int_{(nT)^4} |\Psi(u)\Psi(v)\Psi(w)\Psi(u+v+w)| du dv dw dt.$$

Now, since $\Psi \in L^2$, the Cauchy–Schwarz inequality yields

$$\int_{\mathbb{R}^D} |\Psi(w)\Psi(u+v+w)| dw \le \int_{\mathbb{R}^D} \Psi(w)^2 dw = C < \infty.$$

Thus,

$$J \le \frac{C}{n^{2D}} \int_{(nT)^3} |\Psi(u)\Psi(v)| du dv dt \le \frac{C}{n^D} \int_{(nT)^2} |\Psi(u)\Psi(v)| du dv.$$

The last inequality is due to the fact that $\int_{nT} \mathrm{d}t/n^D = 1$ for all $n \in \mathbb{N}^*$. From the fact that $\Psi \in L^2$ (and that $|\Psi(t)| \le 1$ for all t) it follows also that $\int_{nT} |\Psi(u)| \mathrm{d}u/n^{D/2} \to_n 0$. Indeed,

$$\frac{1}{n^{D/2}} \int_{nT} |\Psi(u)| \mathrm{d}u = \frac{1}{n^{D/2}} \int_{nT \setminus n^{1/4}T} |\Psi(u)| du + o(1)$$

$$\le \frac{\sqrt{n^D - n^{D/4}}}{n^{D/2}} \left[\int_{nT \setminus n^{1/4}T} \Psi(u)^2 \mathrm{d}u \right]^{1/2} + o(1) \le \left[\int_{(n^{1/4}T)^{\complement}} \Psi(u)^2 \mathrm{d}u \right]^{1/2} + o(1),$$

where the first inequality is due to Jensen's inequality. Hence, the convergence to 0 of (4.14) follows.

This concludes the proof. □

To end this section we prove Lemma 4.14.

Proof of Lemma 4.14 Consider two sequences $(a_k)_k$ and $(b_k)_k$ such that $\sum_{k=0}^{\infty} a_k^2 < \infty$ and $\sup_k b_k^2 = \|b\|_\infty^2 < \infty$. For all $k \in \mathbb{N}$ we get

$$\sum_{k=0}^{q} a_k^2 b_{q-k}^2 \leq ||b||_\infty^2 ||a||_2^2. \tag{4.16}$$

Now, coming back to our problem, since $u = 0$, we set $b = (b_j(0))_{j\in\mathbb{N}}$ and we note that for $j \in \mathbb{N}$ we have $b_{2j+1} = 0$ and $b_{2j} = \frac{(-1)^j (2j-1)!!}{\sqrt{2\pi}(2j)!}$. Thus, for $q = 2l$,

$$||g_q||_2^2 = \sum_{|\boldsymbol{\alpha}|+|\boldsymbol{\beta}|=q} b_{\boldsymbol{\alpha}}(0)^2 f_{\boldsymbol{\beta}}^2 \boldsymbol{\alpha}!\boldsymbol{\beta}! = \sum_{|\boldsymbol{\alpha}|+|\boldsymbol{\beta}|=2l} b_{2\alpha_1}^2 \dots b_{2\alpha_{d'}}^2 f_{(\beta_1,\dots,\beta_{d'})}^2 \boldsymbol{\alpha}!\boldsymbol{\beta}!.$$

Direct computations show that $b_{2j}^2 (2j)!$ is decreasing and $b_0^2 = \frac{1}{2\pi} < 1$, then $||b||_\infty^2 < 1$. Write now

$$||g_q||_2^2 = \sum_{k=0}^{l} \sum_{|\beta_2|+\dots+|\beta_{d'}|=2k} \sum_{|\boldsymbol{\alpha}|+|\beta_1|=q-2k} b_{\boldsymbol{\alpha}}(0)^2 f_{\boldsymbol{\beta}}^2 \boldsymbol{\alpha}!\boldsymbol{\beta}!.$$

Set $\alpha_i + \beta_{1i} = l_i$, such that $q - 2k = l_1 + \dots + l_{d'}$. In this form we obtain

$$\sum_{|\boldsymbol{\alpha}|+|\beta_1|=q-2k} b_{\boldsymbol{\alpha}}(0)^2 f_{\boldsymbol{\beta}}^2 \boldsymbol{\alpha}!\boldsymbol{\beta}! = \sum_{l_1+\dots+l_{d'}=q-2k} \prod_{i=1}^{m} \sum_{\beta_{1i}=0}^{l_i} b_{l_i-\beta_{1i}}^2 f_{\boldsymbol{\beta}}^2 (l_i-\beta_{1i})!\beta_{1i}! \cdot \prod_{j=2}^{m} \beta_j!.$$

Using (4.16) it yields $\sum_{\beta_{1i}=0}^{l_i} b_{\alpha_i}^2 f_{\boldsymbol{\beta}}^2 \alpha_i!\beta_{1i}! \leq \sum_{\beta_{1i}=0}^{l_i} f_{\boldsymbol{\beta}}^2 \beta_{1i}!$ and thus

$$\sum_{|\boldsymbol{\alpha}|+|\beta_1|=q-2k} b_{\boldsymbol{\alpha}}(0)^2 f_{\boldsymbol{\beta}}^2 \boldsymbol{\alpha}!\boldsymbol{\beta}! \leq \sum_{l_1+\dots+l_{d'}=q-2k} \prod_{i=1}^{m} \sum_{\beta_{1i}=0}^{l_i} f_{\boldsymbol{\beta}}^2 \boldsymbol{\beta}! = \sum_{|\beta_1|\leq q-2k} f_{\boldsymbol{\beta}}^2 \boldsymbol{\beta}!.$$

Thus,

$$||g_q||_2^2 \leq \sum_{|\boldsymbol{\beta}|\leq q} f_{\boldsymbol{\beta}}^2 \boldsymbol{\beta}! \leq \|\Delta\|_2^2.$$

This concludes the proof. □

4.10 A More General Framework

In this chapter, we chose to follow an elementary approach to chaotic expansions and to limit theorems based upon them. One of the reasons for this choice is that it allows us to give coherence to several examples, see Chap. 8, where we begin with Hermite expansions and then move to chaotic expansions in terms of multiple stochastic integrals. Though the present chapter shows that this procedure is not strictly necessary, it is worth evoking it so as to take advantage of every point of view. There is a considerable variety of newer and deeper methods which can produce impressive results, such as quantitative bounds for the distance from the distributions of the chaotic random variables to their limit, extensions to other frameworks apart from the Gaussian one, etc. For a general overview of this topic, see the books Peccati and Taqqu (2011), Nourdin and Peccati (2012), and references therein.

Let us give here an extremely concise view of the core of the modern approach to chaotic expansions and limit theorems based on them. A natural setting for the study of this subject is that of a Gaussian Hilbert space, that is, a closed Gaussian subspace of $L^2(B)$, where B is a standard Brownian motion. It is well known that Wiener integrals with respect to B provide an isometry between $L^2(B)$ and $L^2(dt)$. Moreover, Itô's formula Major (1981, Th. 4.3.) states that polynomials in simple Wiener integrals can be written in terms of multiple Wiener–Itô integrals, allowing for a definition of Wiener Chaos in terms of multiple Wiener–Itô integrals. The isometry of simple stochastic integrals is easily extended to this framework.

Taking advantage of the stochastic calculus framework, Nualart and Peccati (2005) and Peccati and Tudor (2005) proved the celebrated Fourth Moment Theorem, and in doing so also gave an alternative practical criterion for checking the asymptotic normality in terms of the contractions which is analogous to our condition (4.14).

Later, Malliavin calculus and Stein's method were combined in order to produce quantitative bounds for the distance between the distribution of a chaotic random variable and a given Gaussian one in terms of norms of the Malliavin derivative, which in turn can be bounded in terms of the L^2-norm of the contractions.

Chapter 5
Critical Points of Isotropic Gaussian Fields

Focusing on the geometric landscape of random fields, this chapter provides a rigorous study of critical points. It delivers exact expressions for the mean number of critical points of a given index below a level for isotropic Gaussian random fields defined on both flat spaces and spheres. Some logarithmic asymptotics of this mean number are given in particular cases. This chapter also studies the phenomenon of attraction and repulsion between critical points through second factorial moments. Finally, it establishes a multivariate central limit theorem for all types of critical points above a level.

5.1 Introduction

This chapter is devoted to the study of the number of critical points of different types of a given isotropic Gaussian field (although the isotropy assumption is not necessary in the last two sections of the chapter). The first seminal works concerning critical points of random fields came in the form of papers by Nosko (1969), Lindgren (1972), Belyaev and Piterbarg (1972), Hasofer (1976b), and books by Adler (1981) and by Piterbarg (1996). They mainly focus on local maxima of Gaussian random fields, their positions, their mean number and second factorial moments. The main applications of such results concern the detection of peaks in a random field, since the mean number of local maxima above a level appear in the calculation of the distribution of the test statistic. This mean number can be approximated by the expectation of the Euler characteristic of the appropriate excursion set, as explained in Adler (2000), Delmas (1998, 2001). For example, many papers concern the detection of activation zones in the human brain (Taylor and Worsley 2007; Worsley et al. 2004, 1992), applications in astronomy (Torres 1994; Vogeley et al. 1994), and the detection of genes on a chromosome (Azaïs et al. 2014; Azaïs and Cierco-Ayrolles 2002; Rabier and Delmas 2021). Local maxima of

D. Armentano et al., *Geometry of Level Sets of Random Fields, Kac–Rice Formulas, Hermite Expansions and Applications*, Probability Theory and Stochastic Modelling 111, https://doi.org/10.1007/978-3-032-29144-8_5

random fields are also used to define a crest in the modeling of random sea waves, see Azaïs and Wschebor (2005a), Azaïs and Wschebor (2009, Ch11.) and references therein. In this chapter we give exact expressions for the mean number of critical points of different types above or below a level for random fields defined on the sphere $S^d \subset \mathbb{R}^{d+1}$ (Sect. 5.2.1) or on $\mathbb{R}^d$ (Sects. 5.3.1 and 5.3.2). Such results can be found in existing literature: Auffinger and Ben Arous (2013), Auffinger et al. (2013), Cheng and Schwartzman (2018), Azaïs and Delmas (2022).

The asymptotic behavior of the mean number of local maxima above a level has been studied in Adler (1981), Delmas (1998, 2001), Azaïs and Delmas (2002) when the level tends to infinity. We have chosen to present these results, without proof, in Sect. 3.4.4, before extending them to random fields defined on the sphere in Sect. 3.5.4. More recently, several authors (Auffinger and Ben Arous 2013; Auffinger et al. 2013) have studied the asymptotic behavior of the mean number of critical points of different types below a level for Gaussian random fields defined on the sphere $S^d \subset \mathbb{R}^{d+1}$ when the dimension d tends to infinity. They noted an interesting layering of the different types of critical points. Their results suggest compelling applications in random functions' minimization in large dimension since there is a critical interval in which only minima can be found. Some of their results are given in Sect. 5.2 and extended to flat Gaussian fields defined on $\mathbb{R}^d$ in Sect. 5.3.

More recent works study the spread of the critical points of a Gaussian random field. In particular, it can be interesting to determine if there is attraction or repulsion between these points. The answer can depend on the types of the critical points under consideration. The seminal work by Beliaev et al. (2019) considers the particular case of the *Random Plane Wave*, a popular Gaussian field defined on $\mathbb{R}^2$. They prove, in particular, neutrality between critical points, repulsion between maxima (or minima), repulsion between saddles and also between extrema. Their results have been extended to more general planar Gaussian random fields in Beliaev et al. (2020) and Ladgham and Lachièze-Rey (2023) and to any dimension in Azaïs and Delmas (2022). This is the subject of Sect. 5.4.

A final area of study pertains to the establishing of Central Limit Theorems on a family of growing sets. The first paper on this is by Estrade and León (2016). It considers the Euler characteristic of the excursion set above the level u of a real-valued isotropic Gaussian random field over a set $T \subset \mathbb{R}^d$. The total number of critical points has been considered in Nicolaescu (2017). A multivariate generalization of both papers is given in Azaïs et al. (2024). It is presented in Sect. 5.5.

Additionally Sect. 5.6 presents a small result on bounds for the moments.

5.2 Mean Number: The Spherical Case

Our framework is the same as that presented in Sect. 3.5.3. We recall Assumption **(A.3.1)**: $\{X(t) : t \in S^d \subset \mathbb{R}^{d+1}\}$ is a real-valued centered isotropic Gaussian field defined on the unit sphere $S^d \subset \mathbb{R}^{d+1}$. Its covariance satisfies

$$c(s,t) = \mathbb{E}(X(s)X(t)) = \nu(\langle s,t\rangle),$$

where

$$\nu(\langle s,t\rangle) = \sum_{k=2}^{+\infty} c_k \langle s,t\rangle^k \text{ with } c_k \geq 0, \quad \sum_{k=2}^{+\infty} c_k = 1,$$

$$\sum_{k=2}^{+\infty} k c_k > 2, \quad \sum_{k=2}^{+\infty} k^5 c_k < +\infty.$$

We recall that

$$\nu_0 := \mathrm{Var}(X(t)) = \nu(1) = 1,$$
$$\nu_2 := \mathrm{Var}(\nabla X_i(t)) = \nu'(1) \text{ and}$$
$$\nu_4 := \mathrm{Var}(\nabla^2 X_{ii}(t) = 3\nu''(1) + \nu'(1).$$

It is easy to check that the expression of the covariance implies

$$\nu_4 + 2\nu_2 - 3\nu_2^2 = \sum_{i \leq j} c_i c_j (i-j)^2 \geq 0.$$

This last quantity is zero if and only if $\nu(\langle s,t\rangle) = c_p\langle s,t\rangle^p$, which corresponds to p-spin spherical spin glass models in the physics literature, see Auffinger et al. (2013), Auffinger and Ben Arous (2013).

5.2.1 *Exact Expressions*

The aim of this section is to derive exact expressions for the mean number of critical points with index k below the level u on S^d, denoted by $\mathrm{Crt}^k_{u\downarrow}(S^d)$:

$$\mathrm{Crt}^k_{u\downarrow}(S^d) := \#\{t \in S^d : X(t) \leq u, \nabla X(t) = 0, i\left(\nabla^2 X(t)\right) = k\}.$$

We first give in the following theorem the Kac–Rice formula for $\mathbb{E}\left(\mathrm{Crt}^k_{u\downarrow}(S^d)\right)$. We denote by $p_{\nabla X(t),X(t)}(0,x)$ the joint probability density of $\nabla X(t)$ and $X(t)$ at 0 and x. This theorem is a reminder of what was demonstrated in Chap. 2, namely Theorem 2.27 and equation (2.44).

Theorem 5.1 *Under (A.3.1), we have*

$$\mathbb{E}(\mathrm{Crt}^k_{u\downarrow}(S^d)) = |S^d| \int_{-\infty}^{u} p_{\nabla X(t),X(t)}(0,x)$$
$$\mathbb{E}\left(|\det \nabla^2 X(t)| \mathbb{1}_{\{i(\nabla^2 X(t))=k\}} / \nabla X(t) = 0, X(t) = x\right) \mathrm{d}x. \tag{5.1}$$

We start this section by giving the mean number of maxima and minima below the level u of X on the unit sphere when $d = 1$.

Proposition 5.2 *Under (A.3.1),*

$$\mathbb{E}\left(\mathrm{Crt}^0_{u\downarrow}(S^1)\right) = \sqrt{\frac{\nu_4}{\nu_2}}\Phi\left(u\sqrt{\frac{\nu_4}{\nu_4 - \nu_2^2}}\right) + \sqrt{2\pi}\,\nu_2\phi(u)\bar{\Phi}\left(\frac{u\nu_2}{\sqrt{\nu_4 - \nu_2^2}}\right),$$

$$\mathbb{E}\left(\mathrm{Crt}^1_{u\downarrow}(S^1)\right) = \sqrt{\frac{\nu_4}{\nu_2}}\Phi\left(u\sqrt{\frac{\nu_4}{\nu_4 - \nu_2^2}}\right) - \sqrt{2\pi}\,\nu_2\phi(u)\Phi\left(\frac{u\nu_2}{\sqrt{\nu_4 - \nu_2^2}}\right).$$

Let us now state a lemma that is, with Lemma 3.28, the main tool in the calculation of (5.1).

Lemma 5.3 *Let G be a d-*GOE *matrix, V an independent Gaussian variable with mean m and variance σ^2 and $v \in \mathbb{R}$. Then*

$$\mathbb{E}\left(|\det(G - v\mathrm{Id}_d)| \mathbb{1}_{\{i(G - v\mathrm{Id}_d)=k\}}\right) = \sqrt{2}\Gamma\left(\frac{d+1}{2}\right)\exp\left(\frac{v^2}{2}\right) p_{L_{k+1}}(v),$$

$$\mathbb{E}\left(|\det(G - V\mathrm{Id}_d)| \mathbb{1}_{\{i(G - V\mathrm{Id}_d)=k\}}\right)$$
$$= \frac{1}{\sigma\sqrt{\pi}}\Gamma\left(\frac{d+1}{2}\right)\mathbb{E}\left(\exp\left(\frac{L_{k+1}^2}{2} - \frac{(L_{k+1} - m)^2}{2\sigma^2}\right)\right),$$

*where L_{k+1} is the $(k+1)$th ordered eigenvalue of a $(d+1)$-*GOE *matrix and $p_{L_{k+1}}(\cdot)$ its density.*

Proof of Lemma 5.3 Since the joint density f_d of the eigenvalues $(\mu_1, \ldots, \mu_d)$ of G, see Mehta (2004, Theorem 3.3.1), is given by

$$f_d(\mu_1, \ldots, \mu_d) = k_d \exp\left(-\frac{\sum_{i=1}^d \mu_i^2}{2}\right) \prod_{1 \le i < j \le d} |\mu_j - \mu_i|, \tag{5.2}$$

with

$$k_d := (2\pi)^{-d/2} (\Gamma(3/2))^d \left(\prod_{i=1}^{d} \Gamma\left(1+\frac{i}{2}\right) \right)^{-1}, \tag{5.3}$$

denoting $O_k(v) := \{(\mu_i, i=1,\dots,d) \in \mathbb{R}^d : \mu_1 \le \mu_2 \le \cdots \le v \le \mu_k \le \cdots \le \mu_d\}$, we have

$$\begin{aligned}
&\mathbb{E}\left(|\det(G - v\mathrm{Id}_d)| \mathbb{1}_{\{i(G - v\mathrm{Id}_d)=k\}}\right) \\
&= k_d\, d! \int_{O_k(v)} \prod_{i=1}^{d} |\mu_i - v| \exp\left(-\frac{\sum_{i=1}^{d} \mu_i^2}{2}\right) \prod_{1 \le i < j \le d} |\mu_j - \mu_i| \mathrm{d}\mu_1 \dots \mathrm{d}\mu_d \\
&= \frac{k_d}{k_{d+1}} \frac{1}{d+1} \exp\left(\frac{v^2}{2}\right) p_{L_{k+1}}(v).
\end{aligned}$$

In the same way, when V is a Gaussian variable with mean m and variance σ^2 we have

$$\begin{aligned}
&\mathbb{E}\left(|\det(G - V\mathrm{Id}_d)| \mathbb{1}_{\{i(G - V\mathrm{Id}_d)=k\}}\right) = k_d\, d! \\
&\times \int \int_{O_k(v)} \prod_{i=1}^{d} |\mu_i - v| \exp\left(-\frac{\sum_{i=1}^{d} \mu_i^2}{2}\right) \prod_{1 \le i < j \le d} |\mu_j - \mu_i| \mathrm{d}\mu_1 \dots \mathrm{d}\mu_d \frac{\mathrm{e}^{-\frac{(v-m)^2}{2\sigma^2}}}{\sqrt{2\pi}\sigma} \mathrm{d}v \\
&= \frac{k_d}{k_{d+1}} \frac{1}{(d+1)\sqrt{2\pi}\sigma} \mathbb{E}\left(\exp\left(\frac{L_{k+1}^2}{2} - \frac{(L_{k+1} - m)^2}{2\sigma^2}\right)\right).
\end{aligned}$$

□

Before giving Theorem 5.4, we set

$$K := \frac{\nu_4 - 4\nu_2}{\nu_4 + 2\nu_2}, \quad E_{\star\star} := -\frac{2\sqrt{3}\nu_2\sqrt{\nu_4 - \nu_2}}{\nu_4 + 2\nu_2}, \quad E_{\star} := -\frac{\nu_4 - 4\nu_2 + 3\nu_2^2}{\sqrt{3}\nu_2\sqrt{\nu_4 - \nu_2}}.$$

Theorem 5.4 *Let L_p be the pth ordered eigenvalue of a $(d+1)$-GOE matrix ($L_1 \le L_2 \le \cdots \le L_{d+1}$). Under* **(A.3.1)***, with the notation above, we have for $k = 0, 1, \dots, d$,*

$$\begin{aligned}
\mathbb{E}\left(\mathrm{Crt}_{u\downarrow}^{k}(S^d)\right) &= 2\sqrt{2} \left(\frac{\nu_4 - \nu_2}{3\nu_2}\right)^{\frac{d+1}{2}} \sqrt{\frac{3\nu_2}{\nu_4 + 2\nu_2}} \\
&\quad \mathbb{E}\left[\exp\left(-\frac{L_{k+1}^2}{2} K\right) \Phi\left(\sqrt{\frac{2K}{E_{\star\star}(E_{\star} - E_{\star\star})}} \left(u + L_{k+1} \frac{E_{\star\star}}{\sqrt{2}}\right)\right)\right],
\end{aligned}$$

with the convention that when $\nu_4 + 2\nu_2 - 3\nu_2^2 = 0$,

$$\Phi\left(\sqrt{\frac{2K}{E_{\star\star}(E_\star - E_{\star\star})}}\left(u + L_{k+1}\frac{E_{\star\star}}{\sqrt{2}}\right)\right) = \mathbb{1}_{\{L_{k+1} < u\sqrt{\frac{\nu_2}{2(\nu_2-1)}}\}}.$$

Now we set $q := \dfrac{\nu_2}{\nu_4}$, then

$$\mathbb{E}(\mathrm{Crt}^k(S^d)) = 2\sqrt{2}\left(\frac{1-q}{3q}\right)^{\frac{d+1}{2}}\sqrt{\frac{3q}{1+2q}}\,\mathbb{E}\left(\exp\left(\frac{(4q-1)L_{k+1}^2}{2(1+2q)}\right)\right).$$

Using the exact expressions of the probability densities of the ordered eigenvalues of a $(d+1)$-GOE matrix for $d = 2$ and $d = 3$, obtained in Azaïs and Delmas (2022, Section 3.2.1), by straightforward calculations we obtain Proposition 5.5.

Proposition 5.5 *Under (A.3.1), when* $d = 2$

$$\mathbb{E}(\mathrm{Crt}^0(S^2)) = \mathbb{E}(\mathrm{Crt}^2(S^2)) = 1 + 2\frac{(1-q)^{3/2}}{3^{3/2}q},$$

$$\mathbb{E}(\mathrm{Crt}^1(S^2)) = 4\frac{(1-q)^{3/2}}{3^{3/2}q}.$$

When $d = 3$

$$\mathbb{E}(\mathrm{Crt}^0(S^3)) = \mathbb{E}(\mathrm{Crt}^3(S^3)) = \frac{16q^3 + 24q^2 - 15q + 29}{108q^{3/2}} + \frac{4q^2 + 7q - 2}{6\sqrt{3}q^{3/2}\sqrt{2+q}},$$

$$\mathbb{E}(\mathrm{Crt}^1(S^3)) = \mathbb{E}(\mathrm{Crt}^2(S^3)) = \frac{16q^3 + 24q^2 - 15q + 29}{108q^{3/2}} - \frac{4q^2 + 7q - 2}{6\sqrt{3}q^{3/2}\sqrt{2+q}}.$$

Remark 5.1 The proportions $\dfrac{\mathbb{E}(\mathrm{Crt}^k(S^d))}{\mathbb{E}(\mathrm{Crt}(S^d))}$ depend on $q = \nu_2/\nu_4$. They are not universal.

Remark 5.2 When $q \to 0$, we get for $d = 2$ and $d = 3$,

$$\frac{\mathbb{E}(\mathrm{Crt}^0(S^2))}{\mathbb{E}(\mathrm{Crt}(S^2))} = \frac{\mathbb{E}(\mathrm{Crt}^2(S^2))}{\mathbb{E}(\mathrm{Crt}(S^2))} \simeq \frac{1}{4}, \qquad \frac{\mathbb{E}(\mathrm{Crt}^1(S^2))}{\mathbb{E}(\mathrm{Crt}(S^2))} \simeq \frac{1}{2},$$

$$\frac{\mathbb{E}(\mathrm{Crt}^0(S^3))}{\mathbb{E}(\mathrm{Crt}(S^3))} = \frac{\mathbb{E}(\mathrm{Crt}^3(S^3))}{\mathbb{E}(\mathrm{Crt}(S^3))} \simeq \frac{29 - 6\sqrt{6}}{116},$$

$$\frac{\mathbb{E}(\mathrm{Crt}^1(S^3))}{\mathbb{E}(\mathrm{Crt}(S^3))} = \frac{\mathbb{E}(\mathrm{Crt}^2(S^3))}{\mathbb{E}(\mathrm{Crt}(S^3))} \simeq \frac{29 + 6\sqrt{6}}{116}.$$

These proportions correspond to those obtained in the flat case when the Gaussian field is defined on a subset of $\mathbb{R}^d$; see Sect. 5.3.2 below, Eq. (5.23) and Proposition 5.11. See also Azaïs and Delmas (2022, Proposition 6).

In the following proposition we give the exact expressions of $\mathbb{E}(\mathrm{Crt}^k_{u\downarrow}(S^2))$ for $k = 0, 1, 2$ in the particular case $\nu_4 + 2\nu_2 - 3\nu_2^2 = 0$ that corresponds to $\nu(\langle s, t\rangle) = c_p \langle s, t\rangle^p$.

Proposition 5.6 *We assume that* $\nu_4 + 2\nu_2 - 3\nu_2^2 = 0$ *then, under* ***(A.3.1)***,

$$\mathbb{E}(\mathrm{Crt}^1_{u\downarrow}(S^2)) = \frac{4}{\sqrt{3\nu_2 - 2}} \nu_2^{3/2} \Phi\left(u\sqrt{\frac{3\nu_2 - 2}{2(\nu_2 - 1)}}\right).$$

$$\begin{aligned}\mathbb{E}(\mathrm{Crt}^0_{u\downarrow}(S^2)) &= \frac{2\sqrt{\nu_2 - 1}}{\sqrt{\nu_2}} \int_{-\infty}^{u\sqrt{\frac{\nu_2}{\nu_2 - 1}}} \bar{\Phi}(z)\phi\left(z\sqrt{\frac{\nu_2 - 1}{\nu_2}}\right) \mathrm{d}z \\ &+ \frac{\sqrt{2\nu_2(\nu_2 - 1)}}{\sqrt{\pi}} \phi\left(u\sqrt{\frac{2\nu_2 - 1}{\nu_2 - 1}}\right) - 2u\nu_2\phi(u)\bar{\Phi}\left(u\sqrt{\frac{\nu_2}{\nu_2 - 1}}\right) \\ &\quad + 4(\nu_2 - 1)^{3/2}\sqrt{\frac{1}{\nu_2}} \int_{-\infty}^{u\sqrt{\frac{\nu_2}{2(\nu_2 - 1)}}} \bar{\Phi}(z)\phi\left(z\sqrt{\frac{3\nu_2 - 2}{\nu_2}}\right) \mathrm{d}z.\end{aligned}$$

$$\begin{aligned}\mathbb{E}(\mathrm{Crt}^2_{u\downarrow}(S^2)) &= \frac{2\sqrt{\nu_2 - 1}}{\sqrt{\nu_2}} \int_{-\infty}^{u\sqrt{\frac{\nu_2}{\nu_2 - 1}}} \Phi(z)\phi\left(z\sqrt{\frac{\nu_2 - 1}{\nu_2}}\right) \mathrm{d}z \\ &- \frac{\sqrt{2\nu_2(\nu_2 - 1)}}{\sqrt{\pi}} \phi\left(u\sqrt{\frac{2\nu_2 - 1}{\nu_2 - 1}}\right) - 2u\nu_2\phi(u)\Phi\left(u\sqrt{\frac{\nu_2}{\nu_2 - 1}}\right) \\ &\quad + 4(\nu_2 - 1)^{3/2}\sqrt{\frac{1}{\nu_2}} \int_{-\infty}^{u\sqrt{\frac{\nu_2}{2(\nu_2 - 1)}}} \Phi(z)\phi\left(z\sqrt{\frac{3\nu_2 - 2}{\nu_2}}\right) \mathrm{d}z.\end{aligned}$$

5.2.2 *Logarithmic Asymptotics*

This section presents the results of Auffinger et al. (2013) and Auffinger and Ben Arous (2013) concerning the landscape of isotropic random fields on the sphere. Broadly speaking, under some conditions, there exist two levels u_0 and u_1 between which there are (logarithmically speaking) only local minima and no index 1 (or higher) critical points, see Remark 5.4. This can explain why descent minimization algorithms can fail to find the global minimum in large dimension.

We take the notation $v := \dfrac{u}{\sqrt{d+1}}$ and $\tilde{L}_{k+1} := \dfrac{L_{k+1}}{\sqrt{d+1}}$ for $k = 0, \dots d$. The Large Deviations Principle (LDP) for the $(k+1)$th smallest eigenvalue of a d-GOE matrix obtained in Auffinger et al. (2013, Theorem A.9), formulated below as a lemma, is the main tool for the results of this section.

Lemma 5.7 *Let $L_1 \le L_2 \le \cdots \le L_{d+1}$ be the ordered eigenvalues of a $(d+1)$-*GOE *matrix. We set $\tilde{L}_{k+1} := \dfrac{L_{k+1}}{\sqrt{d+1}}$, then $\tilde{L}_{k+1}$ satisfies the LDP with speed $(d+1)$ and good rate function*

$$J_k(x) = (k+1)I_1(-x, 2^{-1/2}), \tag{5.4}$$

with

$$I_1(x,\sigma) := \begin{cases} \displaystyle\int_{2\sigma}^{x} \sigma^{-1}\sqrt{\left(\frac{z}{2\sigma}\right)^2 - 1}\,\mathrm{d}z, & \text{if} x \ge 2\sigma, \\ +\infty, & \text{otherwise.} \end{cases} \tag{5.5}$$

When $\nu_4 + 2\nu_2 - 3\nu_2^2 = 0$ (p-spin spherical spin glass models, $\nu(\langle s,t\rangle) = c_p\langle s,t\rangle^p$), the following proposition, due to Auffinger et al. (2013, Theorem 2.5), gives the logarithmic limit of the mean number of critical points with index k below a level u. It is a direct consequence of Lemma 5.7 and Varadhan's Integral Lemma (Dembo and Zeitouni 1998, Theorem 4.3.1). We set

$$E_\infty := -2\sqrt{\frac{\nu_2 - 1}{\nu_2}},$$

and we define the function $\Theta_k(v)$

$$\Theta_k(v) := \begin{cases} \dfrac{1}{2}\log(\nu_2 - 1) - \dfrac{\nu_2 - 2}{\nu_2} & \text{if } v \ge E_\infty \\ \dfrac{1}{2}\log(\nu_2 - 1) - \dfrac{v^2}{4}\left(\dfrac{\nu_2 - 2}{\nu_2 - 1}\right) - (k+1)I_1(-v, -E_\infty/2) & \text{if } v \le E_\infty \end{cases}$$

We have the following logarithmic asymptotic for $\mathbb{E}\left(\mathrm{Crt}_{u\downarrow}^k(S^d)\right)$.

Theorem 5.8 *We assume that $\nu_4 + 2\nu_2 - 3\nu_2^2 = 0$. Then, under* ***(A.3.1)****,*

$$\lim_{d\to+\infty} \frac{1}{d}\log \mathbb{E}\left(\mathrm{Crt}_{u\downarrow}^k(S^d)\right) = \Theta_k(v).$$

This theorem is a particular case of Theorem 5.9 below, for which we will give a proof.

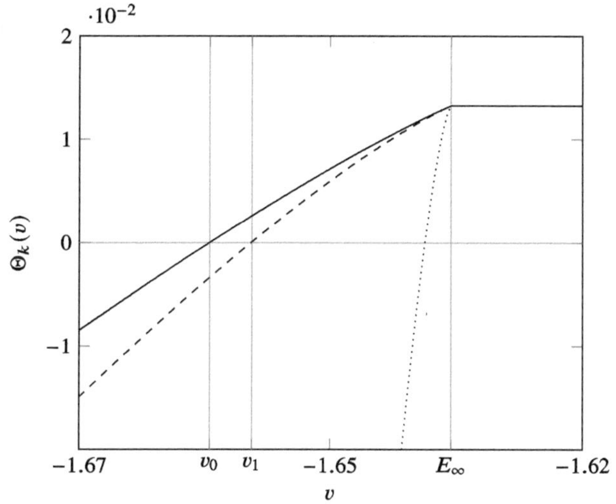

Fig. 5.1 Function $\Theta_k(v)$ for $\nu(\langle s,t\rangle) = \langle s,t\rangle^3$, $k = 0$ (solid), $k = 1$ (dashed) and $k = 100$ (dotted)

Remark 5.3 It is easy to check that for $p > 2$, $\Theta_k(v)$ is a positive constant for $v \geq E_\infty$ and an increasing function of v depending on k for $v \leq E_\infty$. For v fixed, $\Theta_k(v)$ is a decreasing function of k. This function $\Theta_k(v)$ is plotted in Fig. 5.1 for $\nu(\langle s,t\rangle) = \langle s,t\rangle^3$.

Remark 5.4 There exists a level v_0 (defined such that $\Theta_0(v_0) = 0$) below which $\Theta_k(v) < 0$, $\forall k$ and $\forall v$. Since $\mathbb{E}\left(\mathrm{Crt}^k_{u\downarrow}(S^d)\right) \simeq \exp\left(d\Theta_k(v)\right)$, the mean number of critical points below $u_0 = \sqrt{d+1}v_0$ tends to zero when $d \to +\infty$. In the same way there exists a level v_1 (defined as $\Theta_1(v_1) = 0$) such that for $v_0 \leq v \leq v_1$, $\Theta_0(v) > 0$ and $\Theta_k(v) < 0$ $\forall k \geq 1$. So between the corresponding levels u_0 and u_1 we find essentially only critical points with index 0 (minima). And so on; between u_1 and u_2 we find essentially only critical points with indexes 0 and 1. So we have a layered structure for the lowest critical values of X, as described in Auffinger et al. (2013).

Following Auffinger and Ben Arous (2013), we can now extend these results to the more general case $\nu_4 + 2\nu_2 - 3\nu_2^2 > 0$, which corresponds to more general polynomial covariances (3.48). We recall the following notation

$$K := \frac{\nu_4 - 4\nu_2}{\nu_4 + 2\nu_2}, \quad E_{\star\star} := -\frac{2\sqrt{3}\nu_2\sqrt{\nu_4 - \nu_2}}{\nu_4 + 2\nu_2}, \quad E_\star := -\frac{\nu_4 - 4\nu_2 + 3\nu_2^2}{\sqrt{3}\nu_2\sqrt{\nu_4 - \nu_2}}.$$

When $K^2E_\star^2 - (k+1)^2(E_{\star\star} - E_\star)^2 \neq 0$, we define $x_{k,v}$ as the quantity, depending on k and v,

$$x_{k,v} := \frac{-\sqrt{2}vE_\star K^2 + \sqrt{2}(k+1)(E_{\star\star} - E_\star)\sqrt{(k+1)^2(E_{\star\star} - E_\star)^2 + K^2(v^2 - E_\star^2)}}{K^2E_\star^2 - (k+1)^2(E_{\star\star} - E_\star)^2},$$

and when $K^2E_\star^2 - (k+1)^2(E_{\star\star} - E_\star)^2 = 0$,

$$x_{k,v} := -\frac{E_\star^2 + v^2}{\sqrt{2}vE_\star}.$$

We also define the function $\Theta_k^\star(v)$ by

$$\Theta_k^\star(v) := \begin{cases} \frac{1}{2}\log\left(\frac{\nu_4 - \nu_2}{3\nu_2}\right) - K & \text{if } v \geq E_{\star\star} \\ \frac{1}{2}\log\left(\frac{\nu_4 - \nu_2}{3\nu_2}\right) - K\left(1 - \frac{(v - E_{\star\star})^2}{E_{\star\star}(E_{\star\star} - E_\star)}\right) & \text{if } E_\star \leq v \leq E_{\star\star} \\ \frac{1}{2}\log\left(\frac{\nu_4 - \nu_2}{3\nu_2}\right) - K\left(\frac{x_{k,v}^2}{2} - \frac{\left(v + x_{k,v}\frac{E_{\star\star}}{\sqrt{2}}\right)^2}{E_{\star\star}(E_{\star\star} - E_\star)}\right) - J_k(x_{k,v}) & \text{if } v \leq E_\star. \end{cases}$$

We have the following logarithmic asymptotic for $\mathbb{E}\left(\mathrm{Crt}_{u\downarrow}^k(S^d)\right)$.

Theorem 5.9 *We assume that $\nu_4 + 2\nu_2 - 3\nu_2^2 > 0$. Then, under* ***(A.3.1)****,*

$$\lim_{d\to+\infty}\frac{1}{d}\log\mathbb{E}\left(\mathrm{Crt}_{u\downarrow}^k(S^d)\right) = \Theta_k^\star(v).$$

Proof of Theorem 5.9 By Theorem 5.4 we have

$$\lim_{d\to+\infty}\frac{1}{(d+1)}\log\mathbb{E}\left(\mathrm{Crt}_{u\downarrow}^k(S^d)\right) = \frac{1}{2}\log\left(\frac{\nu_4 - \nu_2}{3\nu_2}\right)$$
$$+\lim_{d\to+\infty}\frac{1}{(d+1)}\log\mathbb{E}\left(\Phi\left(\sqrt{\frac{2(d+1)K}{E_{\star\star}(E_\star - E_{\star\star})}}\left(v + \tilde{L}_{k+1}\frac{E_{\star\star}}{\sqrt{2}}\right)\right)\mathrm{e}^{(d+1)g\left(\tilde{L}_{k+1}\right)}\right), \tag{5.6}$$

where

$$g(x) = -K\frac{x^2}{2}.$$

First case: $v \geq E_{\star\star}$. This implies $-\sqrt{2} \leq -\sqrt{2}v/E_{\star\star}$.

(a) We have

$$\frac{1}{(d+1)} \log \mathbb{E}\left(\Phi\left(\sqrt{\frac{2(d+1)K}{E_{\star\star}(E_\star - E_{\star\star})}}\left(v + \tilde{L}_{k+1}\frac{E_{\star\star}}{\sqrt{2}}\right)\right) \mathrm{e}^{(d+1)g\left(\tilde{L}_{k+1}\right)}\right)$$
$$\leq \frac{1}{d+1} \log \mathbb{E}\left(\mathrm{e}^{(d+1)g\left(\tilde{L}_{k+1}\right)}\right). \tag{5.7}$$

Now we use Varadhan's Integral Lemma (Dembo and Zeitouni 1998, Theorem 4.3.1). Since, by Lemma 5.7, $\tilde{L}_{k+1}$ satisfies the LDP with good rate function J_k and since $g(x) \leq 0$, condition (4.3.3) in Dembo and Zeitouni (1998) holds and

$$\lim_{d\to+\infty} \frac{1}{d+1} \log \mathbb{E}\left(\mathrm{e}^{(d+1)g\left(\tilde{L}_{k+1}\right)}\right) = \sup_{x\in\mathbb{R}} (g(x) - J_k(x)) = -K. \tag{5.8}$$

(b) We have

$$\mathbb{E}\left(\Phi\left(\sqrt{\frac{2(d+1)K}{E_{\star\star}(E_\star - E_{\star\star})}}\left(v + \tilde{L}_{k+1}\frac{E_{\star\star}}{\sqrt{2}}\right)\right) \mathrm{e}^{(d+1)g\left(\tilde{L}_{k+1}\right)}\right)$$
$$\geq \mathbb{E}\left(0.5\mathrm{e}^{(d+1)g\left(\tilde{L}_{k+1}\right)} \mathbb{1}_{\tilde{L}_{k+1} \leq -\frac{\sqrt{2}v}{E_{\star\star}}}\right),$$

then

$$\lim_{d\to+\infty} \frac{1}{d+1} \log \mathbb{E}\left(\Phi\left(\sqrt{\frac{2(d+1)K}{E_{\star\star}(E_\star - E_{\star\star})}}\left(v + \tilde{L}_{k+1}\frac{E_{\star\star}}{\sqrt{2}}\right)\right) \mathrm{e}^{(d+1)g\left(\tilde{L}_{k+1}\right)}\right)$$
$$\geq \lim_{d\to+\infty} \frac{1}{d+1} \log \mathbb{E}\left(\mathrm{e}^{(d+1)g\left(\tilde{L}_{k+1}\right)} \mathbb{1}_{\tilde{L}_{k+1} \leq -\frac{\sqrt{2}v}{E_{\star\star}}}\right). \tag{5.9}$$

Since condition (4.3.3) in Dembo and Zeitouni (1998) holds, by Dembo and Zeitouni (1998, Lemma 4.3.8) the tail condition (4.3.2) in Dembo and Zeitouni (1998) also holds and by Dembo and Zeitouni (1998, exercise 4.3.11)

$$\lim_{d\to+\infty} \frac{1}{d+1} \log \mathbb{E}\left(\mathrm{e}^{(d+1)g\left(\tilde{L}_{k+1}\right)} \mathbb{1}_{\tilde{L}_{k+1} \leq -\frac{\sqrt{2}v}{E_{\star\star}}}\right) = \sup_{x \leq -\frac{\sqrt{2}v}{E_{\star\star}}} (g(x) - J_k(x)) = -K. \tag{5.10}$$

By (5.6)–(5.10), we conclude that

$$\lim_{d\to+\infty} \frac{1}{(d+1)} \log \mathbb{E}\left(\mathrm{Crt}^k_{u\downarrow}(T_d)\right) = \frac{1}{2}\log\left(\frac{\nu_4-\nu_2}{3\nu_2}\right) - K.$$

This gives the first line of $\Theta^\star_k(V)$ in Theorem 5.9.

Second case $v < E_{\star\star}$. This implies $-\sqrt{2}v/E_{\star\star} < -\sqrt{2}$.

(a) For $-\dfrac{v\sqrt{2}}{E_{\star\star}} \le x \le -\sqrt{2}$, there exists a constant $C := \sqrt{\pi/2}$ such that

$$\Phi\left(\sqrt{\frac{2(d+1)K}{E_{\star\star}(E_\star - E_{\star\star})}}\left(v + x\frac{E_{\star\star}}{\sqrt{2}}\right)\right) \le C\phi\left(\sqrt{\frac{2(d+1)K}{E_{\star\star}(E_\star - E_{\star\star})}}\left(v + x\frac{E_{\star\star}}{\sqrt{2}}\right)\right),$$

where ϕ is the pdf of a standard Gaussian variable. We deduce

$$\mathbb{E}\left(\Phi\left(\sqrt{\frac{2(d+1)K}{E_{\star\star}(E_\star - E_{\star\star})}}\left(v + \tilde{L}_{k+1}\frac{E_{\star\star}}{\sqrt{2}}\right)\right) \mathrm{e}^{(d+1)g\left(\tilde{L}_{k+1}\right)}\right) \le \mathbb{E}\left(\mathrm{e}^{(d+1)\tilde{g}\left(\tilde{L}_{k+1}\right)}\right),$$

where

$$\tilde{g}(x) = -K\frac{x^2}{2} - \frac{K}{E_{\star\star}(E_\star - E_{\star\star})}\left(v + x\frac{E_{\star\star}}{\sqrt{2}}\right)^2 \mathbb{1}_{\{-\frac{v\sqrt{2}}{E_{\star\star}} \le x \le -\sqrt{2}\}}.$$

Therefore

$$\lim_{d\to+\infty} \frac{1}{(d+1)} \log \mathbb{E}\left(\Phi\left(\sqrt{\frac{2(d+1)K}{E_{\star\star}(E_\star - E_{\star\star})}}\left(v + \tilde{L}_{k+1}\frac{E_{\star\star}}{\sqrt{2}}\right)\right) \mathrm{e}^{(d+1)g\left(\tilde{L}_{k+1}\right)}\right)$$
$$\le \lim_{d\to+\infty} \frac{1}{(d+1)} \log \mathbb{E}\left(\mathrm{e}^{(d+1)\tilde{g}\left(\tilde{L}_{k+1}\right)}\right). \quad (5.11)$$

Since $\tilde{g}(x) \le 0$, condition (4.3.3) in Dembo and Zeitouni (1998) holds and by Dembo and Zeitouni (1998, Theorem 4.3.1), the r.h.s. of (5.11) is equal to

$$\sup_{x\in\mathbb{R}} (\tilde{g}(x) - J_k(x)).$$

(b) For $x \leq 0$, we have $\Phi(x)/\phi(x) \geq f(x)$, where $f(x) := (-1/x + 1/x^3)\mathbb{1}_{x\leq -2} + (\Phi(-2)/\phi(-2))\mathbb{1}_{-2\leq x\leq 0}$ is an increasing function. Therefore

$$\mathbb{E}\left(\Phi\left(\sqrt{\frac{2(d+1)K}{E_{\star\star}(E_\star - E_{\star\star})}}\left(v + \tilde{L}_{k+1}\frac{E_{\star\star}}{\sqrt{2}}\right)\right)\mathrm{e}^{(d+1)g\left(\tilde{L}_{k+1}\right)}\right)$$
$$\geq \mathbb{E}\left(\frac{f\left(\sqrt{\frac{2(d+1)K}{E_{\star\star}(E_\star - E_{\star\star})}}(v - E_{\star\star})\right)}{\sqrt{2\pi}}\mathrm{e}^{(d+1)\tilde{g}\left(\tilde{L}_{k+1}\right)}\mathbb{1}_{-\frac{v\sqrt{2}}{E_{\star\star}}\leq \tilde{L}_{k+1}\leq -\sqrt{2}}\right).$$

We deduce

$$\lim_{d\to+\infty}\frac{1}{(d+1)}\log\mathbb{E}\left(\Phi\left(\sqrt{(d+1)\kappa}\left(v + \tilde{L}_{k+1}\frac{E_{\star\star}}{\sqrt{2}}\right)\right)\mathrm{e}^{(d+1)g\left(\tilde{L}_{k+1}\right)}\right)$$
$$\geq \lim_{d\to+\infty}\frac{1}{(d+1)}\log\mathbb{E}\left(\mathrm{e}^{(d+1)\tilde{g}\left(\tilde{L}_{k+1}\right)}\mathbb{1}_{-\frac{v\sqrt{2}}{E_{\star\star}}\leq \tilde{L}_{k+1}\leq -\sqrt{2}}\right). \tag{5.12}$$

Since condition (4.3.3) in Dembo and Zeitouni (1998) holds, by Dembo and Zeitouni (1998, Lemma 4.3.8) the tail condition (4.3.2) in Dembo and Zeitouni (1998) also holds and by Dembo and Zeitouni (1998, exercise 4.3.11), the r.h.s. of (5.12) is equal to

$$\sup_{-\frac{v\sqrt{2}}{E_{\star\star}}\leq x\leq -\sqrt{2}}(\tilde{g}(x) - J_k(x)).$$

It is easy to check that

$$\sup_{x\in\mathbb{R}}(\tilde{g}(x) - J_k(x)) = \sup_{-\frac{v\sqrt{2}}{E_{\star\star}}\leq x\leq -\sqrt{2}}(\tilde{g}(x) - J_k(x)).$$

In the case $E_\star < v < E_{\star\star}$, we can check that $\tilde{g}'(x) > 0$ and $-J_k'(x) > 0$ $\forall x \in [-\frac{v\sqrt{2}}{E_{\star\star}}; -\sqrt{2}]$. Thus the maximum of $\tilde{g}(x) - J_k(x)$ is obtained for $x = -\sqrt{2}$. This gives the second line of $\Theta_k^\star(v)$ in Theorem 5.9.

When $v < E_\star$ we look for the maximum of $\tilde{g}(x) - J_k(x)$ on $[-\frac{v\sqrt{2}}{E_{\star\star}}; -\sqrt{2}]$. We have $\tilde{g}'(x) > 0$ and $-J_k'(x) > 0$ $\forall x \in [-\frac{v\sqrt{2}}{E_{\star\star}}; -\frac{v\sqrt{2}}{E_\star}]$. Moreover $\tilde{g}'(-\sqrt{2}) - J_k'(-\sqrt{2}) < 0$. So the maximum is reached on $[-\frac{v\sqrt{2}}{E_\star}; -\sqrt{2}]$

$$\tilde{g}'(x) - J_k'(x) = 0 \Leftrightarrow (k+1)\sqrt{x^2 - 2} = -\frac{K(xE_\star + \sqrt{2}v)}{(E_{\star\star} - E_\star)}. \tag{5.13}$$

This implies that

$$\left(\frac{K^2E_\star^2}{(E_{\star\star}-E_\star)^2}-(k+1)^2\right)x^2+\frac{2\sqrt{2}vE_\star K^2}{(E_{\star\star}-E_\star)^2}x+2(k+1)^2+\frac{2v^2K^2}{(E_{\star\star}-E_\star)^2}=0.$$

When $K^2E_\star^2-(k+1)^2(E_{\star\star}-E_\star)^2\neq 0$, this equation admits two roots

$$x_\pm=\frac{-\sqrt{2}vE_\star K^2\pm\sqrt{2}(k+1)(E_{\star\star}-E_\star)\sqrt{(k+1)^2(E_{\star\star}-E_\star)^2+K^2(v^2-E_\star^2)}}{K^2E_\star^2-(k+1)^2(E_{\star\star}-E_\star)^2}.$$

When $K^2E_\star^2-(k+1)^2(E_{\star\star}-E_\star)^2<0$, we have $x_->0$, and the only suitable root is x_+. Otherwise, when $K^2E_\star^2-(k+1)^2(E_{\star\star}-E_\star)^2>0$, we can check that $x_-<-\sqrt{2}v/E_\star$, and the only suitable root is again x_+.

When $K^2E_\star^2-(k+1)^2(E_{\star\star}-E_\star)^2=0$, the root is

$$x:=-\frac{E_\star^2+v^2}{\sqrt{2}vE_\star}.$$

This concludes the proof.

Remark 5.5 We have $E_\star<E_{\star\star}<0$. Moreover $\Theta_k^\star(v)$ is a positive constant for $v\geq E_{\star\star}$, an increasing function of v not depending on k for $E_\star\leq v<E_{\star\star}$ and an increasing function of v depending on k for $v\leq E_\star$. For v fixed, $\Theta_k^\star(v)$ is a decreasing function of k. For more details, see Auffinger and Ben Arous (2013, Proposition 1).

Remark 5.6 In the general case $\nu_4+2\nu_2-3\nu_2^2>0$ we retrieve the layered structure described above for random fields such that $\Theta_k^\star(E_\star)>0$, that is to say

$$\sqrt{\frac{\nu_4-\nu_2}{3\nu_2}}>\exp\left(K\frac{E_\star}{E_{\star\star}}\right).$$

In that case ν is called a pure-like mixture in Auffinger and Ben Arous (2013, Definition 4.1). When $\Theta_k^\star(E_\star)=0$, ν is called critical and when $\Theta_k^\star(E_\star)<0$, ν is called a full mixture. We give an example of a pure-like mixture in Fig. 5.2 and an example of a full mixture in Fig. 5.3. Note that all pure p-spins $\nu(\langle s,t\rangle)=\langle s,t\rangle^p$ with $p\geq 3$ are pure-like while $\nu(\langle s,t\rangle)=\langle s,t\rangle^2$ is critical, as noticed in Remark 5.3 and Auffinger and Ben Arous (2013, Example 1).

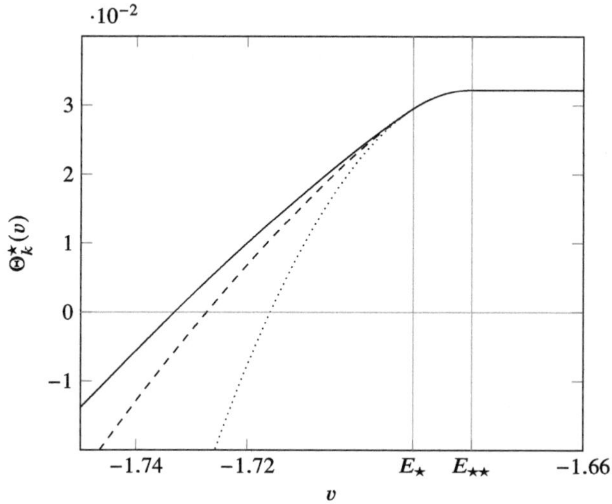

Fig. 5.2 Function $\Theta_k^\star(v)$ for $\nu(\langle s,t\rangle) = \frac{1}{2}\langle s,t\rangle^3 + \frac{1}{2}\langle s,t\rangle^4$, $k = 0$ (solid), $k = 1$ (dashed) and $k = 100$ (dotted)

Fig. 5.3 Function $\Theta_k^\star(v)$ for $\nu(\langle s,t\rangle) = \frac{17}{21}\langle s,t\rangle^2 + \frac{4}{21}\langle s,t\rangle^6$, $k = 0$ (solid), $k = 1$ (dashed) and $k = 100$ (dotted)

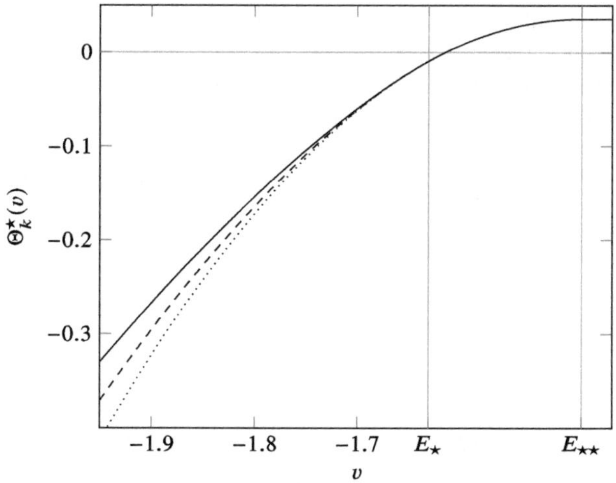

5.3 Mean Number: The Flat Case

In this section $\{X(t) : t \in T \subset \mathbb{R}^d\}$ is a flat stationary and isotropic Gaussian random field with real values. The aim of this section is to show that under some conditions, we observe a behavior which is similar to that of Sects. 5.2.1 and 5.2.2. We first define the following assumption.

Assumption ($\mathbf{A_k}$.5.1):

- $\{X(t) : t \in T \subset \mathbb{R}^d\}$ is a stationary and isotropic Gaussian field with real values. We assume, without loss of generality, that it is centered with variance 1.
- The paths of $X(\cdot)$ are C^2 and $\mathbf{r}$ is C^{2k} with $k = 2, 3$ or 4.
- $\mathbb{E}(X(s)X(t)) = c(s,t) = \mathbf{r}(\|s-t\|^2); \quad \mathbf{r}(0) = 1; \quad \mathbf{r}'(0) < 0.$

Remark 5.7

- Note that when $k \geq 3$, the condition $\mathbf{r}$ is C^{2k} implies that the paths of $X(\cdot)$ are C^2.
- The condition $\mathbf{r}'(0) < 0$ is supposed to avoid the uninteresting case of a constant random field. Indeed $\lambda_2 := \mathrm{Var} X_i(t) = -2\mathbf{r}'(0) > 0$.
- $\forall \ell = 1, \dots, d$ and $n = 1, 2$,

$$1 = \mathrm{Var}(X(t)) = \mathbf{r}(0),$$
$$\lambda_{2n} := \mathrm{Var}\left(\frac{\partial^n X(t)}{\partial t_\ell^n}\right) = (-1)^n \frac{(2n)!}{n!} \mathbf{r}^{(n)}(0). \tag{5.14}$$

In this section we assume that $\{X(t) : t \in T \subset \mathbb{R}^d\}$ is a flat Gaussian random field satisfying Assumption ($\mathbf{A_2}$.5.1). Then, the sample paths are almost surely Morse and this implies that the number of critical points with index k in a Borel set T, $\mathrm{Crt}^k(T)$, is well defined. More precisely,

$$\mathrm{Crt}^k(T) := \#\{t \in T : X'(t) = 0,\ i(X''(t)) = k\}. \tag{5.15}$$

We also define the number of critical points of index k above the level u by

$$\mathrm{Crt}^k_{u\uparrow}(T) := \#\{t \in T : X'(t) = 0,\ i(X''(t)) = k,\ X(t) > u\}, \tag{5.16}$$

and below the level u by

$$\mathrm{Crt}^k_{u\downarrow}(T) := \#\{t \in T : X'(t) = 0,\ i(X''(t)) = k,\ X(t) < u\}. \tag{5.17}$$

First, we derive in Sects. 5.3.1 and 5.3.2 exact expressions for the expectation of (5.15) and (5.16). Then, in Sect. 5.3.3 we give logarithmic asymptotics as $d \to +\infty$ for $\mathbb{E}\left(\mathrm{Crt}^k_{u\downarrow}(T)\right)$, as is done in the spherical case in Sect. 5.2.2.

5.3.1 General Case

By Kac–Rice formulas and Proposition 3.20 we get

$$\mathbb{E}\left(\mathrm{Crt}^k(T)\right) = \frac{|T|}{(2\pi\lambda_2)^{d/2}}\mathbb{E}\left(|\det(X''(t))|\mathbb{1}_{i(X''(t))=k}\right), \tag{5.18}$$

$$\mathbb{E}\left(\mathrm{Crt}^k_{u\uparrow}(T)\right) = \frac{|T|}{\lambda_2^{d/2}(2\pi)^{(d+1)/2}}\int_u^{+\infty}\exp\left(-\frac{x^2}{2}\right) \tag{5.19}$$
$$\times\,\mathbb{E}\left(|\det(X''(t))\mathbb{1}_{i(X''(t))=k}|\,/X(t)=x\right)\mathrm{d}x.$$

Note that isotropy is not needed for the validity of (5.18) and (5.19).

Using Proposition 3.20 we get the following proposition, whose proof can be found in Azaïs and Delmas (2022, Proposition 4).

Proposition 5.10 *Let L_p be the pth ordered eigenvalue of a $(d+1)$-GOE matrix $(L_1 \le L_2 \le \cdots \le L_{d+1})$. For X and T as above, under Assumption*

$$\mathbb{E}\left(\mathrm{Crt}^k(T)\right) = \frac{\sqrt{2}|T|}{\pi^{(d+1)/2}}\left(\frac{\lambda_4}{3\lambda_2}\right)^{d/2}\Gamma\left(\frac{d+1}{2}\right)\mathbb{E}\left(\exp\left(-\frac{L_{k+1}^2}{2}\right)\right). \tag{5.20}$$

When $\lambda_4 - 3\lambda_2^2 > 0$,

$$\mathbb{E}\left(\mathrm{Crt}^k_{u\uparrow}(T)\right) = \frac{\sqrt{2}|T|}{\pi^{(d+1)/2}}\left(\frac{\lambda_4}{3\lambda_2}\right)^{d/2}\Gamma\left(\frac{d+1}{2}\right)$$
$$\times\,\mathbb{E}\left\{\exp\left(-\frac{L_{k+1}^2}{2}\right)\bar{\Phi}\left[\sqrt{\frac{\lambda_4}{\lambda_4-3\lambda_2^2}}\left(u - L_{k+1}\frac{\sqrt{6}\lambda_2}{\sqrt{\lambda_4}}\right)\right]\right\}, \tag{5.21}$$

and when $\lambda_4 - 3\lambda_2^2 = 0$

$$\mathbb{E}\left(\mathrm{Crt}^k_{u\uparrow}(T)\right) = \frac{\sqrt{2}|T|\lambda_2^{d/2}}{\pi^{(d+1)/2}}\Gamma\left(\frac{d+1}{2}\right)\mathbb{E}\left(\exp\left(-\frac{L_{k+1}^2}{2}\right)\mathbb{1}_{L_{k+1}>u/\sqrt{2}}\right). \tag{5.22}$$

Remark 5.8 This kind of result was first obtained in the spherical case by Auffinger et al. (2013) and Auffinger and Ben Arous (2013). This result can also be found in Cheng and Schwartzman (2018, Proposition 3.9), written in a slightly different way.

Remark 5.9 As remarked in Azaïs and Wschebor (2010), the condition $\lambda_4 \ge 3\lambda_2^2$ is met if the covariance function **r** is a "Schoenberg covariance", see Definition 3.2. Note that more general cases have been studied by Cheng and Schwartzman (2018).

Proof of Proposition 5.10 Using (5.18), Proposition 3.20 and the expression (5.2) of the joint density of the eigenvalues of a GOE matrix,

$$\mathbb{E}\left(|\det X''(t)|\mathbb{1}_{i(X''(t))=k}\right)$$
$$= \left(\frac{2\lambda_4}{3}\right)^{d/2} k_d d! \int_{\mu_1<\mu_2<\cdots<\mu_k<\lambda<\mu_{k+1}<\cdots<\mu_d} \prod_{1\le i<k\le d} |\mu_k-\mu_i| \prod_{1\le i\le d} |\lambda-\mu_i|$$
$$\exp\left(-\frac{\sum_{i=1}^d \mu_i^2}{2}\right)\pi^{-1/2}\exp(-\lambda^2/2)\exp(-\lambda^2/2)\mathrm{d}\mu_1\ldots\mathrm{d}\mu_d\mathrm{d}\lambda$$
$$= \left(\frac{2\lambda_4}{3}\right)^{d/2}\pi^{-1/2}\frac{k_d}{k_{d+1}}\frac{d!}{(d+1)!}\mathbb{E}\left(\exp\left(-\frac{L_{k+1}^2}{2}\right)\right),$$

where k_d is given in (5.3). This proves (5.20). Now, using Proposition 3.20, by (5.19) and (5.2) we can write

$$\mathbb{E}\left(\mathrm{Crt}_{u\uparrow}^k(T)\right) = \frac{|T|}{\lambda_2^{d/2}(2\pi)^{(d+1)/2}}\int_u^{+\infty}\exp\left(-\frac{x^2}{2}\right)\left(\frac{2\lambda_4}{3}\right)^{d/2} k_d d!$$
$$\times\int_{\mu_1<\mu_2<\cdots<\mu_k<y<\mu_{k+1}<\cdots<\mu_d}\prod_{1\le i<k\le d}|\mu_k-\mu_i|\prod_{1\le i\le d}|y-\mu_i|\exp\left(-\frac{\sum_{i=1}^d\mu_i^2}{2}\right)$$
$$\times\exp(-y^2/2)\exp(y^2/2)p_Y(y)\mathrm{d}\mu_1\ldots\mathrm{d}\mu_d\,\mathrm{d}y\,\mathrm{d}x.$$

Integrating first with respect to x we obtain (5.21). In the same way, when $\lambda_4-3\lambda_2^2=0$, $X''(t)/X(t)=x$ is distributed as $\sqrt{2\lambda_2^2}G-\lambda_2 x\mathrm{Id}_d$, where G is a d-GOE matrix, and following the same lines as previously we obtain (5.22).

5.3.2 *Particular Cases*

Using the exact expressions of the probability densities of the ordered eigenvalues of a $(d+1)$-GOE matrix for $d\le 4$, given in Azaïs and Delmas (2022, Section 3.2.1), we obtain Propositions 5.11–5.13 and for $d=2$ we can easily check

$$\mathbb{E}\left(\mathrm{Crt}^0(T)\right)=\mathbb{E}\left(\mathrm{Crt}^2(T)\right)=\frac{1}{2}\mathbb{E}\left(\mathrm{Crt}^1(T)\right)=\frac{1}{4}\mathbb{E}\left(\mathrm{Crt}(T)\right), \tag{5.23}$$

with

$$\mathbb{E}\left(\mathrm{Crt}(T)\right) = \frac{2|T|}{\sqrt{3}\pi}\left(\frac{\lambda_4}{3\lambda_2}\right).$$

This result was first obtained, in a different way, by Longuet-Higgins (1957). See also Adler (1981, Theorem 6.2.1 and Lemma 6.1.1).

Proposition 5.11 *Under the conditions of Proposition 5.10 when* $d = 3$

$$\mathbb{E}\left(\mathrm{Crt}^0(T)\right) = \mathbb{E}\left(\mathrm{Crt}^3(T)\right) = \frac{|T|}{\pi^2\sqrt{2}}\left(\frac{\lambda_4}{3\lambda_2}\right)^{3/2}\left(\frac{29-6\sqrt{6}}{12\sqrt{6}}\right),$$

$$\mathbb{E}\left(\mathrm{Crt}^1(T)\right) = \mathbb{E}\left(\mathrm{Crt}^2(T)\right) = \frac{|T|}{\pi^2\sqrt{2}}\left(\frac{\lambda_4}{3\lambda_2}\right)^{3/2}\left(\frac{29+6\sqrt{6}}{12\sqrt{6}}\right).$$

Proposition 5.12 *Set* $\mathcal{I} := \mathbb{E}\left(\Phi(Y)\Phi(\sqrt{2}Y)\right)$, *$Y$ being a Gaussian centered variable with variance* $1/3$. *Under the conditions of Proposition 5.10 when* $d = 4$

$$\mathbb{E}\left(\mathrm{Crt}^1(T)\right) = \mathbb{E}\left(\mathrm{Crt}^3(T)\right) = \frac{|T|}{\pi^2}\left(\frac{\lambda_4}{3\lambda_2}\right)^2\left(\frac{25}{24\sqrt{3}}\right),$$

$$\mathbb{E}\left(\mathrm{Crt}^0(T)\right) = \mathbb{E}\left(\mathrm{Crt}^4(T)\right) = \frac{|T|}{\pi^2}\left(\frac{\lambda_4}{3\lambda_2}\right)^2\left(\frac{\mathcal{I}\times 100\pi - 57}{48\sqrt{3}\pi}\right),$$

$$\mathbb{E}\left(\mathrm{Crt}^2(T)\right) = \frac{|T|}{\pi^2}\left(\frac{\lambda_4}{3\lambda_2}\right)^2\left(\frac{50\pi(1-2\mathcal{I})+57}{24\sqrt{3}\pi}\right).$$

Proposition 5.13 *Under the conditions of Proposition 5.10, for* $u \in \mathbb{R}$, *when* $d = 2$ *and* $\lambda_4 - 3\lambda_2^2 = 0$,

$$\mathbb{E}\left(\mathrm{Crt}^1_{u\uparrow}(T)\right) = \frac{|T|\lambda_2}{\pi}\frac{1}{\sqrt{3}}\bar{\Phi}\left(\sqrt{\frac{3}{2}}u\right),$$

$$\mathbb{E}\left(\mathrm{Crt}^2_{u\uparrow}(T)\right) = \frac{|T|\lambda_2}{\pi}\left(\frac{u\mathrm{e}^{-u^2/2}\Phi(u)}{2\sqrt{2\pi}} + \frac{\mathrm{e}^{-u^2}}{4\pi} + \int_{\frac{u}{\sqrt{2}}}^{+\infty}\frac{\mathrm{e}^{-3\ell^2/2}}{\sqrt{2\pi}}\Phi(\ell)\mathrm{d}\ell\right).$$

$$\mathbb{E}\left(\mathrm{Crt}^0_{u\uparrow}(T)\right) = \frac{|T|\lambda_2}{\pi}\left(\frac{u\mathrm{e}^{-u^2/2}\bar{\Phi}(u)}{2\sqrt{2\pi}} - \frac{\mathrm{e}^{-u^2}}{4\pi} + \int_{\frac{u}{\sqrt{2}}}^{+\infty}\frac{\mathrm{e}^{-3\ell^2/2}}{\sqrt{2\pi}}\bar{\Phi}(\ell)\mathrm{d}\ell\right),$$

$$\mathbb{E}\left(\mathrm{Crt}_{u\uparrow}(T)\right) = \frac{|T|\lambda_2}{\pi}\left(\frac{u\mathrm{e}^{-u^2/2}}{2\sqrt{2\pi}} + \frac{2}{\sqrt{3}}\bar{\Phi}\left(\sqrt{\frac{3}{2}}u\right)\right).$$

5.3.3 Logarithmic Asymptotics

In this section we consider the logarithmic asymptotic behavior of $\mathbb{E}\left(\mathrm{Crt}^k_{u\downarrow}(T)\right)$ as $d \to \infty$. So we consider a covariance function $\mathbf{r}$ which is valid in every dimension, i.e. a Schoenberg covariance (see Definition 3.2). In the same way, the set T will be a function of d and will be denoted T_d. Roughly speaking we obtain, in this section, the same kind of results as in Sect. 5.2.2 when the volume of the set T_d has an appropriate behavior as $d \to +\infty$. More precisely we assume that

$$\frac{1}{d}\log\left(\frac{|T_d|}{\kappa_d}\right) \to A \in \mathbb{R} \text{ when } d \to +\infty, \tag{5.24}$$

where $\kappa_d = \dfrac{2\pi^{(d+1)/2}}{\Gamma\left(\frac{d+1}{2}\right)}$ denotes the volume of the unit sphere S^d in $\mathbb{R}^{d+1}$. Note that (5.24) means that, at the logarithmic scale, T_d has a volume exponentially comparable to that of S^d. The is the case for the ball with radius R in $\mathbb{R}^{d+1}$. Its volume is $|T_d| = \dfrac{\pi^{(d+1)/2}R^{d+1}}{\Gamma\left(\frac{d+1}{2}+1\right)}$. In that case $A = \log R$. It is also the case for the square $[0, L/\sqrt{d}]$ with $0 < L < +\infty$. In that case $A = \log\left(\dfrac{L}{\sqrt{2\pi}}\right) - \dfrac{1}{2}$.

We also assume in a first part that $\lambda_4 - 3\lambda_2^2 = 0$. By Lemma 3.19

$$\mathbf{r}(\|s-t\|^2) = \exp\left(-a^2\|s-t\|^2\right).$$

We set $v := \dfrac{u}{\sqrt{d+1}}$. We define $\tilde{E}_\infty := -2$, and

$$\tilde{\Theta}_k(v) := \begin{cases} A + \dfrac{1}{2}\log(\lambda_2) - 1 & \text{if } v \geq \tilde{E}_\infty \\ A + \dfrac{1}{2}\log(\lambda_2) - \dfrac{v^2}{4} - (k+1)I_1\left(\dfrac{v\sqrt{2}}{\tilde{E}_\infty}, \dfrac{1}{\sqrt{2}}\right) & \text{if } v \leq \tilde{E}_\infty \end{cases}$$

where I_1 is defined in (5.5).

Theorem 5.14 *We assume Assumption ($\mathbf{A_2}$.5.1), that $\mathbf{r}$ is a Schoenberg covariance as defined in Definition 3.2, that $\lambda_4 - 3\lambda_2^2 = 0$ and that the family of sets T_d satisfies* (5.24). *We have*

$$\lim_{d\to+\infty}\frac{1}{d}\log\mathbb{E}\left(\mathrm{Crt}^k_{u\downarrow}(T_d)\right) = \tilde{\Theta}_k(v).$$

The proof of this theorem follows the same lines as that of Theorem 5.9.

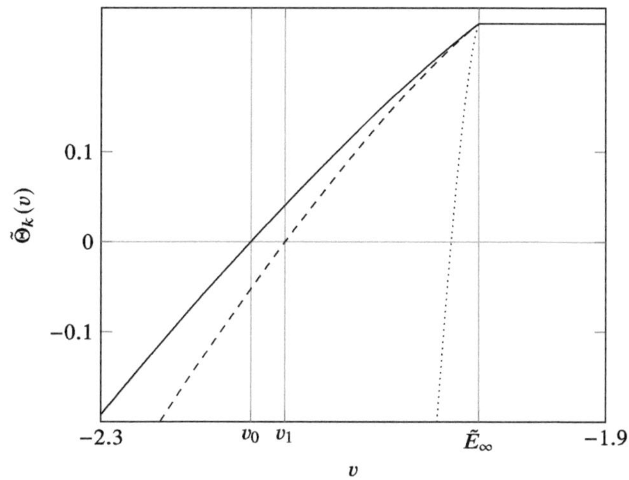

Fig. 5.4 Function $\tilde{\Theta}_k(v)$ for $\mathbf{r}(\|s-t\|^2) = \exp\left(-6\|s-t\|^2\right)$, $A = 0$, $k = 0$ (solid), $k = 1$ (dashed) and $k = 100$ (dotted)

Remark 5.10 When $a^2 > \dfrac{1}{2}\mathrm{e}^{2(1-A)}$, the function $\tilde{\Theta}_k(v)$ is a positive constant for $v \geq \tilde{E}_\infty$ and a strictly increasing function of v depending on k for $v \leq \tilde{E}_\infty$. For v fixed, $\tilde{\Theta}_k(v)$ is a strictly decreasing function of k. We observe the same layered structure for the critical points with finite indexes as observed and described in the spherical case in Sect. 5.2.2 or in Auffinger et al. (2013). We recall that v_i is defined as $\tilde{\Theta}_i(v_i) = 0$ and $u_i := \sqrt{d+1}v_i$. Since $\forall k$ $\tilde{\Theta}_k(v) < 0$ for $v < v_0$, there are essentially no critical points below u_0; since for $v_0 < v < v_1$, $\tilde{\Theta}_0(v) > 0$ and $\tilde{\Theta}_k(v) < 0$ $\forall k \geq 1$, there are essentially only critical points with index 0 between u_0 and u_1. Between v_1 and v_2, only $\tilde{\Theta}_0(v)$ and $\tilde{\Theta}_1(v)$ are positive so we find only critical points with indexes 0 and 1 between the corresponding levels u_1 and u_2 and so on. We give an example in Fig. 5.4.

We now extend these results to the more general case $\lambda_4 - 3\lambda_2^2 > 0$ with also the condition (5.24) on the set T_d. The following results also appear in Azaïs and Delmas (2026). We set

$$v := \frac{u}{\sqrt{d+1}}, \qquad \tilde{E}_\star := -\left(\frac{\lambda_4 + 3\lambda_2^2}{\lambda_2\sqrt{3\lambda_4}}\right), \qquad \tilde{E}_{\star\star} := -\frac{2\sqrt{3}\lambda_2}{\sqrt{\lambda_4}}.$$

When $\tilde{E}_\star^2 - (k+1)^2(\tilde{E}_{\star\star} - \tilde{E}_\star)^2 \neq 0$, we define

$$\tilde{x}_{k,v} := \frac{-\sqrt{2}v\tilde{E}_\star + \sqrt{2}(k+1)(\tilde{E}_{\star\star} - \tilde{E}_\star)\sqrt{(k+1)^2(\tilde{E}_{\star\star} - \tilde{E}_\star)^2 + (v^2 - \tilde{E}_\star^2)}}{\tilde{E}_\star^2 - (k+1)^2(\tilde{E}_{\star\star} - \tilde{E}_\star)^2},$$

and when $\tilde{E}_\star^2 - (k+1)^2(\tilde{E}_{\star\star} - \tilde{E}_\star)^2 = 0$,

$$\tilde{x}_{k,v} := -\frac{\tilde{E}_\star^2 + v^2}{\sqrt{2} v \tilde{E}_\star}.$$

We define

$$\tilde{\Theta}_k^\star(v) := \begin{cases} A + \frac{1}{2}\log\left(\frac{\lambda_4}{3\lambda_2}\right) - 1 \text{ if } v \geq \tilde{E}_{\star\star} \\ A + \frac{1}{2}\log\left(\frac{\lambda_4}{3\lambda_2}\right) - \left(1 - \frac{(v - \tilde{E}_{\star\star})^2}{\tilde{E}_{\star\star}(\tilde{E}_{\star\star} - \tilde{E}_\star)}\right) \text{ if } \tilde{E}_\star \leq v \leq \tilde{E}_{\star\star} \\ A + \frac{1}{2}\log\left(\frac{\lambda_4}{3\lambda_2}\right) - \left(\frac{\tilde{x}_{k,v}^2}{2} - \frac{\left(v + \tilde{x}_{k,v}\frac{\tilde{E}_{\star\star}}{\sqrt{2}}\right)^2}{\tilde{E}_{\star\star}(\tilde{E}_{\star\star} - \tilde{E}_\star)}\right) - J_k(\tilde{x}_{k,v}) \text{ if } v \leq \tilde{E}_\star \end{cases}$$

where the function J_k is defined in (5.4).

Theorem 5.15 *With the same conditions as in Theorem 5.14 except that now* $\lambda_4 - 3\lambda_2^2 > 0$, *we have*

$$\lim_{d\to+\infty} \frac{1}{d} \log \mathbb{E}\left(\mathrm{Crt}_{u\downarrow}^k(T_d)\right) = \tilde{\Theta}_k^\star(v).$$

The proof of this theorem follows the same lines as that of Theorem 5.9.

Remark 5.11 (Azaïs and Delmas 2026, Lemma 5) We can check that $\tilde{E}_\star < -2 < \tilde{E}_{\star\star} < 0$. Moreover, $\tilde{\Theta}_k^\star(v)$ is a constant for v above the level $\tilde{E}_{\star\star}$, a strictly increasing function of v not depending on k between $\tilde{E}_\star$ and $\tilde{E}_{\star\star}$ and a strictly increasing function of v depending on k below $\tilde{E}_\star$. For A and $v < \tilde{E}_\star$ fixed, $\tilde{\Theta}_k^\star(v)$ is a strictly decreasing function of k. We will observe a layered structure when $\tilde{\Theta}_k^\star(\tilde{E}_\star) > 0$, that is to say

$$\sqrt{\frac{\lambda_4}{3\lambda_2}} > \exp\left(\frac{\tilde{E}_\star}{\tilde{E}_{\star\star}} - A\right). \tag{5.25}$$

We give an example in Fig. 5.5. Note that it is always possible to find A such that (5.25) is satisfied.

Remark 5.12 In a more general case, when Condition 5.24 on the set T_d is not satisfied, we have a Taylor expansion when $d \to +\infty$:

$$\frac{1}{d}\log \mathbb{E}\left(\mathrm{Crt}_{u\downarrow}^k(T_d)\right) \simeq \frac{1}{2}\log\left(\frac{d}{2}\right) - \frac{1}{2}(1 + \log\pi) + \frac{\log|T_d|}{d} + \tilde{\Theta}_k^\star(v) + o(1).$$

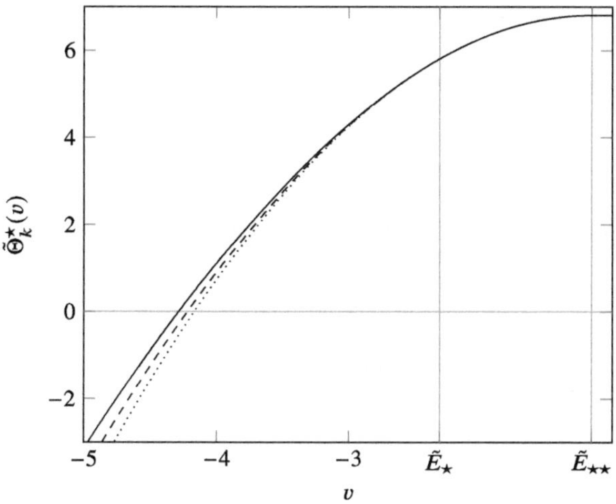

Fig. 5.5 Function $\tilde{\Theta}_k^\star(v)$ for $A = 7$, $k = 0$ (solid), $k = 1$ (dashed), $k = 100$ (dotted) and $\mathbf{r}(\|s - t\|^2)$ is the Matérn covariance function, see Azaïs and Delmas (2026, Section 3.2) for the definition and notation, with $\nu = 5/2$, $l = 1$ and $\sigma^2 = 1$

In particular, when $T_d = [0, a]^d$:

$$\frac{1}{d} \log \mathbb{E}\left(\mathrm{Crt}_{u\downarrow}^k(T_d)\right) \simeq \frac{1}{2} \log\left(\frac{d}{2}\right) - \frac{1}{2}(1 + \log \pi) + \log(a) + \tilde{\Theta}_k^\star(v) + o(1).$$

When $\lambda_4 - 3\lambda_2^2 = 0$, just replace $\tilde{\Theta}_k^\star(v)$ by $\tilde{\Theta}_k(v)$.

5.4 Correlation Functions Between Critical Points

Consider a stationary and isotropic Gaussian field $\{X(t) : t \in \mathbb{R}^d\}$ with real values satisfying Assumption . Its spectral moments $\lambda_{2n} : n = 0, 1, 2, 3$ are defined in (5.14). They satisfy the property given in (3.30) for $n = 2, 3$.

Let us consider two points $s, t \in \mathbb{R}^d$. The correlation function of a point process $\mathcal{P}$ is defined by

$$A(s,t) := \lim_{\varepsilon \to 0} \frac{1}{V^2(\varepsilon)} \mathbb{E}\left(\mathcal{P}(B(s,\varepsilon))\mathcal{P}(B(t,\varepsilon))\right) \text{ when it exists,}$$

where $\mathcal{P}(B(s, \varepsilon))$ is the number of points in the ball with center s and radius ε and $V(\varepsilon)$ is the volume of this ball. When the considered process is stationary and isotropic, this function depends only on the norm $\rho := \|s - t\|$ and by a small abuse of notation we will denote it by $A(\rho)$.

Suppose now that the point process $\mathcal{P}$ is the process of critical points of X. When S and T are disjoint subsets of $\mathbb{R}^d$, sufficiently close and sufficiently small (as $B(s, \varepsilon)$ and $B(t, \varepsilon)$ for $s \neq t$ and $\|s - t\|$ and ε sufficiently small), the Kac–Rice formula of order two yields

$$\mathbb{E}[\mathrm{Crt}(S)\mathrm{Crt}(T)] \tag{5.26}$$

$$= \int_{S\times T} \mathbb{E}\left(|\det(X''(s))\det(X''(t))| \mid X'(s) = X'(t) = 0\right) p_{X'(s),X'(t)}(0,0)\mathrm{d}s\mathrm{d}t,$$

proving that for ρ sufficiently small, $A(\rho)$ is well defined and given by

$$A(\rho) = p_{X'(0),X'(t)}(0,0) \times \mathbb{E}\left(|\det(X''(0))\det(X''(t))| \mid X'(0) = X'(t) = 0\right). \tag{5.27}$$

In the same way, the correlation function between critical points with indexes i_1 and i_2 is defined as

$$A^{i_1,i_2}(\rho) = \tag{5.28}$$

$$p_{X'(0),X'(t)}(0,0)\mathbb{E}\left(|\det(X''(0))\det(X''(t))|\mathbb{1}_{\substack{i(X''(0))=i_1\\ i(X''(t))=i_2}} \mid X'(0) = X'(t) = 0\right).$$

Definition 5.1 There is repulsion, attraction or neutrality, between critical points when $A(\rho) \to 0$, $A(\rho) \to +\infty$ or $A(\rho) \to (\texttt{const})$, as $\rho \to 0$. In the same way there is repulsion, attraction or neutrality between critical points with index i_1 and critical points with index i_2 when $A^{i_1,i_2}(\rho) \to 0, \ +\infty,$ or $(\texttt{const})$ as $\rho \to 0$.

In this section we present the results of Azaïs and Delmas (2022, Section 4).

5.4.1 General Case

Theorem 5.16 below gives the asymptotic expression (as $\rho \to 0$) of the correlation function between critical points (5.27) of any isotropic Gaussian field. It generalizes the result of Beliaev et al. (2019, 2020) to general fields in any dimension.

Theorem 5.16 *Let $\{X(t) : t \in \mathbb{R}^d\}$ be a stationary and isotropic Gaussian field with real values satisfying Assumption ($\mathbf{A}_3$.5.1). Let $A(\rho)$ be the correlation function between critical points* (5.27). *Then as $\rho \to 0$,*

$$A(\rho) \simeq \rho^{2-d} \frac{\gamma_{d-1}}{2^3 3^{(d-1)/2}\pi^d} \left(\sqrt{\frac{\lambda_4}{\lambda_2}}\right)^d \frac{(\lambda_2\lambda_6 - \lambda_4^2)}{\lambda_2\lambda_4}, \tag{5.29}$$

where γ_{d-1} is defined by

$$\gamma_{d-1} := \mathbb{E}\left(\det{}^2\left(G_{d-1} - V\mathrm{Id}_{d-1}\right)\right),$$

with G_{d-1} a $(d-1)$-GOE matrix (as in Definition 3.3) and V an independent centered Gaussian random variable with variance $1/3$. When $d = 1$ we have $\gamma_0 := 1$.

Remark 5.13 We set $\gamma_{d-1,2}(x) := \mathbb{E}\left(\det^2(G_{d-1} - x\mathrm{Id}_{d-1})\right)$. Note that Mehta and Normand (2001, formulas 2.2.16 and 2.2.17) give explicit expressions for the calculation of $\gamma_{d-1,2}(x)$ for $x \in \mathbb{R}$.

Remark 5.14 We have repulsion between critical points when $d = 1$, neutrality when $d = 2$ and attraction between critical points when $d \geq 3$.

Remark 5.15 For more details and results in dimension $d = 1$ we refer to Proposition 3.4.

Proof of Theorem 5.16 For simplicity of exposition we admit in a first step all passages to the limit. They will be justified at the end of the proof. By (5.27)

$$A(\rho) = \mathbb{E}\left(|\det(\xi(0))\det(\xi(t))|\right) p_{X'(0),X'(t)}(0,0), \tag{5.30}$$

with $t = \rho e_1$. We denote by $\xi(0)$ and $\xi(t)$ a representation of the conditional distribution of the Hessian matrix of X at 0 and t. Because of (3.32),

$$p_{X'(0),X'(t)}(0,0) \simeq \rho^{-d} 3^{(d-1)/2} \lambda_2^{-d/2} \lambda_4^{-d/2} (2\pi)^{-d}. \tag{5.31}$$

The expectation in (5.30) remains to be studied. We denote by $\xi_{-1}(t)$ the $(d-1) \times (d-1)$ matrix corresponding to the Hessian matrix $\xi(t)$ without its first row and column.

Let us expand $\det(\xi(t))$ along the first row, for example. The first term is $\xi_{11}(t)\det(\xi_{-1}(t))$.

Consider another term and expand it now along the first column. Each of the $d-1$ terms that appear are products that include a factor $\xi_{1j}\xi_{j'1}$ with $j, j' \neq 1$ so, because of (3.33), they are $O_p(\rho^2)$: divided by ρ^2 they are bounded in probability. As a consequence, we have proved that

$$\det(\xi(t)) = \xi_{11}(t)\det(\xi_{-1}(t)) + O_p(\rho^2).$$

By (3.34) and (3.33) we have

$$\xi_{-1}(t) \simeq \sqrt{\frac{2\lambda_4}{3}}\left(G_{d-1} - V\mathrm{Id}_{d-1}\right), \tag{5.32}$$

where V is an $\mathcal{N}(0, 1/3)$ random variable, G_{d-1} is a size $(d-1)$ GOE matrix, and V and G_{d-1} are independent. So

$$\det(\xi_{-1}(t)) \simeq \left(\frac{2\lambda_4}{3}\right)^{(d-1)/2} \det(G_{d-1} - V\mathrm{Id}_{d-1}). \tag{5.33}$$

The order of magnitude of the first term in the development of the determinant is then ρ.

In conclusion, we have proved that

$$\det(\xi(t)) = O_p(\rho) = \xi_{11}(t)\det(\xi_{-1}(t)) + O_p(\rho^2) \tag{5.34}$$

$$\det(\xi(0))\det(\xi(t)) = O_p(\rho^2) = \xi_{11}(0)\xi_{11}(t)\det(\xi_{-1}(0))\det(\xi_{-1}(t)) + O_p(\rho^3).$$

By (3.37), (3.35) and (3.36):

$$\mathrm{Corr}(\xi_{11}(0), \xi_{11}(t)) \to -1, \ \mathrm{Corr}(\xi_{jk}(t), \xi_{jk}(0)) \to 1, \ \forall j, k \in 2, \dots, d,$$
$$\mathrm{Cov}(\xi_{11}(t), \xi_{jj}(t)) = O(\rho^2), \ \mathrm{Cov}(\xi_{11}(t), \xi_{jk}(t)) = 0 \, \forall j \neq k \in 1, \dots, d.$$

Then,

$$\det(\xi(0))\det(\xi(t)) \simeq -\xi_{11}(t)^2 \det{}^2(\xi_{-1}(t)) \tag{5.35}$$

$$\mathbb{E}\left(|\det(\xi(0))\det(\xi(t))|\right) \simeq \mathrm{Var}(\xi_{11}(t))\mathbb{E}\left(\det{}^2(\xi_{-1}(t))\right).$$

Finally, Eqs. (3.33) and (5.33) with (5.31) and (5.30) give (5.29).

Justification of the passages to the limit: Since we are in the Gaussian space generated by the random field X, all the variables considered above are jointly Gaussian. Their absolute values are thus bounded by the maximum in absolute value of a bounded random field. General results about the maximum of random fields, for example the Borell–Sudakov–Tsirelson theorem (Azaïs and Wschebor 2009, Section 2.4), imply that the maximum of its absolute value has moments of every order, giving all the dominations needed.

5.4.2 Correlation Between Critical Points with Adjacent Indexes

Now we will consider the behavior of the correlation function between critical points with index k and critical points with index $k+1$.

Theorem 5.17 *Let $\{X(t) : t \in \mathbb{R}^d\}$ be a stationary and isotropic Gaussian field with real values satisfying Assumption ($\mathbf{A_3}$.5.1). Let $A^{k,k+1}(\rho)$ be the correlation function* (5.28) *between critical points with index k and critical points with index $k+1$, $k = 0, \dots, d-1$. Then, as $\rho \to 0$,*

$$A^{k,k+1}(\rho) \simeq \rho^{2-d} \frac{\gamma_{d-1}^k}{2^4 3^{(d-1)/2} \pi^d} \left(\sqrt{\frac{\lambda_4}{\lambda_2}}\right)^d \frac{(\lambda_2\lambda_6 - \lambda_4^2)}{\lambda_2\lambda_4}, \tag{5.36}$$

where γ_{d-1}^k is defined by

$$\gamma_{d-1}^k := \mathbb{E}\left(\det^2\left(G_{d-1} - V\mathrm{Id}_{d-1}\right)\mathbb{1}_{\{i(G_{d-1}-V\mathrm{Id}_{d-1})=k\}}\right), \tag{5.37}$$

with G_{d-1} a $(d-1)$-GOE matrix (as in Definition 3.3) and V an independent centered Gaussian random variable with variance $1/3$. When $d = 1$ we have $\gamma_0^0 :=$ 1.

This theorem gives an interpretation of Theorem 5.16: the attraction for $d \geq 3$ is in fact due to the attraction between critical points with adjacent indexes.

Again, in the particular case of the random plane wave ($d = 2$), our result agrees with Beliaev et al. (2019, Formula (9)).

We refer to Azaïs and Delmas (2022) for some remarks on the calculation of (5.37).

Proof of Theorem 5.17 We keep the same notation as in the proof of Theorem 5.16. By (5.28) we have

$$\begin{aligned} A^{k,k+1}(\rho) &= p_{X'(0),X'(t)}(0,0)\mathbb{E}\left(|\det\xi(0)|\mathbb{1}_{\{i(\xi(0))=k\}}|\det\xi(t)|\mathbb{1}_{\{i(\xi(t))=k+1\}}\right) \\ &= -p_{X'(0),X'(t)}(0,0)\mathbb{E}\left(\det(\xi(0))\,\mathbb{1}_{\{i(\xi(0))=k\}}\det\xi(t)\mathbb{1}_{\{i(\xi(t))=k+1\}}\right). \end{aligned}$$

We can modify the computation of the determinant of $\xi(0)$ in (5.34) to compute $\det\left(\xi(0) - \lambda\mathrm{Id}_d\right)$:

$$\det(\xi(0) - \lambda\mathrm{Id}_d) = \left(\xi_{11}(0) - \lambda\right)\det(\xi_{-1}(0) - \lambda\mathrm{Id}_{d-1}) + O_p(\rho^2). \tag{5.38}$$

Note that as $\rho \to 0$, $\xi(0)$ converges in L^2 (or a.s. if we use a Skorohod imbedding argument) to

$$\begin{pmatrix} 0 & 0 & \dots & 0 \\ 0 & & & \\ \vdots & & \xi_{-1}(0) & \\ 0 & & & \end{pmatrix}.$$

Its eigenvectors converge to those of this last matrix. These eigenvectors are associated with different eigenvalues so we can properly define the eigenvalue μ_1 that tends to zero. Because of (5.38) with $\lambda = \mu_1$:

$$(\xi_{11}(0) - \mu_1)\det(\xi_{-1}(0) - \mu_1\mathrm{Id}_{d-1}) = O_p(\rho^2),$$

implying in turn that

$$\mu_1 = \xi_{11}(0) + O_p(\rho^2).$$

On the other hand, the others eigenvalues, say $\mu_2, \ldots, \mu_d$, converge in distribution to that given by the right-hand side of (5.32). This implies in turn that

$$\mathbb{1}_{\{i(\xi(0))=k\}} \simeq \mathbb{1}_{\{\xi_{11}(0)<0; i(\xi_{-1}(0))=k-1\}} + \mathbb{1}_{\{\xi_{11}(0)>0; i(\xi_{-1}(0))=k\}}.$$

As a consequence, when computing $\mathbb{E}\left(\det(\xi(0))\,\mathbb{1}_{\{i(\xi(0))=k\}} \det \xi(t) \mathbb{1}_{\{i(\xi(t))=k+1\}}\right)$, we have four cases to consider depending on the signs of $\xi_{11}(0), \xi_{11}(t)$. First, we consider the case $\xi_{11}(0) > 0, \xi_{11}(t) < 0$. We have the equivalent of (5.35)

$$\begin{aligned}
&\det(\xi(0)) \det(\xi(t))\, \mathbb{1}_{\{\xi_{11}(0)>0, \xi_{11}(t)<0\}} \mathbb{1}_{\{i(\xi_{-1}(t))=i(\xi_{-1}(0))=k\}} \\
&\qquad = -\xi_{11}(t)^2 \mathbb{1}_{\{\xi_{11}(t)<0\}} \det^2(\xi_{-1}(t))\, \mathbb{1}_{\{i(\xi_{-1}(t))=k\}} + O_p(\rho^3),
\end{aligned}$$

giving

$$\begin{aligned}
&\mathbb{E}\left(|\det(\xi(0)) \det(\xi(t))| \mathbb{1}_{\{\xi_{11}(0)>0, \xi_{11}(t)<0\}} \mathbb{1}_{\{i(\xi_{-1}(t))=i(\xi_{-1}(0))=k\}}\right) \\
&\qquad = 1/2 \mathrm{Var}(\xi_{11}(t)) \mathbb{E}\left(\det^2(\xi_{-1}(t))\, \mathbb{1}_{\{i(\xi_{-1}(t))=k\}}\right) + O(\rho^3).
\end{aligned}$$

This gives (5.36) as soon as we have checked that the three other cases give smaller contributions, which is direct.

5.4.3 *Correlation Between Critical Points with Extremes Indexes*

First, we give a bound for the correlation function between maxima and minima (Theorem 5.18), then a bound for the correlation function between maxima (or between minima) (Theorem 5.19).

Theorem 5.18 *Let $\{X(t) : t \in \mathbb{R}^d, d > 1\}$ be a stationary and isotropic Gaussian field with real values satisfying Assumption ($\mathbf{A_3}$.5.1). Let $A^{0,d}(\rho)$ be the correlation function* (5.28) *between minima and maxima. Then, for all $\varepsilon > 0$ there exists a positive constant $K(\varepsilon) > 0$ such that as $\rho \to 0$,*

$$A^{0,d}(\rho) \le K(\varepsilon) \times \rho^{2d-1-\varepsilon}. \tag{5.39}$$

Theorem 5.19 *Let $\{X(t) : t \in \mathbb{R}^d, d > 1\}$ be a stationary and isotropic Gaussian field with real values satisfying Assumption ($\mathbf{A_4}$.5.1). Let $A^{d,d}(\rho)$ (respectively $A^{0,0}(\rho)$) be the correlation function* (5.28) *between maxima (respectively minima). Then, $\forall \varepsilon > 0$ there exists a constant $\tilde{K}(\varepsilon) > 0$ such that as $\rho \to 0$,*

$$A^{0,0}(\rho) = A^{d,d}(\rho) \le \tilde{K}(\varepsilon) \times \rho^{5-d-\varepsilon}. \tag{5.40}$$

Remark 5.16 Upper bound (5.39) proves strong repulsion between maxima and minima. Upper bound (5.40) proves repulsion between maxima (or minima) for $d < 5$. For more details and results in dimension $d = 1$ see Proposition 3.4.

Now, we give the proof of Theorem 5.19. The proof of Theorem 5.18 is not given here. It follows the same lines as that of Theorem 5.19. For more details we refer to Azaïs and Delmas (2022).

Proof of Theorem 5.19 We use Hadamard's inequality. If M is a positive semidefinite matrix of size d

$$\det(M) \leq M_{11} \times \cdots \times M_{dd}.$$

As a consequence, for any symmetric matrix M'

$$\det(M')\mathbb{1}_{\{i(M')=0\}} \leq (M'_{11})^+ \times \cdots \times (M'_{dd})^+ . \tag{5.41}$$

By (5.28):

$$A^{0,0}(\rho) = p_{X'(0),X'(t)}(0,0)\mathbb{E}\left(|\det\xi(0)|\mathbb{1}_{\{i(\xi(0))=0\}}|\det\xi(t)|\mathbb{1}_{\{i(\xi(t))=0\}}\right).$$

Using (5.41), the Cauchy–Schwarz inequality and symmetry of the role of 0 and t we get

$$A^{0,0}(\rho) \leq p_{X'(0),X'(t)}(0,0)\mathbb{E}\left(\xi_{11}^+(0)\xi_{11}^+(t)\xi_{22}^+(0)^2\ldots\xi_{dd}^+(0)^2\right).$$

By (5.31) we have $p_{X'(0),X'(t)}(0,0) = O(\rho^{-d})$. Now let us prove that

$$\breve{I}_\rho := \mathbb{E}\left(\xi_{11}^+(0)\xi_{11}^+(t)\xi_{22}^+(0)^2\ldots\xi_{dd}^+(0)^2\right) = O(\rho^{4+1/p}). \tag{5.42}$$

By Hölder's inequality for every $p > 1$

$$\breve{I}_\rho \leq \left(\mathbb{E}\left(\xi_{11}^+(0)^p\xi_{11}^+(t)^p\right)\right)^{1/p}\left(\mathbb{E}\left(\xi_{22}^+(0)^{2p/(p-1)}\ldots\xi_{dd}^+(0)^{2p/(p-1)}\right)\right)^{(p-1)/p}. \tag{5.43}$$

Since $\mathrm{Cov}(\xi_{11}(0),\xi_{11}(t)) \simeq -\mathrm{Var}(\xi_{11}(t))$ (see Lemma 3.24), by Azaïs and Delmas (2022, Lemma 12)

$$\mathbb{E}\left(\xi_{11}^+(0)^p\xi_{11}^+(t)^p\right) \simeq (\texttt{const})\frac{1}{(\mathrm{Var}(\xi_{11}(t)))^{1+p}}(\det(\mathrm{Var}(\xi_{11}(0),\xi_{11}(t))))^{p+1/2}. \tag{5.44}$$

By Lemma 3.24 we have

$$\mathrm{Var}(\xi_{11}(t)) \simeq \frac{\rho^2}{4}\frac{(\lambda_2\lambda_6 - \lambda_4^2)}{\lambda_2},$$

$$\det(\text{Var}(\xi_{11}(0), \xi_{11}(t))) \simeq \frac{\rho^6}{144} \frac{(\lambda_2\lambda_6 - \lambda_4^2)(\lambda_4\lambda_8 - \lambda_6^2)}{\lambda_2\lambda_4},$$

and by (5.44) we deduce

$$\left(\mathbb{E}\left(\xi_{11}^+(0)^p \xi_{11}^+(t)^p\right)\right)^{1/p} \simeq (\text{const})\rho^{4+1/p}. \tag{5.45}$$

We now consider the term $\left(\mathbb{E}\left(\xi_{22}^+(0)^{2p/(p-1)} \dots \xi_{dd}^+(0)^{2p/(p-1)}\right)\right)^{(p-1)/p}$ in (5.43).

We set $\breve{\xi} := (\xi_{22}(0), \xi_{33}(0), \dots, \xi_{dd}(0))$. $\breve{\xi}$ is a Gaussian vector centered with variance-covariance matrix

$$\text{Var}(\breve{\xi}) = (a_\rho \text{Id}_{d-1} + b_\rho J_{d-1,d-1})$$

explicitly given in Lemma 3.23. According to Lemma 3.24, as $\rho \to 0$,

$$a_\rho \to \frac{2\lambda_4}{3} \text{ and } b_\rho \to \frac{2\lambda_4}{9}.$$

Using Lebesgue's dominated convergence theorem we can deduce that there exists a constant $C(p) > 0$ such that, as $\rho \to 0$

$$\mathbb{E}\left(\xi_{22}^+(0)^{2p/(p-1)} \dots \xi_{dd}^+(0)^{2p/(p-1)}\right) \simeq C(p). \tag{5.46}$$

Then (5.46) and (5.45) give (5.42). Finally, we conclude the proof taking $p = 1/(1-\varepsilon)$.

5.5 Central Limit Theorem

The seminal paper Estrade and León (2016) studied the asymptotic variance and distribution of the Euler characteristic of an isotropic Gaussian random field defined on $\mathbb{R}^d$ and, soon afterwards, Nicolaescu (2017) complemented these results by considering the number of critical points of such a field.

In this section we set out the results of Azaïs et al. (2024), which extend Estrade and León (2016), Nicolaescu (2017), on the limit variance of the number of critical points above a fixed level $u \geq -\infty$ with given index $k = 0, 1, \dots, d$ and their asymptotic distribution. A more detailed picture can be found in Azaïs et al. (2024).

Consider a real-valued random field $X(\cdot)$ defined on $\mathbb{R}^d$, $d \geq 1$. We denote by $X'(\cdot)$ the gradient of $X(\cdot)$ and by $X''(\cdot)$ its Hessian matrix. The covariance function of $X(\cdot)$ is denoted by $r(\cdot)$, that is, $r(t) = \mathbb{E}(X(0)X(t)), t \in \mathbb{R}^d$. We consider the following assumptions.

Assumption (A.5.2) The random field $X(\cdot)$ is centered, stationary and Gaussian with C^2 paths and $\mathrm{Var}(X'(t))$ is N.D.

Assumption (A.5.3) (Geman's Condition) For sufficiently small $B \subset \mathbb{R}^d$, $\mathbb{E}\big((N(v,B))^2\big)$ is continuous as a function of v at zero, where

$$N(v,B) := \#\{t \in B : X'(t) = v\}. \tag{5.47}$$

Assumption (A.5.4) (Arcones' Condition) Defining

$$\Psi(t) := \max\left\{\left|\frac{\partial^j r(t)}{\partial t_m}\right| : m \in \{1,\dots,d\}^j, 0 \le j \le 4\right\}, \tag{5.48}$$

we have $\Psi(t) \underset{\|t\|\to\infty}{\to} 0$ and $\Psi \in L^1(\mathbb{R}^d)$.

Remark 5.17 A few remarks are in order here. Assumption **(A.5.2)** establishes minimal regularity and non-degeneracy for the underlying random field. Compared to previous works, e.g. Nicolaescu (2017), Estrade and León (2016), Azaïs and León (2020b), it is worth pointing out that isotropy is not asked for. Furthermore, note that Assumption **(A.5.2)** implies, by Proposition (2.11), that the sample paths of $X(\cdot)$ are almost surely Morse:

$$\mathbb{P}\big\{\exists t \in \mathbb{R}^d : X'(t) = 0,\ \det(X''(t)) = 0\big\} = 0.$$

Assumption **(A.5.3)** is natural after Geman (1972), Kratz and León (2006b), Azaïs and León (2020b). Assumption **(A.5.4)** provides a simple way to unify the necessary integrability of the covariance function and its derivatives.

We consider the set $\mathcal{T} = [-T,T]^d$ and we define the number of critical points of $X(\cdot)$ above level $u \ge -\infty$ which lie within $\mathcal{T}$ and have index $k = 0,1,\dots,d$:

$$\mathrm{Crt}_u^k(\mathcal{T}) := \#\big\{t \in \mathcal{T} : X(t) > u, X'(t) = 0, i(X''(t)) = k\big\},$$

where $i(X''(t))$ stands for the index of $X''(t)$ as defined in Sect. 3.4.4. We consider the normalized variables:

$$C_u^k(\mathcal{T}) := \frac{\mathrm{Crt}_u^k(\mathcal{T}) - \mathbb{E}(\mathrm{Crt}_u^k(\mathcal{T}))}{\sqrt{T^d}}, \qquad k = 0,\dots,d,$$

and in order to study the joint distribution of these random variables we define for $\boldsymbol{\alpha} = (\alpha_0,\alpha_1,\dots,\alpha_d) \in \mathbb{R}^{d+1}$:

$$\mathrm{Crt}_u^{\boldsymbol{\alpha}}(\mathcal{T}) = \sum_{k=0}^d \alpha_k \mathrm{Crt}_u^k(\mathcal{T}) \ \text{ and } \ C_u^{\boldsymbol{\alpha}}(\mathcal{T}) := \frac{\mathrm{Crt}_u^{\boldsymbol{\alpha}}(\mathcal{T}) - \mathbb{E}(\mathrm{Crt}_u^{\boldsymbol{\alpha}}(\mathcal{T}))}{\sqrt{T^d}}. \tag{5.49}$$

The following theorem is the main result of this section.

Theorem 5.20 *Consider a real-valued random field* $X(\cdot)$ *defined on* $\mathbb{R}^d$ *and satisfying Assumptions* **(A.5.2)–(A.5.4)**. *Let* $u \in \mathbb{R} \cup \{-\infty\}$ *and* $\boldsymbol{\alpha} \in \mathbb{R}^{d+1}$. *Then,*

1. *There exists* $V_{\boldsymbol{\alpha}}(u) < \infty$ *such that*

$$\lim_{T \to \infty} \operatorname{Var}\big(C_u^{\boldsymbol{\alpha}}(\mathcal{T})\big) = V_{\boldsymbol{\alpha}}(u).$$

2. *The distribution of* $C_u^{\boldsymbol{\alpha}}(\mathcal{T})$ *converges, as* $T \to \infty$, *towards the centered normal distribution with variance* $V_{\boldsymbol{\alpha}}(u)$.
3. *The random vector* $\big(C_u^0(\mathcal{T}), \dots, C_u^d(\mathcal{T})\big)$ *converges, as* $T \to \infty$, *to a multivariate Gaussian distribution.*

Remark 5.18 As stated here, the limit distribution may degenerate, as is surely the case when $u = -\infty$, see Azaïs et al. (2024). Besides, Azaïs et al. (2024) compute an explicit expression for the limit variance and discuss its positivity for $\boldsymbol{\alpha} \neq 0$, see also Gass (2025) for related computations and discussions.

5.5.1 Hermite Expansion

We will now give the explicit expression of the chaotic expansion for the numbers of critical points.

The starting point of our analysis is an integral formula for $\mathrm{Crt}_u^k(\mathcal{T})$. Set $\mathcal{A}_k$ for the set of symmetric $d \times d$ matrices with index k, $k = 0, 1, \dots, d$.

Lemma 5.21 *Suppose that* $X(\cdot)$ *is a random field satisfying* **(A.5.2)** *and* **(A.5.3)**. *With the above notation, for* $k = 0, 1, \dots, d$, *and* $u > -\infty$, *we have*

$$\mathrm{Crt}_u^k(\mathcal{T}) = \lim_{\varepsilon \to 0} \frac{1}{W(\varepsilon)} (-1)^k \int_{\mathcal{T}} \det(X''(t)) \mathbb{1}_{\mathcal{A}_k}(X''(t)) \mathbb{1}_{[u,\infty)}(X(t)) \cdot \mathbb{1}_{\|X'(t)\| \leq \varepsilon}\, \mathrm{d}t, \tag{5.50}$$

both a.s. and in the L^2*-sense. Here,* $W(\varepsilon)$ *is the volume of a ball of radius* ε *in* $\mathbb{R}^d$.

When $u = -\infty$, *both a.s. and in the* L^2*-sense, we have*

$$\mathrm{Crt}^k(\mathcal{T}) = \lim_{\varepsilon \to 0} \frac{1}{W(\varepsilon)} (-1)^k \int_{\mathcal{T}} \det(X''(t)) \mathbb{1}_{\mathcal{A}_k}(X''(t)) \cdot \mathbb{1}_{\|X'(t)\| \leq \varepsilon}\, \mathrm{d}t. \tag{5.51}$$

Note that in the second case the integrand is a product of two independent factors.

The Hessian $X''(t)$ can be viewed either as a symmetric $d \times d$ matrix or as a size $d(d+1)/2$ vector. Set $D = d + \frac{d(d+1)}{2} + 1$,

$$\boldsymbol{X}(t) = \big(X'(t), X''(t), X(t)\big) \in \mathbb{R}^D, \tag{5.52}$$

and let Ξ be its variance-covariance matrix, i.e. $\Xi = \mathbb{E}(\boldsymbol{X}(t)\boldsymbol{X}(t)^\top)$. Let Λ be such that $\Lambda\Lambda^\top = \Xi$. Note that Λ can be chosen as a block matrix $\Lambda = \begin{pmatrix} \Lambda_1 & 0 \\ 0 & \Lambda_2 \end{pmatrix}$ due to the independence of $X'(t)$ from $(X''(t), X(t))$. Thus, we can write

$$\boldsymbol{X}(t) = \Lambda \boldsymbol{Y}(t), \tag{5.53}$$

with $\boldsymbol{Y}(t)$ a standard Gaussian random vector in $\mathbb{R}^D$ for each fixed t.

For $\underline{y} \in \mathbb{R}^d$ and $\overline{y} = (x, z) \in \mathbb{R}^{\frac{1}{2}d(d+1)} \times \mathbb{R}$, set $\mathbf{y} = (\underline{y}, \overline{y}) \in \mathbb{R}^D$ and define

$$\tilde{G}_\varepsilon(\mathbf{y}) = G_\varepsilon(\Lambda \mathbf{y}) = \tilde{\delta}_\varepsilon(\underline{y}) \cdot \tilde{f}_k(\overline{y}),$$

with $\delta_\varepsilon(\cdot) := \frac{1}{W(\varepsilon)} \mathbb{1}_{\|\cdot\| \le \varepsilon}$, $\tilde{\delta}_\varepsilon = \delta_\varepsilon \circ \Lambda_1$, $f_k(x, z) = (-1)^k \det(x) \mathbb{1}_{\mathcal{A}_k}(x) \mathbb{1}_{[u,\infty)}(z)$ and $\tilde{f}_k = f_k \circ \Lambda_2$. Since these functions are in $L^2(\phi_d)$ and $L^2(\phi_{D-d})$ respectively, we can consider their Hermite expansions. Define for $p \in \mathbb{N}$, $\mathbf{m} = (m_1, \dots, m_p) \in \mathbb{N}^p$ and $\mathbf{y} = (y_1, \dots, y_p)$:

$$H_{\otimes \mathbf{m}}(\mathbf{y}) := \prod_{i=1}^{p} H_{m_i}(y_i),$$

where H_m is the Hermite polynomial of degree m. Fix $k = 0, \dots, d$, for $\mathbf{n} = (\underline{n}, \overline{n}) \in \mathbb{N}^d \times \mathbb{N}^{D-d}$, set

$$a_{k,\varepsilon}(\mathbf{n}) := c(\tilde{\delta}_\varepsilon, \underline{n}) c(\tilde{f}_k, \overline{n}),$$

where $c(\tilde{\delta}_\varepsilon, \underline{n})$ and $c(\tilde{f}_k, \overline{n})$ are respectively the $\underline{n}$, $\overline{n}$-th Hermite coefficients of $\tilde{\delta}_\varepsilon$ and $\tilde{f}_k$, that is:

$$c(\tilde{\delta}_\varepsilon, \underline{n}) = \frac{1}{\underline{n}!} \int_{\mathbb{R}^d} \tilde{\delta}_\varepsilon(\underline{y}) H_{\otimes \underline{n}}(\underline{y}) \phi_d(\underline{y}) \, \mathrm{d}\underline{y};$$

$$c(\tilde{f}_k, \overline{n}) = \frac{1}{\overline{n}!} \int_{\mathbb{R}^{D-d}} \tilde{f}_k(\overline{y}) H_{\otimes \overline{n}}(\overline{y}) \phi_{D-d}(\overline{y}) \, \mathrm{d}\overline{y}.$$

It is easy to see that $c(\tilde{\delta}_\varepsilon, \underline{n})$ converges, so we set

$$d(\underline{n}) = \frac{1}{\underline{n}!} \frac{H_{\otimes \underline{n}}(\underline{0})}{(2\pi)^{d/2}} = \lim_{\varepsilon \to 0} c(\tilde{\delta}_\varepsilon, \underline{n}).$$

The L^2 convergence in (5.50) implies the following proposition. For $q \in \mathbb{N}$, let $\mathcal{J}_q := \{\mathbf{n} \in \mathbb{N}^D : |\mathbf{n}| = q\}$ and, for $\mathbf{n} \in \mathcal{J}_\mathbf{q}$ set $\mathcal{H}_\mathbf{n}(\mathcal{T}) = \int_\mathcal{T} H_{\otimes \mathbf{n}}(\boldsymbol{Y}(t)) \mathrm{d}t$ and $a_{\boldsymbol{\alpha}}(\mathbf{n}) = \sum_{k=0}^{d} \alpha_k a_k(\mathbf{n})$ with $\boldsymbol{\alpha} \in \mathbb{R}^{d+1}$.

Proposition 5.22 *Let $X(\cdot)$ be a random field that satisfies* **(A.5.2)** *and* **(A.5.3)**. *Then, for $\boldsymbol{\alpha} \in \mathbb{R}^{d+1}$, the following expansion holds in $L^2(\Omega)$:*

$$\mathrm{Crt}_u^{\boldsymbol{\alpha}}(\mathcal{T}) = \sum_{q=0}^{\infty} \sum_{\mathbf{n}\in\mathcal{J}_q} a_{\boldsymbol{\alpha}}(\mathbf{n})\mathcal{H}_{\mathbf{n}}(\mathcal{T}).$$

For fixed $q \geq 1$ we define

$$s_{q,T}^{\boldsymbol{\alpha}} := \frac{1}{(2T)^{d/2}} \sum_{\mathbf{n}\in\mathcal{J}_q} a_{\boldsymbol{\alpha}}(\mathbf{n})\mathcal{H}_{\mathbf{n}}(\mathcal{T}),$$

so that $C_u^{\boldsymbol{\alpha}}(\mathcal{T}) = \sum_{q=1}^{\infty} s_{q,T}^{\boldsymbol{\alpha}}$. That is, $s_{q,T}^{\boldsymbol{\alpha}}$ is the q-th chaotic component of $C_u^{\boldsymbol{\alpha}}(\mathcal{T})$.

5.5.2 *Proof of Theorem 5.20*

We proceed as in the proof of Theorem 4.13.

5.5.2.1 Step 1: Finiteness of the Limit Variance

In a first step we check the finiteness of the limit variance of a fixed chaotic component $s_{q,T}^{\boldsymbol{\alpha}}$ as $T \to \infty$. We have

$$\mathrm{Var}\big(s_{q,T}^{\boldsymbol{\alpha}}\big) = \sum_{\mathbf{m},\mathbf{n}\in\mathcal{J}_q} a_{\boldsymbol{\alpha}}(\mathbf{m})a_{\boldsymbol{\alpha}}(\mathbf{n})\frac{\mathbb{E}\big[\mathcal{H}_{\mathbf{m}}(\mathcal{T})\mathcal{H}_{\mathbf{n}}(\mathcal{T})\big]}{(2T)^d}. \tag{5.54}$$

For each fixed q, this sum has a finite number of terms, thus, it converges if the terms

$$\frac{1}{(2T)^d}\mathbb{E}\big[\mathcal{H}_{\mathbf{m}}(\mathcal{T})\mathcal{H}_{\mathbf{n}}(\mathcal{T})\big]$$

do. We have, for fixed $\mathbf{m}, \mathbf{n} \in \mathcal{J}_q$

$$\begin{aligned}\frac{\mathbb{E}\big[\mathcal{H}_{\mathbf{m}}(\mathcal{T})\mathcal{H}_{\mathbf{n}}(\mathcal{T})\big]}{(2T)^d} &= \frac{1}{(2T)^d}\iint_{\mathcal{T}^2} \mathbb{E}[H_{\bar{\otimes}_m}(\mathbf{Y}(s))H_{\bar{\otimes}_n}(\mathbf{Y}(t))]\, \mathrm{d}s\mathrm{d}t \\ &\underset{T\to\infty}{\longrightarrow} \int_{\mathbb{R}^d} \mathbb{E}[H_{\bar{\otimes}_m}(\mathbf{Y}(0))H_{\bar{\otimes}_n}(\mathbf{Y}(t))]\, \mathrm{d}t,\end{aligned}$$

because of stationarity. Now, Mehler's formula (4.4) gives

$$\mathbb{E}[H_{\bar{\otimes}_m}(\mathbf{Y}(0))H_{\bar{\otimes}_n}(\mathbf{Y}(t))] \leq (\texttt{const})\Psi_Y(t)^q,$$

where $\Psi_Y(t)$ is defined as $\Psi(t)$ replacing $r(t)$ by $\Gamma(t) := \mathbb{E}(Y(0)Y(t))$ in (5.48). Recall that $\boldsymbol{X}(t) = \Lambda\mathbf{Y}(t)$ for every $t \in \mathbb{R}^d$ by (5.53), thus, there exists $K > 0$ such that $\Psi_Y(t) \leq K\Psi(t)$. Hence, $\mathbb{E}[H_{\bar{\otimes}_m}(\mathbf{Y}(0))H_{\bar{\otimes}_n}(\mathbf{Y}(t))] \leq (\texttt{const})\Psi(t)^q$, and provided that $\Psi \in L^1(\mathbb{R}^d)$ by **(A.5.4)**, from Lebesgue's dominated convergence theorem we deduce that

$$V_{\boldsymbol{\alpha}}^q(\mathcal{T}) := \sum_{\mathbf{m},\mathbf{n}\in\mathcal{J}_q} a_{\boldsymbol{\alpha}}(\mathbf{m})a_{\boldsymbol{\alpha}}(\mathbf{n})\frac{\mathbb{E}\big[\mathcal{H}_{\mathbf{m}}(\mathcal{T})\mathcal{H}_{\mathbf{n}}(\mathcal{T})\big]}{(2T)^d} \underset{T\to\infty}{\to} V_{\boldsymbol{\alpha}}^q(u).$$

Now, as a second step, we proceed to the bound for the variance of the tail of the expansion:

$$\sup_T \sum_{q>Q} V_{\boldsymbol{\alpha}}^q(\mathcal{T}) = \sup_T \operatorname{Var}\Big(\pi_Q(C_u^{\boldsymbol{\alpha}}(\mathcal{T}))\Big) \underset{Q\to\infty}{\to} 0.$$

Here π_Q stands for the orthogonal projection onto $\bigoplus_{q>Q} C_q$. Note that, together with the previous step, this fact implies the convergence in (5.54). Set $I_T := [-T, T)^d \cap \mathbb{Z}^d$ and $s[0, 1)^d := s + [0, 1)^d$. Thus,

$$C_u^{\boldsymbol{\alpha}}(\mathcal{T}) = \frac{1}{(2T)^{d/2}} \sum_{s\in I_T} C_u^{\boldsymbol{\alpha}}(s[0, 1)^d).$$

Using the stationarity of $\boldsymbol{Y}$, we get

$$\begin{aligned} V_{T,Q} &:= \operatorname{Var}\big(\pi_Q(C_u^{\boldsymbol{\alpha}}(\mathcal{T}))\big) \\ &= \sum_{s\in I_T} \frac{\# I_T \cap (I_T - s)}{(2T)^d}\, \mathbb{E}\big(\pi_Q(C_u^{\boldsymbol{\alpha}}([0, 1)^d)) \cdot \pi_Q(C_u^{\boldsymbol{\alpha}}(s[0, 1)^d))\big). \end{aligned}$$

Now, Assumption **(A.5.4)** gives us the right to choose $a > 0$ such that $K\Psi(s) \leq \rho < 1$ for $s : ||s||_\infty \geq a$. Thus, we split $V_{T,Q} = V_{T,Q}^1 + V_{T,Q}^2$, where $V_{T,Q}^1, V_{T,Q}^2$ stand respectively for the sums over $s : ||s||_\infty \leq a + 1$ and over $s : ||s||_\infty > a + 1$.

Consider $V_{T,Q}^1$ first. Clearly there are less than $(2a + 2)^d$ terms in the sum and for each one of them, assuming without loss of generality that $2T > a+1$, we have $\# I_T \cap (I_T - s) \leq (2T)^d$. Thus, from the Cauchy–Schwarz inequality and stationarity, we have

$$|V_{T,Q}^1| \leq (2a + 2)^d\, \mathbb{E}\Big(\pi_Q(C_u^{\boldsymbol{\alpha}}([0, 1)^d))^2\Big),$$

which tends to 0 as $Q \to \infty$ uniformly with respect to T.

Now, we move to $V_{T,Q}^2$. We have,

$$\mathbb{E}\big(\pi_Q(C_u^{\alpha}([0,1)^d)) \cdot \pi_Q(C_u^{\alpha}(s[0,1)^d))\big)$$
$$= \sum_{q>Q} \iint_{([0,1)^d)^2} \mathbb{E}\big[F_q(\boldsymbol{Y}(t))F_q(\boldsymbol{Y}(s+u))\big]\, \mathrm{d}t\mathrm{d}u,$$

where $F_q(\cdot) = \sum_{\mathbf{n}\in\mathcal{J}_q} a_{\boldsymbol{\alpha}}(\mathbf{n}) H_{\bar{\otimes}_{\mathbf{n}}}(\boldsymbol{Y}(\cdot))$. Arcones' inequality, Lemma 4.10, implies that

$$\mathbb{E}\big[F_q(\boldsymbol{Y}(t))F_q(\boldsymbol{Y}(s+u))\big] \le K^d \Psi^d(s+u-t)||F_q||_2^2.$$

Since $\|s+u-t\|_\infty \ge a$, we have $K\Psi(s+u-t) \le \rho$. Moreover, analogously to Lemma 4.14, we can check that $||F_q||_2^2 \le ||\sum_{k=0}^d \alpha_k \tilde{f}_k||_2^2 = (\texttt{const}) < \infty$. Thus,

$$|V_{T,Q}^2| \le (\texttt{const}) \sum_{s\in I_T, ||s||_\infty > a} \sum_{q>Q} \rho^{q-1} \iint_{([0,1)^d)^2} \Psi(s+u-t)\, \mathrm{d}t\mathrm{d}u \le (\texttt{const})\rho^Q,$$

which tends to 0 as $Q \to \infty$, uniformly with respect to T. This proves that the limit variance is finite.

5.5.2.2 Step 2: Central Limit Theorem

As in Step 3 in the proof of Theorem 4.13, it suffices to set the convergence to 0 of integrals of the form (4.14). Fixing $q \ge 1$, forgetting about the coefficients, and proceeding as in the above mentioned proof, it suffices check the convergence of

$$\frac{1}{(2T)^{2d}} \int_{(\mathcal{T})^4} \Psi(t-s)\Psi(t'-s')\Psi(t'-t)\Psi(s-s')\, \mathrm{d}s\mathrm{d}t\mathrm{d}s'\mathrm{d}t'$$
$$\le \frac{(\texttt{const})}{(2T)^d} \int_{(\mathbb{R}^d)^3} \Psi(v_1)\Psi(v_2)\Psi(v_3)\, \mathrm{d}v_1\mathrm{d}v_2\mathrm{d}v_3 = \frac{(\texttt{const})}{(2T)^d}\Big[\int_{\mathbb{R}^d} \Psi(v)\, \mathrm{d}v\Big]^3.$$

Note that in the last display, to get the last bound we used an isometric change of variables and the fact that Ψ is bounded. We then enlarged the domain $\mathcal{T}$ to the whole space $\mathbb{R}^d$. Assumption **(A.5.4)** easily shows that the latter expression is bounded by $(\texttt{const})/(2T)^d$. The CLT follows.

5.6 Bounds for the Moments

In this section we restrict ourselves to the simple case of two-dimensional parameter space $d = 2$ and C^3 paths in order to illustrate the application of the tool of Armentano et al. (2023) to the moment of the number of critical points. This section is a continuation of Sects. 1.9.2 and 2.10.

Let $X(\cdot)$ be a random field defined on $T := [-1, 1]^2 \to \mathbb{R}$ and denote its gradient by $X' = (X'_1, X'_2)$.

Theorem 5.23 *Let $X(\cdot)$ be a random field defined on $T := [-1, 1]^2 \to \mathbb{R}$ such that*

Assumption (A.5.5)

(i) X has C^3 paths.

(ii) For all integers m, $\sup_{|v|=1} \mathbb{E} \left\| \frac{\partial^3 X}{\partial v^3} \right\|_\infty^m = C_m < \infty.$

(iii) For any direction $v \in S^1$, the joint density of $D_v(t) := \left(X'_1(t), X'_2(t), \frac{\partial X'_1(t)}{\partial v} \right)$ exists for $t \in T$ and is continuous and uniformly bounded in $T \times [-\varepsilon, \varepsilon]^2 \times \mathbb{R}$ for some $\varepsilon > 0$.

(iv) For any direction $v \in S^1$,

$$\int_{\mathbb{R}} |x''_1| p_{D_v(t)}(x'_1, x'_2, x''_1) \mathrm{d}x''_1$$

is uniformly bounded in t, x'_1 and x'_2.

Then, for all real $p < 2$

$$\mathbb{E}\big((\mathrm{Crt}(T))^p\big) < \infty.$$

Note that these conditions are easy to obtain for stationary Gaussian processes.

The proof follows the same lines as those of Theorems 1.22–2.37. We first prove some auxiliary lemmas to adapt the bound in Lemma 1.25.

Lemma 5.24 *Under assumption **(A.5.5)**, let I be a segment of T, then*

$$\mathbb{P}\{\min_{t \in I} \|X'(t)\|_\infty \leq h\} \leq (\mathtt{const})\big(|I|h + h^2\big). \tag{5.55}$$

Proof We use a parameterization $\eta : [0, |I|] \to [-1, 1]^2$ for I by its arc length, and set $\bar{X}(t) = (\bar{X}_1, \bar{X}_2) = X(\eta(t))$, accordingly $\bar{X}'_i(t) = X'_i(\eta(t))$, for $i = 1, 2$. By (iii), we have

$$\mathbb{P}(\|\bar{X}'(0)\|_\infty \leq h) \leq K_1 \cdot h^2.$$

On the other hand, set

$$\mu := \min_{t \in [0,|I|]} \|\bar{X}'(t)\|_\infty.$$

Then,

$$\mathbb{P}(\mu \le h) \le \mathbb{P}(\|\bar{X}'(0)\|_\infty \le h) + \mathbb{P}(\exists t \in [0, |I|] : \|\bar{X}'(t)\|_\infty = h).$$

The event $\{\|X'(t)\|_\infty = h\}$ can be decomposed into four sub-cases that are all equivalent, so we consider only one which is

$$\{X_1'(t) = h; |X_2'(t)| \le h\}.$$

By the Kac Lemma, Lemma 1.4, and Fatou's lemma we obtain the upper bound part of the Kac–Rice formula. Setting $M = \#\{t \in I : \bar{X}_1'(t) = h, |\bar{X}_2'(t)| \le h\}$, we have

$$\begin{aligned}\mathbb{P}\{\exists t \in I : X_1'(t) = h, |X_2'(t)| \le h\} &\le \mathbb{E}(M)\\ &\le \int_0^{|I|} \int_{-h}^{h} \int_{\mathbb{R}} |x_1''| p_{D_v(t)}(h, x_2', x_1'') \mathrm{d}x_1'' \mathrm{d}x_2' \mathrm{d}t.\end{aligned}$$

Here, $v \in S^2$ has the direction of I. Under the third and fourth hypotheses

$$\mathbb{P}\{\exists t \in I : X_1'(t) = h, |X_2'(t)| \le h\} \le (\mathtt{const}) \cdot |I| h.$$

□

Lemma 5.25 *Under the conditions in Theorem 5.23, let* $\ell = a^2 + r$ *with* $0 < r \le (a+1)^2 - a^2$, *then*

$$\mathbb{P}\{\mathrm{Crt}(T) > \ell; \|X^{(3)}(\cdot)\|_\infty \le B\} \le (\mathtt{const}) B a^{-3}.$$

Proof We partition T into a^2 sub-squares of size $2/a$, one of which, S say, contains at least two critical points C_1 and C_2. Consider the two central axes A_1 and A_2 of S of length $\frac{2\sqrt{2}}{a}$. The line passing through C_1, C_2 crosses at least one axis. We consider the case in which it intersects A_1, the other one is strictly equivalent (see Fig. 5.6). Let P be the intersection of L with A_1. Applying the tools of Lemma 1.23 along L to X_1' and X_2' (we use the first hypothesis), we get

$$|X_i'(P)| \le (\mathtt{const}) \cdot a^{-2} B, \quad i = 1, 2.$$

This implies that

$$m := \inf_{t \in A_1} \|X'(t)\|_\infty \le (\mathtt{const}) a^{-2} B.$$

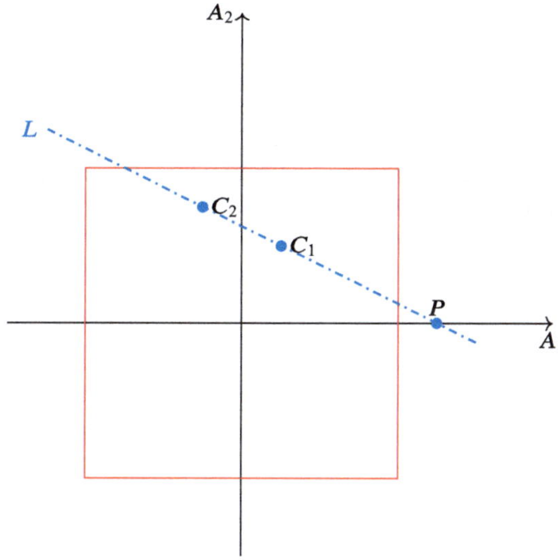

Fig. 5.6 The square S and the two axes

Applying Lemma 5.24 and recalling that $|A_1| = (\texttt{const})/a$, the probability of this event is bounded by $(\texttt{const}) \cdot a^{-3}B$.

Proof of Theorem 5.23 We write C for $\mathrm{Crt}(T)$. Using Lemma 2.7 and setting $\ell = a^2 + r, 0 < r \le (a+1)^2 - a^2$, we have

$$\begin{aligned}\mathbb{E}(C^p) &= p\mathbb{P}\{C \ge 1\} + p\sum_{\ell=2}^{\infty} \ell^{p-1}\mathbb{P}\{C \ge \ell\}\\ &\le p + p\sum_{a=1}^{\infty}(a+1)^{2(p-1)}\mathbb{P}\{C \ge \ell\}.\end{aligned}$$

We split $\mathbb{P}\{C \ge \ell\}$ as in the proof of Theorem 1.22 using Lemma 5.25 with $B = a^{\alpha}$, $\alpha > 0$, and Chebyshev's inequality (we use the second hypothesis), we get

$$\mathbb{E}(C^p) \le p + (\texttt{const}) \cdot p\sum_{a=1}^{\infty}(a+1)^{2(p-1)}\left[a^{\alpha-3} + a^{-m\alpha}\right].$$

Thus, for convergence we need:

$$5 - 2p - \alpha > 1, \qquad 2 - 2p + m\alpha > 1.$$

Namely,

$$\frac{1-2p}{m} < \alpha < 4 - 2p.$$

Taking $m \to \infty$, we can take $\alpha \to 0^+$ and deduce that the p-th moment of C is finite for $p < 2$.

Chapter 6
Second Maximum and Sparse Models

This chapter establishes a novel statistical framework for detecting sparse signals embedded within continuous spatial domains, bridging the topological geometry of random fields with high-dimensional statistical inference. By modeling observations as Gaussian random fields, we leverage Reproducing Kernel Hilbert Spaces alongside Mercer and Karhunen–Loève expansions to formalize the concept of continuous sparse regression. A central focus of the analysis is the topological relationship between the global maximum and the *second maximum* of the field. To overcome classical limitations in computing the joint distributions of these extrema, an ad-hoc Kac–Rice formula is developed, yielding the exact conditional density of the first maximum given the second maximum and the residual Hessian. This geometric derivation forms the basis of the *Spacing Test*, an exact statistical procedure for hypothesis testing on the mean of the field against unspecified sparse alternatives. Recognizing practical observational constraints, the chapter further derives a Karhunen–Loève variance estimator, culminating in the *t-Spacing test* for scenarios with unknown background variance. These theoretical mechanics are extensively validated through modern applications, including identifying phase transitions in spiked tensor Principal Component Analysis, resolving point sources in super-resolution deconvolution, and mapping the highly non-convex loss surfaces of two-layer neural networks.

6.1 Introduction

Sparse deconvolution over the space of measures has recently attracted a lot of attention in signal processing (Candès and Fernandez-Granda 2014; Duval and Peyré 2015) among the super-resolution community; in machine learning (De Castro et al. 2021); and in Optimization (Chizat 2022). The super-resolution framework aims to recover fine scale details of an image from just a few low

D. Armentano et al., *Geometry of Level Sets of Random Fields, Kac–Rice Formulas, Hermite Expansions and Applications*, Probability Theory and Stochastic Modelling 111, https://doi.org/10.1007/978-3-032-29144-8_6

frequency measurements, where ideally the observation is given by a low-pass filter. It has applications in astronomy, medical imaging, and microscopy.

The novel aspects of this body of work rely on new statistical and optimization guarantees of sparse regression over the space of measures. Initiated by the work presented in Azaïs et al. (2020) and generalized by the paper (Azaïs et al. 2024), a line of research investigates the possibility of detecting a sparse object (say a celestial object in astronomy, or a rank one tensor in machine learning) from a test on the mean $m(\cdot)$ of a Gaussian random field under a sparsity-type assumption. In this field, we consider a random field $Z(\cdot) := \{Z(t) \ : \ t \in M\}$ on a C^2 compact manifold M of dimension d and without boundary. It can be decomposed as follows

$$Z(t) = m(t) + \sigma X(t), \tag{6.1}$$

where $m(\cdot)$ is a deterministic C^2 function, $\sigma > 0$ is the standard deviation and $X(\cdot)$ a centered Gaussian random field with constant variance 1.

We would like to construct a test of the null hypothesis $m(\cdot) = 0$ when σ might be unknown. This, for the moment, against an unspecified alternative. The test statistic is, in some sense, the quantile $\hat{\alpha}$ of the maximum of $Z(\cdot)$ under the null. More precisely,

$$\lambda_1 = \max_{t \in M} \{\sigma X(t)\} \tag{6.2}$$

and we denote by $t_1 \in M$ its argument maximum. This quantile is built from the distribution of λ_1 conditional on the values of the so-called second maximum λ_2 and so-called residual part of the Riemannian Hessian Ω, under the null. Suppose in a first step that σ is known. In such a case, we will demonstrate that the cumulative distribution of this conditional law can be expressed as a ratio, resulting in the following expression derived from an ad-hoc Kac–Rice formula,

$$\hat{\alpha} := \frac{\displaystyle\int_{\frac{\lambda_1}{\sigma}}^{+\infty} \det(u\mathrm{Id} - \Omega/\sigma)\phi(u)\,\mathrm{d}u}{\displaystyle\int_{\frac{\lambda_2}{\sigma}}^{+\infty} \det(u\mathrm{Id} - \Omega/\sigma)\phi(u)\,\mathrm{d}u}, \tag{6.3}$$

where ϕ is the standard Gaussian density. We will show the exactness of this test, meaning that $\hat{\alpha}$ is uniformly distributed on the interval $(0, 1)$ under the null hypothesis. It demonstrates that the right-hand side of (6.3) is one minus the quantile of the distribution of λ_1 conditional on λ_2 and Ω, under the null hypothesis. Hence, it represents its survival function. Small values of $\hat{\alpha}$ indicate that the maximum λ_1 is abnormally large compared to the value of the second maximum λ_2, making $\hat{\alpha}$ interpretable as a p-value. This test is referred to as the "spacing test".

The second maximum is defined as

$$\lambda_2 = \max_{s \neq t_1 \in M} \left\{ \sigma X^{|t_1}(s) \right\},$$

$$\text{with, for } s \neq t,\ X^{|t}(s) := \frac{X(s) - c(s,t)\,X(t) - \nabla_t c(s,t)^\top \Lambda_2^{-1}(t) \nabla X(t)}{1 - c(s,t)}, \tag{6.4}$$

where $c(\cdot,\cdot)$ is the covariance function of $X(\cdot)$ and $\Lambda_2(t) := \mathrm{Var}[\nabla X(t)]$, the formal definition will be given in Sect. 6.4.2. Note that $X^{|t}(s)$ is a normalization of the remainder of the regression of $X(s)$ w.r.t. $(X(t), \nabla X(t))$. An example of the second maximum in the rank-one tensor detection case, referred to as Spiked tensor PCA, is illustrated in Fig. 6.3 and developed in Sect. 6.7.2.

Section 6.2 provides a brief overview of Reproducing Kernel Hilbert Spaces (RKHS). Section 6.3 uses this framework to describe an interesting class of alternatives. The second maximum and the residual part of the Hessian are discussed in Sect. 6.4. A first test is proposed in Sect. 6.5. A generalization to unknown variance is given in Sect. 6.6. Finally, Sect. 6.7 presents examples of spacing tests for several problems of kernel regression in statistical learning. The kernel regression framework is presented in Sect. 6.7.1.

Note that the spacing test presented in the introduction can be used against any alternative, it is an "omnibus test". The mathematical study of the power of the spacing test for kernel regression problems is an ongoing work but it is thought that this test would be even more effective under specified sparse alternatives. To define them, we need to define Reproducing Kernel Hilbert Spaces (RKHS). These spaces are defined by the covariance kernel $c(\cdot,\cdot)$, and some useful properties that could be used later.

6.2 Reproducing Kernel Hilbert Spaces and Gaussian Random Fields

6.2.1 *Pre-Hilbert Space and Isometries*

First, consider a covariance function $K(\cdot,\cdot)$ of some Gaussian random field or field $Y(\cdot)$ which belongs to some space L^2 of square integrable functions. Consider also the set of functions $\mathcal{H}_0$ given by

$$t \in M \mapsto \sum_{i=1}^{n} a_i\, K(s_i, t) \in \mathbb{R}, \tag{6.5}$$

for all $n \geq 1$, for all $(a_i, s_i)_i \in (\mathbb{R} \times M)^n$, equipped with the inner product $\langle \cdot, \cdot \rangle_{\mathcal{H}}$ defined by

$$\Big\langle \sum_{i=1}^{k} a_i K(s_i, \cdot), \sum_{j=1}^{\ell} b_j K(t_j, \cdot) \Big\rangle_{\mathcal{H}} := \sum_{i=1}^{k} \sum_{j=1}^{\ell} a_i b_j K(s_i, t_j)$$

for any $k, \ell \geq 1$, any reals $a_i, b_j \in \mathbb{R}$ and any points $s_i, t_j \in M$. We can check that this definition is independent of the representation of the functions on the left, and that this defines a valid inner product. The RKHS $\mathcal{H}$ is the completion of $\mathcal{H}_0$ using the metric induced by the inner product $\langle \cdot, \cdot \rangle_{\mathcal{H}}$ only. As such, the completion is not a space of functions, yet it can be identified with a space of functions using the so-called *reproducing formula*

$$\forall h \in \mathcal{H}, \ \forall t \in M, \quad h(t) = \langle K(t, \cdot), h \rangle_{\mathcal{H}}, \tag{6.6}$$

where $s \mapsto K(t, s)$ is the so-called *canonical feature map*, which is a function from M to $\mathcal{H}$. This construction extends the identity $\langle K(t, \cdot), K(s, \cdot) \rangle_{\mathcal{H}} = K(s, t)$ to the Hilbert space generated by the functions $\{K(t, \cdot) \ : \ t \in M\}$.

The construction that we have presented above is the proof of the Moore–Aronszajn Theorem: there is a unique Hilbert space $\mathcal{H}$ of functions on M for which $K(\cdot, \cdot)$ is a reproducing kernel (i.e. it satisfies the reproducing formula (6.6)). The functions in (6.5) can be written as

$$t \mapsto \mathbb{E}[Y(t) H], \quad H = \sum_{i=1}^{n} a_i \, Y(s_i).$$

We denote this function by $\mathbb{E}[Y \cdot H]$. The inner product is equal to

$$\langle \mathbb{E}[Y \cdot H_1], \mathbb{E}[Y \cdot H_2] \rangle_{\mathcal{H}}. = \mathbb{E}[H_1 H_2] \tag{6.7}$$

Hence, the map $H \mapsto \mathbb{E}[Y \cdot H]$ is an isometry for the norm of the L^2-space on which $Y(\cdot)$ is defined and the RKHS norm. The RKHS $\mathcal{H}$ is the set of functions $t \mapsto \mathbb{E}[Y(t) \cdot H]$ with H ranging over the closure of the pre-Hilbert space of functions $H = \sum_{i=1}^{n} a_i \, Y(s_i)$. Using the Borell–Sudakov–Tsirelson inequality, it is standard to prove that the paths of $Y(\cdot)$ belongs to $\mathcal{H}$ almost surely. The interested reader can consult Steinwart and Christmann (2008) for further details.

6.2.2 *Mercer and Karhunen–Loève Expansions*

Assume that the covariance $K : M \times M \to \mathbb{R}$ is a continuous function and M is a compact metric space. We will prove below that what is known as Mercer's representation holds. Given a probability measure ρ on M, we consider the integral operator on $L^2(M, \rho)$ defined by

$$\forall g \in L^2(M, \rho), \quad \mathbb{H}_K(g) : t \mapsto \int_M K(t, s)g(s)\mathrm{d}\rho(s).$$

Since $K(\cdot, \cdot)$ is symmetric, one can check that $\mathbb{H}_K$ is self-adjoint in $L^2(M, \rho)$. Moreover, $K(\cdot, \cdot)$ being continuous on a compact set, it is bounded and one can deduce that $\mathbb{H}_K$ is compact. By the spectral theorem, there exists an at most countable $L^2(M, \rho)$-orthonormal basis $(e_i)_{i=1}^m$, with m being the dimension of $\mathrm{Im}(\mathbb{H}_K)$ (possibly infinite), such that

$$\mathbb{H}_K = \sum_{i=1}^{m} \nu_i e_i \otimes e_i, \tag{6.8}$$

where $\nu_i > 0$ are the positive eigenvalues of $\mathbb{H}_K$, with 0 their only accumulation point when $m = +\infty$, and $e_i \otimes e_i$ is the rank one operator $g \mapsto (\int_M e_i g \mathrm{d}\rho)e_i$, see for instance Steinwart and Christmann (2008). Using the relation

$$\nu_i e_i(t) = \int_M e_i(s)K(t, s)\mathrm{d}\rho(s),$$

one can deduce that e_i is continuous, from the continuity of $K(\cdot, \cdot)$. Using the Cauchy–Schwarz inequality and (6.8), we have

$$\sum_{i=1}^{m} \nu_i |e_i(s)e_i(t)| \leq \sup_{t \in M} K(t, t) < +\infty,$$

which shows that the sequence $\sum_{i=1}^m \nu_i e_i(s)e_i(t)$ converges absolutely and uniformly to a kernel which is easily seen to define the same operator as the kernel $K(\cdot, \cdot)$.

This proves the Mercer's theorem: there exists an at most countable orthonormal basis $(e_i)_{i=1}^m$ such that the corresponding sequence of eigenvalues ν_i are positive. The eigenfunctions are continuous on M and $K(\cdot, \cdot)$ has the representation

$$K(s, t) = \sum_{i=1}^{m} f_i(s)\, f_i(t), \tag{6.9}$$

where the convergence is absolute and uniform and $f_i := \sqrt{\nu_i} e_i$. The functions in the pre-Hilbert space (6.5) can be written as

$$t \mapsto \sum_{i=1}^{m} h_i\, f_i(t), \quad h_i = \sum_{j=1}^{n} a_j f_i(s_j).$$

We denote this function by $F \cdot H$, where $F := (f_1, \ldots, f_m)$ and $H = (h_1, \ldots, h_m)$. The inner product is equal to

$$\langle F \cdot H_1, F \cdot H_2 \rangle_{\mathcal{H}} = \langle H_1, H_2 \rangle_{\ell_2}. \tag{6.10}$$

Hence, the map $H \mapsto F \cdot H$ is an isometry for the norm of the standard ℓ_2-space of dimension m and the RKHS norm. This shows that m is the dimension of $\mathcal{H}$.

The so-called Karhunen–Loève expansion almost surely holds:

$$Y(\cdot) = \sum_{i=1}^{m} \zeta_i f_i(\cdot) \text{ with } \zeta_i \text{ i.i.d. } \mathcal{N}(0,1) \text{ and } \forall t \in M,\ \sum_{i=1}^{m} f_i(t)^2 = K(t,t). \tag{6.11}$$

This property is a consequence of the Borell–Sudakov–Tsirelson inequality. Indeed, $\sup_{t \in M} K(t,t) < +\infty$ and $\mathbb{E}[Y(t)^2] = \sum_{i=1}^{m} f_i(t)^2 = K(t,t)$.

A second useful identity is the following. Since $K(\cdot,\cdot)$ is continuous on a compact space M, its RKHS $\mathcal{H}$ is separable, see Steinwart and Christmann (2008, Lemma 4.3.3). Hence there exists ε drawn with respect to the standard Gaussian law on $\mathcal{H}$ and one can check that, in distribution,

$$Y(t) = \langle K(\cdot,t), \varepsilon \rangle_{\mathcal{H}} \tag{6.12}$$

using the reproducing property (6.6).

6.3 Sparse Models for Random Fields

In this section we present the sparse model, and draw a parallel with high-dimensional statistics. We give a simple example on the Euclidean sphere, related to the so-called *spiked Wigner model*. Since covariance kernels are of positive type, they define Reproducing Kernel Hilbert Spaces (RKHS). On this space, our definition of sparsity takes a simple form which is coherent with the high-dimensional statistics framework. Our testing procedure might be particularly appropriate when the expectation $m(\cdot)$ can be decomposed in a sparse manner, which amounts to considering the following assumption.

Assumption (A.6.1) (Sparse Expectation) We assume that there exist $s \geq 0$, points $t_k \in M$, and non-zero reals $w_k \in \mathbb{R}$ such that

$$\forall t \in M, \quad m(t) = \sum_{k=1}^{s} w_k \, c(t, t_k),$$

where $c(\cdot,\cdot)$ is the covariance function of $X(\cdot)$ and equality (6.1) holds.

This assumption means that $m(\cdot)$ belongs to the s-dimensional subspace of the RKHS generated by the functions $c(\cdot, t_k)$. The set of all the functions $m(\cdot)$ satisfying **(A.6.1)** is a union of subspaces, each of them of dimension s and generated by functions of the form $c(\cdot, t)$ with fixed t. The same phenomenon occurs in high-dimensional statistics, the s-sparsity assumption is equivalent to the mean value belonging to the union of s-dimensional subspaces generated by s vectors of predictors. The difficulty is that we do not know the support and selecting them by an exhaustive search is impossible.

6.3.1 Overture: The RKHS Point of View

The sparsity assumption **(A.6.1)** boils down to the assumption that $m(\cdot)$ is an element of $\mathcal{H}_0$, where $n \leq s$ in (6.5). We define $\beta := \sum_{k=1}^{s} w_k \delta_{t_k}$, where δ_t is the Dirac mass at point $t \in M$, and the function $\Theta\beta = \sum_{k=1}^{s} w_k\, c(\cdot, t_k)$, which is a vector of $\mathcal{H}$. We get that

$$m(\cdot) = \Theta\beta = \sum_{k=1}^{s} w_k\, c(\cdot, t_k). \tag{6.13}$$

Observe that, using the reproducing property (6.6),

$$\begin{aligned} \forall t \in M, \quad Z(t) &= m(t) + \sigma X(t) \\ &= \left\langle c(\cdot, t), m(\cdot) + \sigma\varepsilon \right\rangle_{\mathcal{H}} \\ &= \left\langle c(t, \cdot), \Theta\beta + \sigma\varepsilon \right\rangle_{\mathcal{H}}, \end{aligned} \tag{6.14}$$

where ε is drawn with respect to the standard Gaussian law on $\mathcal{H}$ and $X(\cdot) = \langle c(\cdot, t), \varepsilon\rangle_{\mathcal{H}}$, where equality holds in distribution, see (6.12). We recognize that $Z(t)$ is the correlation between the feature $c(\cdot, t)$ and the Gaussian regression model $\Theta\beta + \sigma\varepsilon$ in $\mathcal{H}$, whose design is given by features $c(\cdot, t_k)$, see (6.13). In the next section, we will see that the expression (6.17) of High-Dimensional Statistics is exactly (6.14) when the index M is equal to $[p] := \{1, \ldots, p\}$.

6.3.2 Scherzo: High-Dimensional Models

The parallel with high-dimensional statistics can be illustrated as follows. Suppose that $M = [p]$, with $p \geq 1$, and denote by C the matrix with entries $c(i, j)$, i.e., the variance-covariance matrix of the Gaussian vector X. We assume that $c(i, i) = 1$ and we denote by $n \leq p$ its rank and by C_t its t-th column. Suppose that the expectation of the size p vector Z is $m = C\beta$ with

$$\beta = \sum_{k=1}^{s} w_k e_{t_k}, \quad t_k \in M, \quad m = C\beta = \sum_{k=1}^{s} w_k C_{t_k}, \tag{6.15}$$

where e_k is the k-th element of the canonical basis of $\mathbb{R}^p$. The vector β is called a s-sparse vector of dimension p. Observe that

$$\forall t \in M, \quad Z = C\beta + \sigma X. \tag{6.16}$$

As a variance-covariance matrix, the matrix C is symmetric and can be decomposed as $C = \Theta^\top \Theta$ for some (non-unique) matrix Θ of size $n \times p$. Let $\Theta^\dagger$ be the Moore–Penrose inverse of Θ. It is a $p \times n$ matrix such that $\Theta\Theta^\dagger\Theta = \Theta$ and can be defined as

$$\Theta^\dagger = \lim_{\delta \downarrow 0} (\Theta^\top \Theta + \delta \mathrm{Id})^{-1} \Theta^\top.$$

We get

$$Z = C\beta + \sigma X = \Theta^\top(\Theta\beta + \sigma\varepsilon) \quad \text{and define } Y := \Theta^\dagger Z = \Theta\beta + \sigma\varepsilon, \tag{6.17}$$

giving that the n vector Y obeys a sparse regression model with design Θ, under standard Gaussian noise ε.

Under a sparsity assumption, it is possible to invert this system and uncover β. Many statistical procedures have been proposed to fit high-dimensional linear models under such sparsity assumptions. Most of these procedures, if not all of them, are based on the so-called residual $Z = \Theta^\top Y$ and some Gaussian regressions of this residual. This parallel has scarcely been investigated in the existing literature, and we hope that this chapter will help the reader to find the tools to build efficient statistical procedures for the estimation of parsimonious functions. An attentive reader will recognize the main features of greedy algorithms such as LARS in our forthcoming analysis.

6.3.3 Finale: Spiked Wigner Matrix

Modern statistical models fit the sparsity assumption (**A.6.1**). We present here a simple illustration for a continuous index M given by the Euclidean sphere. More examples will be presented in Sect. 6.5. The spiked Wigner matrix model is the distribution of a $n \times n$ random matrix V such that

$$V = wuu^\top + \sigma W,$$

where u is an unknown unit norm vector referred to as the *spike*, $w > 0$ the unknown strength of the spike, and W has distribution (n − GOE): it is a symmetric matrix with independent (up to symmetries) centered Gaussian entries with variances such that

$$\mathbb{E}(\langle Ws, s\rangle\langle Wt, t\rangle) = \langle s, t\rangle^2.$$

Consider the Euclidean sphere M in which antipodal points are identified. Note that M is the standard projective space of $\mathbb{R}^n$, which is a compact smooth manifold. Given $t \in M$, one can consider

$$Z(t) := \langle Vt, t\rangle = w\langle u, t\rangle^2 + \sigma X(t),$$

where $X(\cdot)$ is a centered and normalized Gaussian random field on M, with covariance $c(s,t) = \langle s, t\rangle^2$ which is well defined (note that all the aforementioned quantities do not depend on the sign of t and s). The maximum λ_1 of the Gaussian random field $Z(\cdot)$ is exactly the top eigenvalue of the Gaussian random matrix V.

The situation can be more intricate when one tackles an s-spikes model, considering

$$Z(t) := \sum_{k=1}^{s} w_k\langle t, t_k\rangle^2 + \sigma X(t),$$

where t_k are unit norm vectors, $X(\cdot)$ has the same distribution as above, and σ is unknown. The mean of $Z(\cdot)$ is s-sparse is the sense of **(A.6.1)**, indeed $m(t) = \sum_{k=1}^{s} w_k\langle t, t_k\rangle^2$ and it corresponds to the matrix model

$$\mathbb{E}(V) = \sum_{k=1}^{s} w_k t_k t_k^\top. \tag{6.18}$$

As detailed in this chapter, it is possible to characterize the law of the maximum λ_1 given by (6.3) and estimate σ, using an ad-hoc Kac–Rice formula and introducing the notion of second maximum λ_2. Based on this analysis, we build a testing procedure on the mean which, in this case, is the rank s deterministic matrix (6.18), from the observation of a random additive Wigner perturbation of V.

6.4 Framework

6.4.1 The Four Conditions

We focus on the centered and normalized part, namely $X(\cdot)$. It is a real-valued Gaussian random field defined on a C^2 compact Riemannian manifold M of dimension d without boundary. We denote by $c(\cdot,\cdot) : M \times M \to \mathbb{R}$ its covariance function.

Except when we explicitly state otherwise, we assume the following four conditions throughout this chapter.

Assumption (A.6.2) (The Four Conditions)

(i) The paths of $X(\cdot)$ are C^2 almost surely;
(ii) $\mathbb{E}[X(t)] = 0$, $\mathrm{Var}[X(t)] = 1$ for every $t \in M$;
(iii) $\forall s \neq t$, $c(s,t) < 1$;
(iv) For every $t \in M$ the gradient $\nabla X(t)$ has a non-degenerate Gaussian distribution.

In addition, the expression of the test may be simpler under a particular hypothesis of second-order marginal stationarity, see Sect. 6.4.3 for details. The variance-covariance matrix of the gradient vector $\nabla X(t)$ is denoted by $\Lambda_2(t)$, which is invertible by Condition *(iv)*. Some useful properties are presented in the next lemma.

Lemma 6.1 *Under Assumption* ***(A.6.2)*** *one has*

$$\circ\ c(t,t) = 1\,; \tag{6.19}$$

$$\circ\ X(t) \textit{ and } \nabla X(t) \textit{ are independent}\,; \tag{6.20}$$

$$\circ\ \Lambda_2(t) = \mathrm{Var}[\nabla X(t)] = -\mathrm{Cov}[\nabla^2 X(t), X(t)]\,; \tag{6.21}$$

$$\circ\ \textit{the argument maximum of } X(\cdot) \textit{ is unique.} \tag{6.22}$$

Proof The first property (6.19) is equivalent to *(ii)*. The other properties can be deduced by differentiating $c(t,t) = 1$. The last statement is a consequence of Tsirelson's Theorem, see Proposition 6.2 hereunder.

The following result is borrowed from Lifshits (1983, Theorem 3) and Tsirelson (1976).

Proposition 6.2 *Let* $\{Y(t)\,;\ t \in M\}$ *be a Gaussian field with a.s. continuous sample paths defined on a compact metric space* M. *Suppose that*

$$\forall s \neq t, \quad \mathbb{P}\{Y(s) = Y(t)\} = 0.$$

Then, almost surely, the maximum of $X(\cdot)$ *on* M *is attained at a single point.*

6.4.2 *The Two Gaussian Regression Remainders*

6.4.2.1 The Gaussian Random Field $X^{|t}(\cdot)$

We know that $X(t)$ and $\nabla X(t)$ are independent by (6.20). For a fixed $t \in M$, consider the remainder of the Gaussian regression of $X(s)$ with respect to $(X(t), \nabla X(t))$ given by the Gaussian random field

$$s \in M \mapsto X(s) - c(s,t)\, X(t) - \nabla_t c(s,t)^\top \Lambda_2^{-1}(t) \nabla X(t) \in \mathbb{R},$$

where $\nabla_t c(s,t)$ is the Riemannian gradient of $t \mapsto c(s,t)$. Recall that $\Lambda_2(t)$ is invertible by Condition *(iv)*. This Gaussian random field is well defined on M and independent of $(X(t), \nabla X(t))$. Now, for $s \neq t$, set

$$X^{|t}(s) := \frac{X(s) - c(s,t)\, X(t) - \nabla_t c(s,t)^\top \Lambda_2^{-1}(t) \nabla X(t)}{1 - c(s,t)}, \tag{6.23}$$

that is, $X^{|t}(s)$ is a normalization of the remainder of the regression of $X(s)$ w.r.t. $(X(t), \nabla X(t))$.

6.4.2.2 The Residual Part of the Hessian

We refer to Sect. 2.2 for further details on the notion of Hessian of Riemannian manifolds. Now, consider the following regression in the space of Gaussian symmetric matrices

$$\nabla^2 X(t) = -\Lambda_2(t)\, X(t) - \Lambda_3(t) \nabla X(t) + \tilde{R}(t), \tag{6.24}$$

for some well-defined 3-way tensor $\Lambda_3(t)$. It will not be necessary to give the explicit expression of $\Lambda_3(t)$ for our purposes, this tensor is well defined by Gaussian regression formulas. Thus, one can identify the symmetric matrix $\tilde{R}(t)$ as the remainder of the regression of $\nabla^2 X(t)$ on $(X(t), \nabla X(t))$. For future use, it is convenient to set

$$R(t) := \Lambda_2^{-\frac{1}{2}}(t)\, \tilde{R}(t)\, \Lambda_2^{-\frac{1}{2}}(t). \tag{6.25}$$

Note that $\tilde{R}(t)$ and $R(t)$ are both symmetric. The following result is straightforward.

Lemma 6.3 *For any $t \in M$, the Gaussian random field $X^{|t}(\cdot)$ and the Gaussian random matrix $R(t)$ are independent of $(X(t), \nabla X(t))$.*

6.4.3 Second-Order Marginal Stationarity

The equations above are simpler in the following particular case. We consider an additional hypothesis referred to as the second-order marginal stationarity.

Assumption (A.6.3) (Second-Order Marginal Stationarity) The law of

$$(X(t), \nabla X(t), \nabla^2 X(t))$$

does not depend on t.

Under this assumption $\Lambda_3(t) = 0$, $\Lambda_2(t)$ is constant, and through scaling this last quantity can be chosen to be equal to the identity. In such a case

$$\forall t \in M, \quad R(t) = \nabla^2 X(t) + X(t)\mathrm{Id}_d.$$

Homogeneity up to the second order is of course met when the random field $X(\cdot)$ is stationary with respect to a simply transitive group operating on M, such as the group of isometries on the sphere. The following example shows that the assumption of second order marginal stationarity is a little more general.

6.4.3.1 A Non-stationary but Second-Order Marginal Stationary Random Process Using a Monge's Curve

Let $Z(\cdot) : \mathbb{R}^3 \to \mathbb{R}$ be a stationary centered Gaussian random field with C^3 paths such that $\mathrm{Var}(Z(m)) = \mathrm{Id}_3$, for every $m \in \mathbb{R}^3$. We assume that $Z(\cdot)$ is isotropic and we denote by λ_4 its fourth spectral moment, see Proposition 3.20 for further details on fourth spectral moments in the isotropic case.

Let $M(\cdot) : \mathbb{R} \to \mathbb{R}^3$ be a Monge's curve (Monge 1785), also called the skew circle (Cesàro 1896), that is not an Helix. Let us parametrize it by its curvilinear abscissa. This curve has constant curvature equal to 1 and a non-constant torsion. We define the real stochastic process

$$X(t) := Z(M(t)).$$

It is clear that

$$\begin{cases} \|M'(t)\| = \|M''(t)\| = 1 \\ M'(t) \perp M''(t) \\ \langle M'(t), M^{(3)}(t)\rangle = -1 \end{cases} \tag{6.26}$$

$$\begin{cases} X'(t) = \langle Z'(M(t)), M'(t)\rangle \\ X''(t) = \langle Z'(M(t)), M''(t)\rangle + \langle M'(t)Z''(M(t)), M'(t)\rangle \end{cases}$$

$$X^{(3)}(t) = \langle Z'(M(t)), M^{(3)}(t)\rangle + 3\langle M''(t)Z''(M(t)), M'(t)\rangle$$
$$+ Z^{(3)}[M'(t), M'(t), M'(t)] \tag{6.27}$$

Direct calculations prove that the variance-covariance matrix of $(X(t), X'(t), X''(t))$ is

$$\begin{pmatrix} 1 & 0 & -1 \\ 0 & 1 & 0 \\ -1 & 0 & 1+\lambda_4 \end{pmatrix}.$$

This proves the second-order marginal stationarity **(A.6.3)**.

The difficult part is proving that there exist values of λ_4 such that $X(\cdot)$ is not stationary. For this goal we first give intermediate results.

- Using Formula (3.30), note that λ_4 varies freely in $[3, +\infty)$.
- The torsion $\mathcal{T}(t)$ of the Monge curve, because of (6.26), takes the particular expression

$$\mathcal{T}(t) = \det\big(M'(t), M''(t), M^{(3)}(t)\big).$$

Using Gram matrices and because of (6.26)

$$\mathcal{T}^2(t) = \det\begin{pmatrix} 1 & 0 & -1 \\ 0 & 1 & 0 \\ 1 & 0 & \|M^{(3)}(t)\|^2 \end{pmatrix} = \|M^{(3)}(t)\|^2 - 1. \tag{6.28}$$

- To prove non-stationarity we consider the variance of $X^{(3)}(\cdot)$. Using isotropy (6.26) and (6.27),

$$\mathrm{Var}(X^{(3)}(t)) = \|M^{(3)}(t)\|^2 + (\texttt{const}) + \lambda_4 f(\|M^{(3)}(t)\|^2), \tag{6.29}$$

where $f(\cdot)$ is a continuous function that comes from the covariance between the first and the third terms in (6.27). This covariance depends, a priori, on the angle between $M'(t)$ and $M^{(3)}(t)$. Because $\langle M'(t), M^{(3)}(t)\rangle = -1$ this angle is a function of $\|M^{(3)}(t)\|^2$.

We return now to (6.29). Suppose that $X(\cdot)$ is stationary for all values of λ_4, choose two such values, say $\lambda_{4,1}$ and $\lambda_{4,2}$. By differentiating this would imply:

$$\big(\lambda_{4,1} - \lambda_{4,2}\big) f(\|M^{(3)}(t)\|^2) = (\texttt{const}).$$

In other words $f(\|M^{(3)}(t)\|^2) = (\texttt{const})$ for all t. This would imply subsequently that

$$\mathrm{Var}(X^{(3)}(t)) = \|M^{(3)}(t)\|^2 + (\texttt{const}).$$

The torsion is not constant so that, by (6.28), $\|M^{(3)}(t)\|^2$ is not constant and $\mathrm{Var}(X^{(3)}(t))$ is not constant, giving a contradiction.

6.4.4 The First and Second Maxima and the Residual Part of the Hessian

We define the (first) maximum λ_1 of $\sigma X(\cdot)$ and its argument maximum t_1 by

$$\lambda_1 := \sigma X(t_1) \quad \text{and} \quad t_1 := \arg\max_{t\in M}\{\sigma X(t)\} = \arg\max_{t\in M}\{X(t)\}. \tag{6.30}$$

The argument maximum is almost surely a singleton by (6.22).

We define the second maximum λ_2 of $\sigma X(\cdot)$ by

$$\forall t \in M, \quad \lambda_2^t := \sup_{s\in M\setminus\{t\}} \{\sigma X^{|t}(s)\} \quad \text{and} \quad \lambda_2 := \lambda_2^{t_1}. \tag{6.31}$$

We define the residual part of the Hessian Ω as

$$\Omega := \sigma R(t_1) = \sigma \Lambda_2^{-\frac{1}{2}}(t_1)\, \tilde{R}(t_1)\, \Lambda_2^{-\frac{1}{2}}(t_1). \tag{6.32}$$

At this stage, it is not clear that $\lambda_2^t < \infty$ a.s. and how $X^{|t}(s)$ is shaped around point t. The next lemma gives a description of $s \mapsto X^{|t}(s)$ around point t, which proves that $\lambda_2^t < \infty$ almost surely.

Lemma 6.4 *Let h be a non-zero tangent vector at point t. For all $\varepsilon \neq 0$, let $s(\varepsilon) := \exp_t(\varepsilon h) \in M$ be the exponential map at t given by the tangent vector εh. Then, under Assumption* **(A.6.2)***,*

$$\lim_{\varepsilon\downarrow 0} X^{|t}(s(\varepsilon)) = \frac{h^\top \tilde{R}(t) h}{h^\top \Lambda_2(t) h} = \frac{h^\top}{\|h\|_2} R(t) \frac{h}{\|h\|_2}, \quad \textit{almost surely.} \tag{6.33}$$

Furthermore, there exists a unit Euclidean norm tangent vector h_0 at point t such that

$$\limsup_{s\to t} X^{|t}(s) = h_0^\top R(t)\, h_0 = \lambda_{\max}(R(t)) < \infty, \quad \textit{almost surely.}$$

Remark 6.1 The aforementioned lemma shows that λ_2^t varies in $(-\infty, \infty)$ almost surely. Also, it shows that $X^{|t}(\cdot)$ is a `helix` random field (Azaïs and Wschebor 2005a, Lemma 4.1) with pole t: the paths of the random field need not extend to a continuous function at the point t; however, the paths have radial limits at t and the random field takes the form of a helix around t, see Fig. 6.1. The shape of the

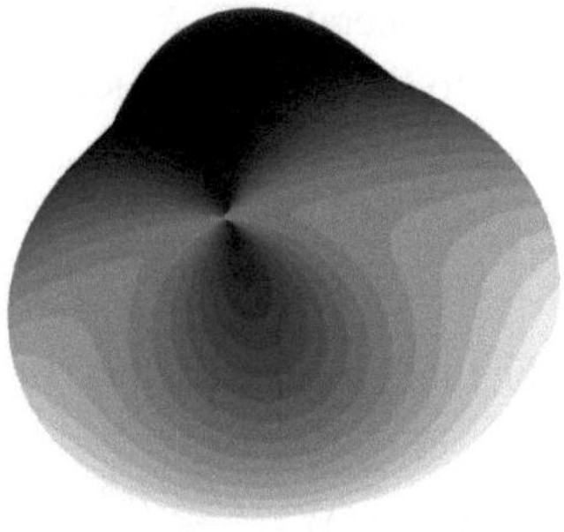

Fig. 6.1 Representation of a "Helix random field" where the paths have radial limits at a singularity point and the random field takes the form of a helix around the singularity. The shape of the helix locally around the singularity is described by the eigen-decomposition of the residual part of the Hessian, as shown by (6.33)

helix locally around the singularity is described by the eigen-decomposition of the residual part of the Hessian, as shown by (6.33).

Proof of Lemma 6.4 This argument is inspired by Azaïs and Wschebor (2005a, Lemma 4.1). Let $t \in M$ be a fixed point. On $L^2(\Omega)$, let Π be the projector on the orthogonal complement to the subspace generated by $(X(t), \nabla X(t))$. Consider a vector h of the tangent space at point t. Consider the exponential map $\varepsilon \mapsto \exp_t(\varepsilon h)$. This function is well defined on a neighborhood of 0. Hence there exists $\varepsilon_0 > 0$ such that for all $\varepsilon \in (-\varepsilon_0, \varepsilon_0)$, the point $s(\varepsilon) := \exp_t(\varepsilon h) \in M$ exists. The function $\varepsilon \mapsto s(\varepsilon)$ is a parametrization of the geodesic starting at point t with velocity h. We denote by $\varepsilon \mapsto h(\varepsilon)$ the parallel transport of h along this geodesic, see for instance do Carmo (2016, Chapter 4.4) for a definition of parallel transport along geodesics. By Condition *(i)* of **(A.6.2)**, the Taylor formula of order two gives

$$X(s(\varepsilon)) = X(t) + \varepsilon\langle \nabla X(t), h\rangle + \frac{\varepsilon^2}{2}\nabla^2 X(s(\varepsilon'))[h(\varepsilon'), h(\varepsilon')], \tag{6.34}$$

where the hessian $\nabla^2 X(\cdot)$ is represented as a bi-linear form, for some $\varepsilon' \in [0, \varepsilon]$, and

$$c(s(\varepsilon), t) = 1 + \frac{\varepsilon^2}{2}\Lambda_2(t)[h, h] + o(\varepsilon^2), \tag{6.35}$$

by Condition *(ii)* of **(A.6.2)**. From (6.34), we have

$$\Pi(X(s(\varepsilon))) = \frac{\varepsilon^2}{2}\tilde{R}(s(\varepsilon'))[h(\varepsilon'), h(\varepsilon')].$$

Note that $-\Pi(X(s(\varepsilon)))$ is the numerator of $X^{|t}(s(\varepsilon))$ while the denominator is $-\frac{\varepsilon^2}{2}\Lambda_2(t)[h,h]+o(\varepsilon^2)$ thanks to (6.35). We deduce that

$$X^{|t}(s(\varepsilon)) = \frac{\tilde{R}(s(\varepsilon'))[h(\varepsilon'),h(\varepsilon')]}{\Lambda_2(t)[h,h]+o(1)}.$$

From (6.24) and passing to the limit, we deduce that

$$\lim_{\varepsilon\to 0} X^{|t}(s(\varepsilon)) = \frac{\tilde{R}(t)[h,h]}{\Lambda_2(t)[h,h]},$$

invoking that $\tilde{R}$ is continuous by regression formulas, Condition *(ii)* of **(A.6.2)**, and that $\Lambda_2(t)[h,h]$ is positive by Condition *(iv)* of **(A.6.2)**.

For the second and last statement, observe that

$$\begin{aligned}\limsup_{s\to t} X^{|t}(s) &\geq \sup_{\|h\|_2=1}\lim_{\varepsilon\to 0} X^{|t}(\exp_t(\varepsilon h))\\ &= \max_{\|h\|_2=1}\frac{\tilde{R}(t)[h,h]}{\Lambda_2(t)[h,h]} = \frac{\tilde{R}(t)[h_0,h_0]}{\Lambda_2(t)[h_0,h_0]},\end{aligned}$$

where the vector h_0 exists by continuity of $h\mapsto \tilde{R}(t)[h,h]/\Lambda_2(t)[h,h]$ on the Euclidean sphere, which is compact. Now, let $\delta>0$ and let s_n be a sequence such that $s_n\to t$ and

$$\lim_{n\to\infty} X^{|t}(s_n) \geq \limsup_{s\to t} X^{|t}(s) - \delta.$$

Note that the exponential map is a local diffeomorphism on a neighborhood of the point t. Hence, there exist a sequence of positive reals ε_n converging to zero and a sequence h_n of unit norm tangent vectors such that $s_n = \exp_t(\varepsilon_n h_n)$. Since the Euclidean sphere is compact, we can extract a sequence such that h_n converges to a unit norm tangent vector h. The Taylor formula gives that

$$\begin{aligned}X(s_n) &= X(t) + \varepsilon_n\langle\nabla X(t), h_n\rangle + \frac{\varepsilon_n^2}{2}\nabla^2 X(\tilde{s}_n)[\tilde{h}_n,\tilde{h}_n],\\ c(s_n,t) &= 1+\frac{\varepsilon_n^2}{2}\Lambda_2(t)[h_n,h_n]+o(\varepsilon_n^2),\end{aligned}$$

for some $\tilde{s}_n$ on the geodesic between t and s_n and $\tilde{h}_n$ the parallel transport at point $\tilde{s}_n$ of the tangent vector h_n along this geodesic. By the Π-projection argument above, we deduce that

$$X^{|t}(s_n) = \frac{\tilde{R}(\tilde{s}_n)[\tilde{h}_n,\tilde{h}_n]}{\Lambda_2(t)[h_n,h_n]+o_n(1)}.$$

Passing to the limit by continuity yields

$$\max_{\|h\|_2=1} \frac{\tilde{R}(t)[h,h]}{\Lambda_2(t)[h,h]} \geq \frac{\tilde{R}(t)[h,h]}{\Lambda_2(t)[h,h]} = \lim_{n\to\infty} X^{|t}(s_n) \geq \limsup_{s\to t} X^{|t}(s) - \delta.$$

Note that the term the furthest to the left above does not depend on $\delta > 0$, hence

$$\max_{\|h\|_2=1} \frac{\tilde{R}(t)[h,h]}{\Lambda_2(t)[h,h]} \geq \limsup_{s\to t} X^{|t}(s),$$

which concludes the proof.

Furthermore, for every t such that $\nabla X(t) = 0$, it follows that

$$X^{|t}(s) = \frac{X(s) - c(s,t)X(t)}{1 - c(s,t)} =: X^t(s). \tag{6.36}$$

In particular, we have the identity $X^{|\hat{t}}(\cdot) = X^{\hat{t}}(\cdot)$ between random fields, which may not be Gaussian (due to the randomness of $\hat{t}$). The following lemma is straightforward.

Lemma 6.5 *For a fixed $t \in M$, consider the following indicator functions:*

- $\imath_1 := \mathbb{1}\{t = \hat{t}\}$;
- $\imath_2 := \mathbb{1}\{\nabla X(t) = 0\}\mathbb{1}\{\lambda_2^t \leq X(t)\}$;

then $\imath_1 = \imath_2$ and $-\infty < \lambda_2 \leq \lambda_1 < \infty$, almost surely.

6.4.5 The Distribution of the Maximum Conditional on the Second Maximum and Ω

We have the following key result giving the distribution of λ_1 conditional on (λ_2, Ω): $\mathcal{D}(\lambda_1 \mid \lambda_2, \Omega)$, where Ω is the residual part of the Hessian of $\sigma X(\cdot)$, given by (6.25).

Theorem 6.6 *Under Assumption **(A.6.2)**, the distribution of the maximum λ_1 conditional on (λ_2, R) has a density with respect to the Lebesgue measure* Leb *and this conditional density at point $\ell \in \mathbb{R}$ is given by*

$$\frac{\mathrm{d}\mathcal{D}(\lambda_1 \mid \lambda_2, \Omega)}{\mathrm{dLeb}}(\ell) = \frac{\det(\ell \mathrm{Id} - \Omega)\phi(\ell/\sigma)}{G_\Omega(\lambda_2)} \mathbb{1}_{\lambda_2 \leq \ell},$$

where ϕ is the standard Gaussian density

$$G_\Omega(\lambda_2) := \int_{\lambda_2}^{+\infty} \det(u\mathrm{Id} - \Omega)\phi(u/\sigma)\, du. \tag{6.37}$$

Proof Consider the set of parameters

$$\mathcal{B} := \Big\{(\ell_1, \ell_2, R) \in \mathbb{R}^2 \times \mathbb{S}_d \;:\; \ell_2 \le \ell_1\Big\},$$

recall that $\mathbb{S}_d$ is the set of $d \times d$ symmetric matrices. Let B_1 be an open set of $\mathbb{R} \times \mathbb{S}_d$ and $b \in \mathbb{R}$, define

$$B = \Big\{(\ell_1, \ell_2, R) \in \mathbb{R}^2 \times \mathbb{S}_d \;:\; \ell_2 < \ell_1,\ \ell_2 < b,\ (\ell_1, R) \in B_1\Big\}.$$

Note that B is an open set in $\mathcal{B}$ and this class of open sets generates the Borel sigma algebra of $\mathcal{B}$. Hence, in order to derive the joint law of the triplet $(\lambda_1, \lambda_2, \Omega)$, we compute

$$
\begin{aligned}
&\mathbb{P}\{(\lambda_1,\lambda_2, \Omega) \in B\} \\
&\quad = \mathbb{P}\{\exists t \in M \;:\; \nabla X(t) = 0,\ (\sigma X(t), \lambda_2^t, \sigma R(t)) \in B\} && (6.38)\\
&\quad = \mathbb{E}\big[\#\{t \in M \;:\; \nabla X(t) = 0,\ (\sigma X(t), \lambda_2^t, \sigma R(t)) \in B\}\big] && (6.39)\\
&\quad = \int_M \mathbb{E}\Big[\big|\det\big(-\nabla^2 X(t)\big)\big| \mathbb{1}_{(\sigma X(t), \lambda_2^t, \sigma R(t)) \in B} \Big| \nabla X(t) = 0\Big]\, p_{\nabla X(t)}(0)\, d\nu(t) && (6.40)\\
&\quad = \int_M \mathbb{E}\Big[\det\big(-\nabla^2 X(t)\big) \mathbb{1}_{(\sigma X(t), \lambda_2^t, \sigma R(t)) \in B} \Big| \nabla X(t) = 0\Big]\, p_{\nabla X(t)}(0)\, d\nu(t) && (6.41)\\
&\quad = \int_M \mathbb{E}\Big[\det(X(t)\Lambda_2(t) - \tilde{R}(t)) \mathbb{1}_{(\sigma X(t), \lambda_2^t, \sigma R(t)) \in B}\Big]\, p_{\nabla X(t)}(0)\, d\nu(t) && (6.42)\\
&\quad = \int_M \mathbb{E}\Big[\det(X(t)\mathrm{Id} - R(t)) \mathbb{1}_{(\sigma X(t), \lambda_2^t, \sigma R(t)) \in B}\Big]\, p_{\nabla X(t)}(0) \det \Lambda_2(t)\, d\nu(t), && (6.43)
\end{aligned}
$$

where ν is the surface area measure on M and

- we used that $(\sigma X(t), \lambda_2^t, \sigma R(t)) \in B$ implies $\lambda_2^t \le \sigma X(t)$, and invoked Lemma 6.5 in (6.38);
- we used (6.22) in (6.39);
- the third equality (6.40) is an ad-hoc Kac–Rice formula, a proof is given in Sect. 6.4.6;

- in (6.41), we used that $\iota_1 = \iota_2$ (see Lemma 6.5) and that the Hessian $-\nabla^2 X(t)$ is positive semi-definite at a maximum;
- in (6.42), we used the regression Formula (6.24) with $\nabla X(t) = 0$ and we invoked the independence of $\nabla X(t)$ from $(\sigma X(t), \lambda_2^t, \sigma \tilde{R}(t))$;
- the last equality is straightforward.

Now, we introduce the measures (defined up to normalizing constants)

$$d\bar{\nu}(t) := p_{\nabla X(t)}(0) \det \Lambda_2(t)\, d\nu(t), \tag{6.44}$$

$$\mu^\star(\cdot) := \int_T \mu_t(\cdot)\, d\bar{\nu}(t) \quad \text{with} \quad \mu_t := \mathcal{D}(\lambda_2^t, \sigma R(t)), \tag{6.45}$$

where $\mathcal{D}(\lambda_2^t, \sigma R(t))$ is the distribution of $(\lambda_2^t, \sigma R(t))$. Since $X(t)$ is independent of $(\lambda_2^t, \sigma R(t))$,

$$\mathcal{D}(\sigma X(t), \lambda_2^t, \sigma R(t)) = \mathcal{N}(0, \sigma^2) \otimes \mu_t,$$

by Condition *(ii)* of Assumption **(A.6.2)**.
Coming back to (6.43) we have

$$\begin{aligned}
\mathbb{P}\{(\lambda_1, \lambda_2, \Omega) \in B\} &= \int_M \mathbb{E}\Big[\det(X(t)\mathrm{Id} - R(t))\mathbb{1}_{(\sigma X(t), \lambda_2^t, \sigma R(t)) \in B}\Big]\, d\bar{\nu}(t) \\
&= \int_M \int_{\mathbb{R}^D} \det(\ell_1 \mathrm{Id}/\sigma - \omega/\sigma)\mathbb{1}_{(\ell_1, \ell_2, \omega) \in B}\phi(\ell_1/\sigma)\, d\ell_1\, d\mu_t(\ell_2, \omega)\, d\bar{\nu}(t) \\
&= \int_{\mathbb{R}^D} \int_M \det(\ell_1 \mathrm{Id}/\sigma - \omega/\sigma)\mathbb{1}_{(\ell_1, \ell_2, \omega) \in B}\phi(\ell_1/\sigma)\, d\ell_1\, d\mu_t(\ell_2, \omega)\, d\bar{\nu}(t) \\
&= \int_{\mathbb{R}^D} \det(\ell_1 \mathrm{Id}/\sigma - \omega/\sigma)\mathbb{1}_{(\ell_1, \ell_2, \omega) \in B}\phi(\ell_1/\sigma)\, d\ell_1\, d\mu^\star(\ell_2, \omega),
\end{aligned}$$

were $D = (d^2 + d + 4)/2$ is the dimension of $\mathbb{R}^2 \times \mathbb{S}_d$.

This implies that the joint density of $(\lambda_1, \lambda_2, \Omega)$ at point (ℓ_1, ℓ_2, ω) on the set $\mathcal{B}$ has a density proportional to $\det(\ell_1 \mathrm{Id}/\sigma - \omega/\sigma)\phi(\ell_1/\sigma)$ with respect to $\mathrm{Leb} \otimes \mu^\star$. This implies in turn that the density of the maximum λ_1 conditional on (λ_2, Ω) is proportional to $\det(\ell_1 \mathrm{Id} - \omega)\phi(\ell_1/\sigma)\mathbb{1}_{\ell_2 \le \ell_1}$.

6.4.6 Ad-Hoc Kac–Rice Formula

The work Azaïs and Wschebor (2009, Theorem 7.2) concerns the weighted sum of the number of roots when the weight is a continuous function of time and level. This has been extended, by a monotone convergence argument, to the case of lower-semicontinuous weights in Theorem 2.23. However this is not sufficient for our purposes, mainly because the regularity of λ_2^t as a function of t is difficult to control. Thus, we must use the following tailored argument. Recall that $\mathrm{dist}(s,t)$ is the geodesic distance between points s and t. Define

$$\forall t \in M, \quad \lambda_{2,\varepsilon}^t := \sup\left\{\sigma X^{|t}(s) \ : s \in M \text{ such that } \mathrm{dist}(s,t) > \varepsilon\right\}$$
$$\text{and} \quad \lambda_{2,\varepsilon} := \lambda_{2,\varepsilon}^{t_1}, \tag{6.46}$$

where t_1 is the argument maximum of $\sigma X(\cdot)$ as defined in (6.30). Note that $\lambda_{2,\varepsilon}^t$ is a continuous function of t. Define a monotone approximation $\xi_n(\cdot)$ of the indicator function $\mathbb{1}_B$. Then

$$\xi_n(\sigma X(t), \lambda_{2,1/n}^t, \sigma R(t)) \uparrow \mathbb{1}_{(\sigma X(t), \lambda_2^t, \sigma R(t)) \in B}. \tag{6.47}$$

So we can use Theorem 2.23, for instance, to compute the expectation of the left side of (6.47). Indeed, all conditions are clear except Condition c) for $\lambda_{2,\varepsilon}$, which is detailed below. We can subsequently use the monotone convergence theorem to obtain the result.

Checking Condition (iii) *of Assumption* **(A.2.5)** *in Theorem 2.23:* Using regression formulas, the distribution of $X(\cdot)$ under the condition $X(t_0) = u$ admits the representation

$$X(t) = \tilde{X}(t) + uf(t),$$

where $\tilde{X}(t)$ corresponds to the distribution conditional on $X(t_0) = 0$. Our goal is to show the continuity of the distribution of λ_2 in this representation. The conditioned random field $\tilde{X}(\cdot)$ satisfies

$$\forall\{s \neq t\} : \mathrm{Corr}(X(s), X(t)) < 1,$$

so that, by Tsirelson's Theorem, the maximum of $\tilde{X}(\cdot) + uf(\cdot)$ is unique. The first consequence is the continuity, as a function of u, of $\hat{t}$. For t fixed, under $X(t_0) = u$,

$$X^{|t}(s) = \tilde{X}^{|t}(s) + ug^{|t}(s), \quad \mathrm{dist}(s,t) > \varepsilon,$$

with obvious notation. The proof of Lemma 6.3 shows that $g^{|t}(s)$ is bounded uniformly in s and t. This gives the desired continuity, as a function of u, of the distribution of $\lambda_{2,\varepsilon}^t$ and then of $\lambda_{2,\varepsilon}$.

6.5 Spacing Test for the Mean, the Known Variance Case

In this section we consider the case where σ^2 is known and without loss of generality we can assume that $\sigma^2 = 1$. As a consequence $Z(\cdot)$ and $X(\cdot)$ coincide. We consider the following test paradigm:

- Null hypothesis H_0: the Gaussian random field $X(\cdot)$ is centered ($m = 0$);
- Alternatives: the Gaussian random field $X(\cdot)$ is not centered.

As usual we will just check the behavior of our testing statistics under the null so that we can directly use the results of Sect. 6.4. We can now state our main result. Recall that, in our particular case, (6.37) can be written as

$$G_\Omega(\lambda_2) = \int_{\lambda_2}^{+\infty} \det(u\mathrm{Id} - \Omega)\phi(u)\,du.$$

Theorem 6.7 *Let $X(\cdot)$ be a Gaussian random field satisfying Assumption* **(A.6.2)***. Assume that σ^2 is known and without loss of generality equal to* 1*. Under $\mathbb{H}_0$, the test statistic*

$$\mathcal{S}(\lambda_1, \lambda_2, \Omega) := \frac{G_\Omega(\lambda_1)}{G_\Omega(\lambda_2)} \sim \mathcal{U}(0, 1),$$

where $\mathcal{U}(0, 1)$ is the uniform distribution on $(0, 1)$.

Remark 6.2 It is an exercise to check that if $\sigma \neq 1$ we obtain the expression (6.3).

Proof It is well known that, if a random variable V has a *continuous* cumulative distribution function F, then $F(V)$ follows a uniform distribution on $(0, 1)$. This yields that, conditional on $(\lambda_2, \Omega) = (\ell_2, r)$,

$$\frac{G_r(\lambda_1)}{G_r(\ell_2)} \sim \mathcal{U}(0, 1).$$

Since the conditional distribution does not depend on (ℓ_2, r), it is also the non-conditional distribution, as claimed.

6.6 Spacing Test for the Mean, the Unknown Variance Case

6.6.1 Existence of Non-degenerate Systems

We consider the order m Karhunen–Loève expansion, which is little more general than (6.11),

$$\sigma X(t) = \sum_{i=1}^{m} \zeta_i f_i(t) \text{ with } \mathrm{Var}(\zeta_i) = \sigma^2 \text{ and } \forall t \in M, \ \sum_{i=1}^{m} |f_i(t)|^2 = 1, \tag{KL(m)}$$

where the equality holds in L^2, uniformly in t, and $(f_1, \dots, f_m)$ is a system of non-zero functions orthogonal on $L^2(M)$. We say that $X(\cdot)$ satisfies Condition (KL(m)) if it admits an order m Karhunen–Loève expansion. Through our analysis, we need to consider the following non-degeneracy condition:

$$\begin{aligned}
&X(\cdot) \text{ is a.s. differentiable, it holds that } m > d+1 \text{ and for all } t_1 \in M,\\
&\exists (t_{d+2}, \dots, t_m) \in (M \setminus \{t_1\})^{m-d-1} \text{ pairwise distinct s.t.}\\
&(X(t_1), \nabla X(t_1), X(t_{d+2}), \dots, X(t_m)) \text{ is non-degenerate.}
\end{aligned} \tag{ND(m)}$$

When $X(\cdot)$ admits an infinite order Karhunen–Loève expansion in the sense that

$$\sigma X(t) = \sum_{i=1}^{\infty} \zeta_i f_i(t) \text{ with } \mathrm{Var}(\zeta_i) = \sigma^2 \text{ and } \forall t \in M, \ \sum_{i=1}^{\infty} |f_i(t)|^2 = 1, \tag{KL(∞)}$$

we say that $X(\cdot)$ satisfies Condition (KL(∞)). In this case, the non-degeneracy condition reads:

$$\begin{aligned}
&X(\cdot) \text{ is a.s. differentiable, for all } p > d+1 \text{ and for all } t_1 \in M,\\
&\exists (t_{d+2}, \dots, t_p) \in (M \setminus \{t_1\})^{p-d-1} \text{ pairwise distinct s.t.}\\
&(X(t_1), \nabla X(t_1), X(t_{d+2}), \dots, X(t_p)) \text{ is non-degenerate.}
\end{aligned} \tag{ND(∞)}$$

A standard result shows that if the covariance function of X is $C^0(M \times M)$ on compact M then the Karhunen–Loève expansion exists (of finite or infinite order). The following key result shows that if X has C^1 paths almost surely, and satisfies Conditions (**A.6.2**), then the non-degeneracy condition also holds.

Proposition 6.8 *Let $X(\cdot)$ be a real-valued Gaussian random field satisfying Assumption (**A.6.2**), then, for all $m > d+1$, Condition* (KL(m)) *implies Condition* (ND(m))*, and also Condition* (KL(∞)) *implies Condition* (ND(∞)).

Proof Since X has C^1 paths almost surely, the covariance function $c(\cdot, \cdot)$ of the random field $X(\cdot)$ has continuous partial derivatives.

Consider the centered Gaussian vector $V := (\nabla X(t_1), X(t_1), X(t_{d+2}), \dots, X(t_p))$ with variance-covariance matrix $\Sigma(V) = \mathbb{E}[V V^\top]$ given by

$$
\begin{pmatrix}
\overbrace{\begin{matrix}\partial_1\partial_1 c(t_1,t_1) & \partial_1\partial_2 c(t_1,t_1) & \dots & \partial_1\partial_d c(t_1,t_1)\end{matrix}}^{\nabla X(t_1)} & \overbrace{\begin{matrix}\partial_1 c(t_1,t_1) & \partial_1 c(t_1,t_{d+2}) & \dots & \partial_1 c(t_1,t_p)\end{matrix}}^{(X(t_1),X(t_{d+2}),\dots,X(t_p))} \\
\begin{matrix}\partial_2\partial_1 c(t_1,t_1) & \partial_2\partial_2 c(t_1,t_1) & \dots & \partial_2\partial_d c(t_1,t_1) \\ \vdots & \vdots & \ddots & \vdots \\ \partial_d\partial_1 c(t_1,t_1) & \partial_d\partial_2 c(t_1,t_1) & \dots & \partial_d\partial_d c(t_1,t_1)\end{matrix} & \begin{matrix}\partial_2 c(t_1,t_1) & \partial_2 c(t_1,t_{d+2}) & \dots & \partial_2 c(t_1,t_p) \\ \vdots & \vdots & \ddots & \vdots \\ \partial_d c(t_1,t_1) & \partial_d c(t_1,t_{d+2}) & \dots & \partial_d c(t_1,t_p)\end{matrix} \\
\star & \begin{matrix} c(t_1,t_1) & c(t_1,t_{d+2}) & \dots & c(t_1,t_p) \\ c(t_{d+2},t_1) & c(t_{d+2},t_{d+2}) & \dots & c(t_{d+2},t_p) \\ \vdots & \vdots & \ddots & \vdots \\ c(t_p,t_1) & c(t_p,t_{d+2}) & \dots & c(t_p,t_p)\end{matrix}
\end{pmatrix}
$$

where $\partial_i\partial_j c(x,x) := \mathbb{E}[(\partial X/\partial x_i)(x)(\partial X/\partial x_j)(x)]$, $\partial_i c(x,y) := \mathbb{E}[(\partial X/\partial x_i)(x)\, X(y)]$ with $(\partial X/\partial x_i)(x)$ are the partial derivatives with respect to x_i at point $x = (x_1,\dots,x_d) \in M$ in Riemannian normal coordinates, and $y \in$ M. We refer to Sect. 2.2 for further details on differential geometry.

Denote by $\mathbb{H}$ the RKHS defined by $c(\cdot,\cdot)$ and let $\varphi : t \in M \mapsto c(\cdot,t) \in \mathbb{H}$ be the canonical feature map of $(\mathbb{H}, \langle\cdot,\cdot\rangle_{\mathbb{H}})$. By a standard result, see for instance Steinwart and Christmann (2008, Corollary 4.36), it holds that $\partial_i\varphi(t_j)$ belongs to $\mathbb{H}$ and $c(t_i,t_j) = \langle\varphi(t_i),\varphi(t_j)\rangle_{\mathbb{H}}$, whose derivatives are

$$
\begin{aligned}
\partial_i c(t_1,t_j) &= \langle\partial_i\varphi(t_1),\varphi(t_j)\rangle_{\mathbb{H}}\,;\\
\partial_i\partial_j c(t_1,t_1) &= \langle\partial_i\varphi(t_1),\partial_j\varphi(t_1)\rangle_{\mathbb{H}},
\end{aligned}
$$

where φ is the canonical feature map of $(\mathbb{H}, \langle\cdot,\cdot\rangle_{\mathbb{H}})$.

We recognize that $\Sigma(V)$ is the Gram matrix, for the scalar product $\langle\cdot,\cdot\rangle_{\mathbb{H}}$, of

$$
\Psi := \left[\partial_1\varphi(t_1)\,\partial_2\varphi(t_1)\,\cdots\,\partial_d\varphi(t_1)\,\varphi(t_1)\,\varphi(t_{d+2})\,\cdots\,\varphi(t_p)\right] \in \mathbb{H}^p,
$$

and $\Sigma(V) = \Psi^\star\Psi$ where $\Psi^\star : h \in \mathbb{H} \mapsto (\langle\partial_1\varphi(t_1),h\rangle_{\mathbb{H}},\cdots,\cdots,\cdots,\langle\varphi(t_p),h\rangle_{\mathbb{H}}) \in \mathbb{R}^p$ is the adjoint operator of Ψ.

Using Conditions (ii), (iv) and (6.20), we ascertain that $W = (X(t_1),\nabla X(t_1))$ is non-degenerate, and its variance-covariance matrix $\Sigma(W)$ has full rank. This matrix is also the Gram matrix of the system of vectors in $\mathbb{H}$ given by $\tilde{\Psi} := \left[\varphi(t_1)\partial_1\varphi(t_1)\cdots\partial_d\varphi(t_1)\right] \in \mathbb{H}^{d+1}$. We deduce that it is a free system (i.e., it spans a vector space of dimension $d+1$). We also know that the dimension of $\mathbb{H}$ is exactly the order of the Karhunen–Loève expansion: in fact, it is standard to prove the Karhunen–Loève expansion from Mercer's theorem. Hence, it is always possible to complete $\tilde{\Psi}$ by $p-d-1$ vectors of the form $\varphi(t)$ to get a p-dimensional free system Ψ, otherwise $\mathbb{H}$ would be of dimension less than p.

We give some examples below:

- The normalized Brownian motion $W_t/\sqrt{t}$ satisfies Condition (KL(∞));

- Any C^k Gaussian stationary field with a spectrum that admits an accumulation point satisfies Conditions (KL(∞)) and (ND(∞)), see Proposition 1.17;
- Any Gaussian random field satisfying the conditions of Proposition 2.22 satisfies (KL(∞)) and (ND(∞));
- Note that in the applications of Sect. 6.7 all the Gaussian random fields satisfy (KL(m)) with m finite and explicitly given.

6.6.2 The Karhunen–Loève Estimator of the Variance

In practical applications, the assumption that the variance is known is too restrictive. In this section, we assume that we observe $\sigma X(\cdot)$ with $X(\cdot)$ satisfying Assumption **(A.6.2)**, $m = 0$ and $\sigma > 0$ is unknown. In particular, $X(\cdot)$ admits a Karhunen–Loève expansion of order, denoted $m_{\mathrm{KL}}(X)$, possibly infinite. We assume the following additional condition:

Assumption (A.6.4) (Degrees of Freedom)

$$\kappa := m_{\mathrm{KL}}(X) - d - 1 \geq 1.$$

We will make use of the notation $m := m_{\mathrm{KL}}(X)$ when there is no ambiguity.

Remark 6.3 Using Condition (iv) of Assumption **(A.6.2)** and the independence between $X(t)$ and $\nabla X(t)$, see (6.20), it is possible to prove that $m_{\mathrm{KL}}(X) \geq d + 1$ by an argument similar to the one used in the proof of Proposition 6.8. Assumption **(A.6.4)** asks for at least one further degree of freedom. This extra freedom is the key to estimate the variance.

6.6.2.1 The Infinite Order Case

When $X(\cdot)$ satisfies (KL(∞)), note that, for every integer $p \geq 2$, from the observation of $(X(t_1), \ldots, X(t_p))$ for conveniently chosen points $t_1, \ldots, t_p$, we can construct an estimator, say $\hat{\sigma}^2_{(p)}$, of the variance σ^2 with distribution $\sigma^2\chi^2(p-1)/(p-1)$ under $\mathbb{H}_0$. Making p tend to infinity, classical concentration inequalities and the Borel–Cantelli lemma prove that $\hat{\sigma}^2_{(p)}$ converges almost surely to σ^2 under $\mathbb{H}_0$. Thus the variance σ^2 is theoretically directly observable from the entire path of X (though in practical applications we will estimate it by a χ^2 with a large number of degrees of freedom). We still denote this observation by $\hat{\sigma}^2_t$.

6.6.2.2 The Finite Order Case

First note that by Proposition 6.8 and Assumption **(A.6.2)** (i)–(ii)–(iv), Condition (ND(m)) holds with $m > d + 1$ by (iv). Let $t \in \mathrm{M}$ and let $(t_{d+2}, \ldots, t_m) \in M^{m-(d+1)}$ be as in Condition (ND(m)), then $(X^{|t}(t_{d+2}), \ldots, X^{|t}(t_m))$ has for variance-covariance matrix $\sigma^2\Sigma$, where Σ is some known matrix and an estimator of σ^2 is

$$\hat{\sigma}_t^2 := \left\| \Sigma^{-1/2}\big(\sigma X^{|t}(t_{d+2}), \ldots, \sigma X^{|t}(t_m)\big) \right\|^2 / \kappa. \tag{6.48}$$

Direct algebra shows that $X^{|t}(\cdot)$ inherits an order κ Karhunen–Loève expansion from the order m Karhunen–Loève expansion of $X(\cdot)$. More precisely, under H_0,

$$\forall s \neq t \in M, \quad X^{|t}(s) = \sum_{i=1}^{\kappa} \bar{\zeta}_i \bar{f}_i(s) \text{ with } \mathrm{Var}(\bar{\zeta}_i) = \sigma^2, \tag{6.49}$$

where equality holds in L^2, uniformly in s, and

$$\hat{\sigma}_t^2 = \frac{1}{\kappa} \sum_{i=1}^{\kappa} \bar{\zeta}_i^2. \tag{6.50}$$

This shows that $\hat{\sigma}_t^2$ does not depend on the choice of $(t_{d+2}, \ldots, t_m)$ in (ND(m)).

Proposition 6.9 *Let $X(\cdot)$ satisfy Assumptions* **(A.6.2)** *and* **(A.6.4)**. *Let $t \in M$, then the following claims are true under H_0.*

(i) *$\hat{\sigma}_t^2$ is well defined and follows a $\sigma^2 \frac{\chi^2(\kappa)}{\kappa}$ distribution;*
(ii) *$\hat{\sigma}_t^2$ is independent of $(X(t), \nabla X(t))$;*
(iii) *the random field $X^{|t}(\cdot)/\hat{\sigma}_t$ is independent of the random variable $\hat{\sigma}_t$.*

Proof Statement (i) follows from (6.50) and Statement (ii) follows from (6.48) and the independence of $X^{|t}(\cdot)$ from $(X(t), \nabla X(t))$. The last statement is a direct consequence of the independence between the angle and the norm of $\bar{\zeta} = (\bar{\zeta}_1, \cdots, \bar{\zeta}_\kappa)$ and (6.49).

6.6.3 t-Spacing Test, an Exact Test Procedure for the Unknown Variance Case

Below, we show how to combine the variance estimate $\hat{\sigma}_t^2$ with the spacing test strategy of proof to derive a test that does not require us to know the true variance σ^2, referred to as the t-spacing test. For fixed $t \in M$, by Proposition 6.9, we know that $X(t)$, $\nabla X(t)$, $X^{|t}(\cdot)/\hat{\sigma}_t$, $\hat{\sigma}_t$ are mutually independent. We know the covariance function of $X(\cdot)$ and hence we know $\Lambda_2(t)$ and $\Lambda_3(t)$. As in (6.25), consider the residual part of the Hessian of $X(\cdot)$, defined by

$$R(t) := \Lambda_2^{-\frac{1}{2}}(t) \left[\nabla^2 X(t) + \Lambda_2(t)\, X(t) + \Lambda_3(t) \nabla X(t)\right] \Lambda_2^{-\frac{1}{2}}(t),$$

and note that $R(t) := \sigma R(t)$ and $X^{|t}(\cdot) = \sigma X^{|t}(\cdot)$. As Lemma 6.4 shows that $R(t)$ can be expressed as radial limits of $X^{|t}(\cdot)$ at point t, we get that

$$\left\{X(t), \nabla X(t), \left(X^{|t}(\cdot)/\hat{\sigma}_t, \ R(t)/\hat{\sigma}_t\right), \ \hat{\sigma}_t\right\}$$

are mutually independent and, by consequence, the variables

$$X(t), \nabla X(t), \left[\frac{\lambda_2^t}{\hat{\sigma}_t}, \frac{R(t)}{\hat{\sigma}_t}\right], \ \hat{\sigma}_t \ \text{are mutually independent,}$$

where, λ_2^t is defined by (6.31). We consider a fixed point $t \in M$ which is a putative value for t_1,

$$T_{1,t} := \frac{\sigma X(t)}{\hat{\sigma}_t}, \ T_{2,t} := \frac{\lambda_2^t}{\hat{\sigma}_t}, \hat{\sigma} := \hat{\sigma}_t \text{ and } R := R(t), \tag{6.51}$$

and let $\bar{\mu}_t$ be the joint law of the pair of random variables $(T_{2,t}, R(t)/\hat{\sigma}_t)$.

Under H_0 note that the variable $\sigma X(t)$ is a centered Gaussian with variance σ^2 and $\sqrt{\kappa}\,\hat{\sigma}_t/\sigma$ is distributed as a χ-distribution with κ degrees of freedom. Hence, the variable $\hat{\sigma}_t/\sigma$ has density

$$f_{\frac{\chi_\kappa}{\sqrt{\kappa}}}(s) = \frac{2^{1-\frac{\kappa}{2}}}{\Gamma\left(\frac{\kappa}{2}\right)}\sqrt{\kappa}\left(s\sqrt{\kappa}\right)^{\kappa-1}\exp\left(-(\kappa s^2/2)\right).$$

Then the quadruplet $(X(t), \hat{\sigma}_t/\sigma, T_{2,t}, R(t)/\hat{\sigma}_t)$ has a density with respect to $\mathrm{Leb}_2 \otimes \bar{\mu}_t$ at point $(\ell, s, t_2, r) \in \mathbb{R}^3 \times \mathbb{S}_d$ equal to

$$(\mathtt{const})\, s^{\kappa-1}\exp\left(-(\kappa s^2/2)\right)\phi(\sigma^{-1}\ell_1),$$

where the constant may depend on κ or σ. Using the same method as in the proof of Theorem 6.6 we have the following proposition, where we define

$$T_1 := T_{1,t_1}, \ T_2 := T_{2,t_1}, \ \Omega := \sigma R(t_1) \tag{6.52}$$

the values of the statistics at the argument maximum point t_1.

Proposition 6.10 *Let $X(\cdot)$ be a Gaussian random field satisfying* **(A.6.2)** *and* **(A.6.4)** *then, under H_0, the joint distribution of $\left(\lambda_1, \hat{\sigma}/\sigma, T_2, \Omega/\hat{\sigma}\right)$ has a density with respect to* $\mathrm{Leb}_2 \otimes \mu^\star$ *at point $(\ell_1, s, t_2, r) \in \mathbb{R}^3 \times \mathbb{S}_d$ equal to*

$$(\text{const})\det(-\ell_1\mathrm{Id} + \sigma s r)\, s^{\kappa-1}\exp\left(-\kappa s^2/2\right)\phi(\sigma^{-1}\ell_1)\mathbb{1}_{\{0<\sigma s t_2<\ell_1\}},$$

where $\bar{\mu}^\star$ is defined by $\bar{\mu}^\star(\cdot) := \int_M \bar{\mu}_t(\cdot)\,\mathrm{d}\bar{\nu}(t)$ and $\bar{\nu}(t)$ by (6.44).

From this proposition, we derive the following result.

Theorem 6.11 *Let $X(\cdot)$ be a Gaussian random field satisfying Assumptions* **(A.6.2)** *and* **(A.6.4)***. For all $r \in \mathbb{S}_d$, define $H_r(\cdot)$ as*

$$\forall \ell \in \mathbb{R}, \quad H_r(\ell) := \int_\ell^{+\infty} \det(-u\mathrm{Id} + r)\, f_{m-1}\Big(u\sqrt{(m-1)/\kappa}\Big)\,\mathrm{d}u, \tag{6.53}$$

where f_{m-1} is the density of the Student law with $m-1$ degrees of freedom and $m = m_{\mathrm{KL}}(X)$. Under the null H_0, the test statistic

$$\mathcal{T}(T_1, T_2, \Omega) := \frac{H_{\Omega/\hat{\sigma}}(T_1)}{H_{\Omega/\hat{\sigma}}(T_2)} \sim \mathcal{U}(0,1),$$

where T_1, T_2, Ω are given by (6.52).

Proof First, using Proposition 6.10 and the change of variable $t_1 = \ell_1/\sigma s$, the joint density of the quadruplet $(T_1, \hat{\sigma}/\sigma, T_2, \Omega/\hat{\sigma}))$ at point (t_1, s, t_2, r) is given by

$$\begin{aligned}
&(\text{const}) \det\big(\sigma s(-t_1\mathrm{Id} + r)\big)\, s^\kappa \exp\big(-(\kappa s^2/2)\big)\, \phi(st_1)\mathbb{1}_{\{0<t_2<t_1\}}\\
&= (\text{const}) \det(-t_1\mathrm{Id} + r)\, s^{m-1} \exp\Bigg[-\bigg(s\sqrt{\frac{m}{m-1}}\bigg)^2 \frac{m-1}{2}\Bigg]\phi(st_1)\mathbb{1}_{\{0<t_2<t_1\}}.
\end{aligned}$$

Second, note that if U and V are two independent random variables of density f_U and f_V then the density of $W := \frac{U}{V}$ satisfies

$$f_W(\omega) = \int_{\mathbb{R}} f_U(\omega v) v f_V(v)\mathrm{d}v.$$

In our case, integrating over s and with the change of variable $s \leftarrow s\sqrt{(m-1)/\kappa}$, we have

$$\begin{aligned}
&\int_{R^+} \phi(st_1)\, s^{m-1} \exp\Bigg[-\bigg(s\sqrt{\frac{\kappa}{m-1}}\bigg)^2 \frac{m-1}{2}\Bigg]\mathrm{d}s\\
&= (\text{const}) \int_{R^+} \phi\bigg(st_1\sqrt{\frac{m-1}{\kappa}}\bigg)\, s\, s^{m-2} \exp\bigg[-\frac{s^2(m-1)}{2}\bigg]\mathrm{d}s\\
&= (\text{const}) \int_{R^+} \phi\bigg(st_1\sqrt{\frac{m-1}{\kappa}}\bigg)\, s\, f_{\frac{\chi_{m-1}}{\sqrt{m-1}}}(s)\,\mathrm{d}s\\
&= (\text{const})\, f_{m-1}\bigg(t_1\sqrt{\frac{m-1}{\kappa}}\bigg).
\end{aligned}$$

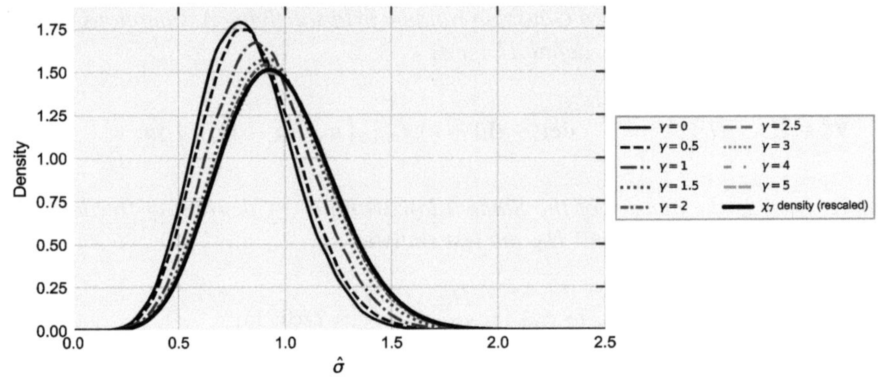

Fig. 6.2 The variance estimation is neither unbiased nor distributed according to a $\chi^2(7)$; it underestimates the variance. The dashed black line is the distribution of a $\chi(7)/\sqrt{7}$, and $\sigma = 1$ in these experiments. The probability distribution function is estimated over 250,000 Monte-Carlo samples for each value of γ

Putting this all together, the density of $(T_1, T_2, \Omega/\hat{\sigma})$ at point (t_1, t_2, r) is now given by

$$(\text{const}) \det(-t_1 \mathrm{Id}_d + r)\ f_{m-1}\left(t_1\sqrt{\frac{m-1}{\kappa}}\right) \mathbb{1}_{\{0<t_2<t_1\}},$$

and we conclude using the same trick as the one in Theorem 6.7.

Remark 6.4 (On the Variance Estimator) Formula (6.53) involves a Student law with $m-1$ degrees of freedom, which necessitates several comments. First, this formula does not show that $\kappa\,\hat{\sigma}^2/\sigma^2$ (resp. $(m-1)\,\hat{\sigma}^2/\sigma^2$) is distributed according to a χ^2-distribution with κ (resp. $m-1$) degrees of freedom.

Figure 6.2 illustrates that $\hat{\sigma}^2$ under-estimates the variance σ^2 under the null ($\gamma = 0$) or moderate alternatives ($\gamma \simeq 1$). The reason for this is clear: the chi-squared distribution assumes that the model is pre-specified, not chosen on the basis of $\sigma X(\cdot)$. However, our procedure has deliberately chosen $t_1 \in M$ as the argument maximum of $Z(\cdot)$, among all of the available choices, so it is not surprising that it yields a drop in the value of the variance estimate on the residuals $Z^{|t_1}(\cdot)$.

Figure 6.2 shows that $\kappa\,\hat{\sigma}^2/\sigma^2$ has an almost χ^2-distribution with κ degrees of freedom for large alternatives ($\gamma = 5$). The reason for this is clear: the mean $m(\cdot)$ is much larger than the noise $\sigma X(\cdot)$, hence $t_1 \simeq t_0$ and $\kappa\hat{\sigma}^2/\sigma^2 \simeq \kappa\hat{\sigma}_{t_1}/\sigma^2$, which has a χ^2-distribution with κ degrees of freedom.

6.7 Examples

6.7.1 *A Continuous Kernel Regression Framework for the Spacing Test*

6.7.1.1 The General Framework

Let $(E, \langle \cdot, \cdot \rangle_E)$ be a Euclidean space. Consider $\psi \ : \ t \in \mathrm{M} \mapsto \psi_t \in E$ satisfying the following conditions:

Assumption (A.6.5) (Sparse Kernel Regression)

(i) $\psi(\cdot)$ is a C^2 function ;
(ii) $\forall t \in M, \ \ \|\psi_t\|_E^2 = 1$;
(iii) $\forall t \in M, \ \forall s \in M \setminus \{t\}, \ \langle \psi_s, \psi_t \rangle_E < 1$;
(iv) $\forall t \in M, \ \psi_t' \psi_t'^{\star}$ has full rank d ;
(v) the span of $\{\psi_t \ : \ t \in M\}$ has dimension m such that $m > d + 1$;
(vi) $\forall t \in M, \ \exists s \in M$ s.t. $\psi_t = -\psi_s$;

where ψ_t' is the Jacobian of ψ_t and $\psi_t'^{\star}$ its adjoint.

6.7.1.2 The MLE of the Regression Model

Let $\lambda_0 \in \mathbb{R}$ and $t_0 \in \mathrm{M}$. Let $W \in E$ be a centered Gaussian vector with variance-covariance matrix Id_E. We consider the regression problem given by

$$Y = \lambda_0 \psi_{t_0} + \sigma W, \tag{6.54}$$

where λ_0, t_0, σ^2 are unknown parameters that one aims to estimate. The Maximum Likelihood Estimator (MLE) of λ_0, t_0 is given by

$$\lambda^{\mathrm{MLE}}, t^{\mathrm{MLE}} = \underset{\lambda, t}{\arg\min} \left\{ \frac{1}{2} \| Y - \lambda \psi_t \|_E^2 \right\}.$$

The minimization with respect to λ is straightforward under Conditions (ii) and (vi), leading to

$$\underset{t}{\arg\min} \min_{\lambda} \left\{ \frac{1}{2} \| Y - \lambda \psi_t \|_E^2 \right\} = \underset{t}{\arg\max} \, \langle Y, \psi_t \rangle_E . \tag{6.55}$$

Now consider the Gaussian random field $Z(t) := \langle Y, \psi_t \rangle_E$, referred to as the profile likelihood. The argument maximum (resp. the maximum) of $Z(\cdot)$ is exactly the MLE of t_0 (resp. of λ_0) by (6.55), namely

$$\lambda^{\mathrm{MLE}} = \max Z \quad \text{and} \quad t^{\mathrm{MLE}} = \arg\max Z .$$

By (6.54), note that we uncover the decomposition (6.1) and the alternative hypothesis as

$$Z(t) = \langle Y, \psi_t \rangle_E = \lambda_0 \langle \psi_{t_0}, \psi_t \rangle_E + \sigma \langle W, \psi_t \rangle_E = \lambda_0 c(t, t_0) + \sigma X(t). \tag{6.56}$$

Remark 6.5 As $X(t) = \langle W, \psi_t \rangle_E$ with $W \in E$ a centered Gaussian vector with variance-covariance matrix Id_E, one can check that Assumptions **(A.6.2)** and **(A.6.4)** are met.

6.7.2 The Spiked Tensor Model

We considered a simple example of the detection of a rank one 3-way symmetric tensor. We consider the sphere $M = S^2$ with $d = 2$. The Euclidean space of 3-way symmetric tensors of size $3 \times 3 \times 3$ is denoted by $(E, \langle \cdot, \cdot \rangle_E)$ with dimension $m = 10$. Then

$$\psi_t := t^{\otimes 3} = (t_i t_j t_k)_{1 \le i,j,k \le 3}$$

$$X(t) := \langle W, \psi_t \rangle_E = \sum_{ijk} t_i t_j t_k W_{ijk},$$

where W is an order 3 symmetric tensor defined using symmetry and the following independent terms,

$$W_{iii} \text{ of variance } 1\,; \text{ there are 3 principal diagonal terms, } i \in [3]$$

$$W_{iij} \text{ of variance } 1/3\,; \text{ there are 6 sub-diagonals terms, } i \neq j \in [3]$$

$$W_{123} \text{ of variance } 1/6\,; \text{ there is 1 off-diagonal term.}$$

Note that the regression (6.54) is given by $Y = \lambda_0 t_0^{\otimes 3} + \sigma W$ and the observed random field is given by $Z(t) = \langle Y, t^{\otimes 3} \rangle_E = t\lambda_0 \langle t_0, t \rangle^3 + \sigma X(t)$. Recall that the mathematical analysis has been developed by studying the random field $\sigma X(\cdot)$ which corresponds to the null hypothesis. The test can be deduced substituting $\sigma X(\cdot)$ by $Z(\cdot)$ in the following expressions.

6.7.2.1 Maxima

The first and second maximum (and their arguments) are given by (6.30). In our experiments, we have used a projected gradient descent algorithm on M to compute these parameters. An illustration is given in Fig. 6.3.

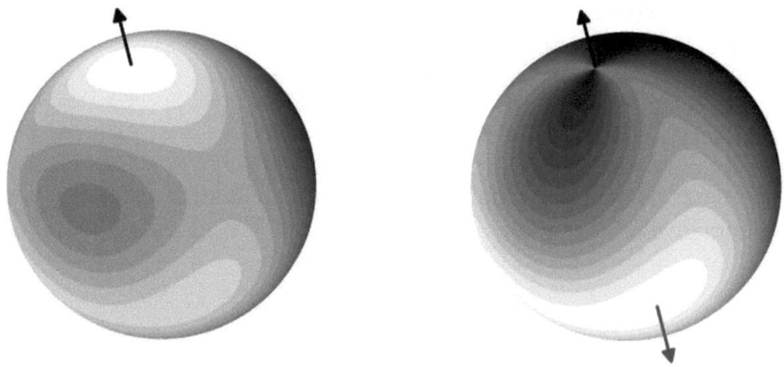

Fig. 6.3 The first eigenvector (resp. first eigenvalue) of a Gaussian symmetric tensor (arrow in the left panel) is the argument maximum (resp. maximum, λ_1) of the Gaussian homogeneous polynomial $X(\cdot)$. The second eigenvector (resp. second maximum, λ_2), represented by an arrow in the right panel, is the argument maximum of $X^{|t_1}(\cdot)$

6.7.2.2 Residual Part of the Hessian

To compute the gradient and Hessian we can take advantage of the isometry of the problem and consider, without loss of generality, the case where $t \in M$ is the so-called "north pole": (0, 0, 1). We have

$$X'(0,0,1) = \begin{pmatrix} 3W_{133} \\ 3W_{233} \\ 3W_{333} \end{pmatrix}.$$

This shows that $\Lambda_2 = 3\,\mathrm{Id}_2$ (on the tangent space). Since the second fundamental form of the unit sphere is $-\mathrm{Id}$, the Riemannian Hessian is equal to the Euclidean Hessian $X''(t)$ restricted to the tangent space minus the normal derivative (oriented outwards) times Id. The Euclidean Hessian $X''(t)$ restricted to the tangent space is given by

$$\begin{pmatrix} 2W_{113} & W_{123} \\ W_{123} & 2W_{223} \end{pmatrix}.$$

This shows that the gradient is independent of the Hessian. To compute the residual R we can use

$$X(0,0,1) = X'_3(0,0,1) = W_{333} \quad \text{and} \quad X'_3(0,0,1) = W_{333}.$$

This shows that $R(t)$ is always equal to the Euclidean Hessian restricted to the tangent space. By (6.32), this yields

$$\Omega = \frac{\sigma}{3}\tilde{R}(t_1) = \frac{\sigma}{3}(\mathrm{Id} - \Pi_{t_1})\, X''(t_1)\, (\mathrm{Id} - \Pi_{t_1}),$$

where $\Pi_t := tt^\top$ is the orthogonal projection onto the normal space at point t, and direct algebra gives

$$\frac{\partial^2 X}{\partial t^i \partial t^j}(t) = 6\sum_{k=1}^{3} W_{ijk} t^k. \tag{6.57}$$

6.7.2.3 Test Statistics

One can compute the statistics (t_1, λ_1) and (t_2, λ_2) using a gradient descent. The expression of Ω has an explicit form by (6.57). Because we are dealing with 2×2 matrices, we have for the spacing test the following identities,

$$\begin{aligned} G_{\Omega/\sigma}(\ell) &= \int_\ell^{+\infty} \det(u\mathrm{Id}_2 - \Omega/\sigma)\phi(u)\,\mathrm{d}u \\ &= \int_\ell^{+\infty} \big[(u^2 - 1) - \mathrm{Tr}(\Omega/\sigma)u + \det(\Omega/\sigma) + 1\big]\phi(u)\,\mathrm{d}u \\ &= \ell\phi(\ell) - \mathrm{Tr}(\Omega/\sigma)\phi(\ell) + (\det(\Omega/\sigma) + 1)(1 - \Phi(\ell)), \end{aligned}$$

with

$$\begin{aligned} \det(\Omega/\sigma) &= (1/9\sigma^2)\det\big[(\mathrm{Id}_3 - \Pi_{t_1})Z''(t_1)(\mathrm{Id}_3 - \Pi_{t_1}) + \Pi_{t_1}\big] \\ \mathrm{Tr}(\Omega/\sigma) &= (1/3\sigma)\big(\mathrm{Tr}(Z''(t_1)) - t_1^\top Z''(t_1)t_1\big). \end{aligned}$$

As for the t-spacing test, recall that we observe $\sigma X(\cdot)$ under the null and we can check that

$$\begin{aligned} H_{\Omega/\hat{\sigma}}(\ell) &= \int_\ell^{+\infty} \det(u\mathrm{Id}_2 - \Omega/\hat{\sigma})\, f_{m-1}\Big(u\sqrt{(m-1)/\kappa}\Big)\,\mathrm{d}u \\ &= \int_\ell^{+\infty} \big[u^2 - \mathrm{Tr}(\Omega/\hat{\sigma})u + \det(\Omega/\hat{\sigma})\big]\, f_{m-1}\Big(u\sqrt{(m-1)/\kappa}\Big)\,\mathrm{d}u \\ &= \frac{\kappa\sqrt{m-3}}{(m-2)\sqrt{m-1}}\frac{\Gamma(\frac{m}{2})\Gamma(\frac{m-3}{2})}{\Gamma(\frac{m-1}{2})\Gamma(\frac{m-2}{2})}\sqrt{\frac{\kappa}{m-3}} \end{aligned}$$

$$
\cdot \left[\ell\sqrt{\frac{m-3}{\kappa}} f_{m-3}(\ell\sqrt{\frac{m-3}{\kappa}}) + 1 - F_{m-3}(\ell\sqrt{\frac{m-3}{\kappa}}) \right]
$$

$$
- \operatorname{Tr}(\Omega/\hat{\sigma}) \frac{\kappa\sqrt{m-3}}{(m-2)\sqrt{m-1}} \frac{\Gamma(\frac{m}{2})\Gamma(\frac{m-3}{2})}{\Gamma(\frac{m-1}{2})\Gamma(\frac{m-2}{2})} f_{m-3}(\ell\sqrt{\frac{m-3}{\kappa}})
$$

$$
+ \det(\Omega/\hat{\sigma})\sqrt{\frac{\kappa}{m-1}}\Big(1 - F_{m-1}(\ell\sqrt{\frac{m-1}{\kappa}})\Big),
$$

with $m = 10$, $\kappa = 7$, F_α (resp. f_α) the cumulative distribution function (resp. density function) of the Student t-distribution with α degrees of freedom and

$$
\det(\Omega/\hat{\sigma}) = (1/9\hat{\sigma}^2) \det\big[(\mathrm{Id}_3 - \Pi_{t_1})\sigma X''(t_1)(\mathrm{Id}_3 - \Pi_{t_1}) + \Pi_{t_1}\big]
$$

$$
\operatorname{Tr}(\Omega/\hat{\sigma}) = (1/3\hat{\sigma})\big(\operatorname{Tr}(\sigma X''(t_1)) - t_1^\top \sigma X''(t_1) t_1\big).
$$

6.7.2.4 The Variance Estimate

Draw κ independent points $(t_{d+2}, \ldots, t_m)$ uniformly distributed on M. They generically satisfy Condition (ND(m)). For a putative point $t \in M$, the Gaussian vector $(X^{|t}(t_{d+2}), \ldots, X^{|t}(t_m))$ has for variance-covariance matrix $\sigma^2\Sigma$, where Σ is some known matrix and an estimator of σ^2 is given by (6.48) with $t = t_1$ as in (6.51). In our experiments, we have drawn 5 independent samples on κ points and we found the same value for $\hat{\sigma}$, as shown by the theory (6.50).

6.7.2.5 The t-Spacing Test

In Figs. 6.5 and 6.4, the alternative is given by $\lambda_0 = \gamma \times \sigma\sqrt{3\log 3 + 3\log\log 3}$, where $\gamma = 1$ corresponds to the so-called phase transition in Spiked tensor PCA, as presented in Perry et al. (2020, Theorem 1.3). In Fig. 6.5, the top right panel shows that λ_1 is stochastically increasing as γ increases while the top right panel shows that the distribution of λ_2 remains unchanged for moderate values of γ, say $\gamma \geq 2$. It illustrates that the spacing between λ_1 and λ_2 grows linearly with γ. In the moderate regime, the bottom left panel shows that λ_1 is distributed as a Gaussian with mean λ_0 and variance σ^2 (dashed black line), which is the distribution of $\sigma X(t_0)$ under $\mathbb{H}_1(t_0, \lambda_0)$. The bottom right panel shows that $t_1 \simeq t_0$, for moderate values of γ. It illustrates that (t_1, λ_1) is weakly close to the distribution of $(t_0, \sigma X(t_0))$; consequently, the (t-)spacing tests detect the alternative $\mathbb{H}_1(t_0, \lambda_0)$.

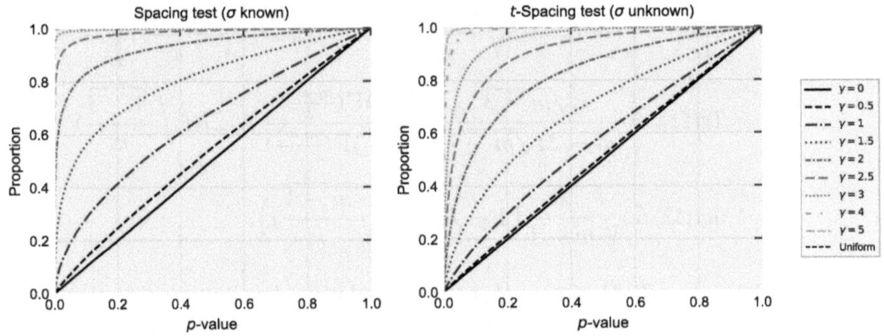

Fig. 6.4 CDFs of the p-value of (t-)spacing tests over 250,000 Monte-Carlo samples for each value of γ. Note that these tests are perfectly calibrated under the null hypothesis ($\gamma = 0$). The parameter γ is a scaling factor of the eigenvalue of the rank-one tensor to be detected. The value $\gamma = 1$ corresponds to the so-called phase transition in Spiked tensor PCA, as presented in Perry et al. (2020, Theorem 1.3). In the t-spacing test, the variance σ has been estimated on $X^{|t_1}(\cdot)$

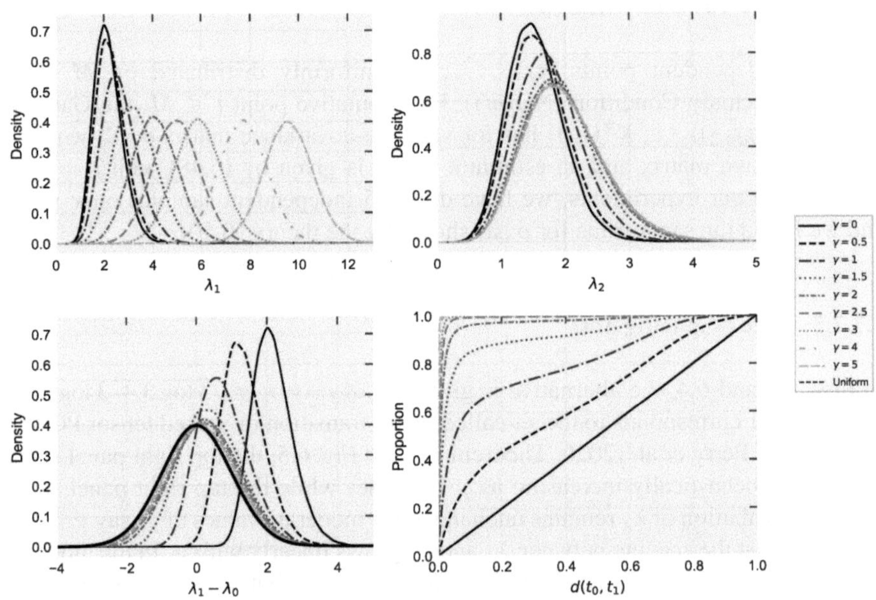

Fig. 6.5 PDFs of the maxima λ_1, λ_2 and CDF of the distance $\mathrm{d}(t_0, t_1)$ over 250,000 Monte-Carlo samples for each value of γ. The alternative is given by t_0 fixed and $\lambda_0 = \gamma \times \sigma\sqrt{3\log 3 + 3\log\log 3} \simeq 0.684\gamma\sigma$, where $\gamma = 1$ corresponds to the so-called phase transition in Spiked tensor PCA as presented in Perry et al. (2020, Theorem 1.3). The distance $\mathrm{d}(t_0, t_1)$ is normalized so that it is uniformly distributed on $(0, 1)$ if t_1 is uniformly distributed on the sphere (e.g., $\gamma = 0$)

6.7.3 The Super-Resolution Random Field

In signal processing, the super-resolution phenomenon refers to the ability to distinguish two close sources from noisy low frequency measurements given by an optical system. This issue can be tackled using sparse regularization on the space of measures using the so-called Beurling-LASSO, introduced by De Castro and Gamboa (2012), Azaïs et al. (2015), Candès and Fernandez-Granda (2014), Duval and Peyré (2015).

We consider the framework of the paper Azaïs et al. (2020), where we observe $N := 2f + 1$ low frequencies between $[-f, f]$ with $f \geq 1$ or equivalently the convolution of some signal on the torus $[0, 2\pi)$ with the Dirichlet kernel $D_N(u) := \sin(Nu/2)/N\sin(u/2)$. Since we want to compute the maximum modulus of a trigonometric polynomial $P_N(\cdot)$ we consider

$$X(t) := \mathrm{Re}(\mathrm{e}^{-\imath\theta} P_N(u)),$$

where $t = (\theta, u)$ is a point of $M := [0, 2\pi)^2$, the bi-dimensional torus. Following the paper Azaïs et al. (2020), the trigonometric polynomial P_N is given by

$$P_N(u) := \frac{1}{\sqrt{N}} \sum_{k=-f}^{f} \xi_k \mathrm{e}^{\imath k u} \quad \text{with } \xi_k := \xi_{k,1} + \imath \xi_{k,2},$$

where $\xi_{k,\ell}$ are independent standard Gaussian variables. An illustration of the super-resolution random field is given in Fig. 6.6, and the covariance of $X(\cdot)$ is given by

$$c(t, t') = \cos(\theta - \theta') D_N(u - u').$$

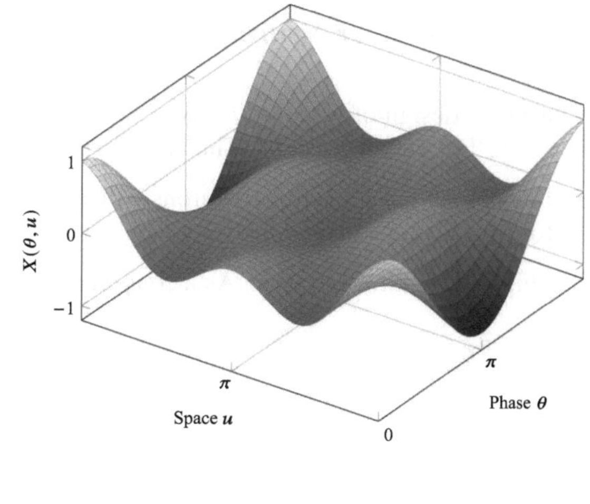

Fig. 6.6 The super-resolution random field $X(\cdot)$ depicts the observation of illuminating point sources on the interval $[0, 2\pi)$ (spacial domain on the x-axis) with phases given on the y-axis. The z-axis corresponds to the observed signal $X(\theta, u)$ at point u with phase θ

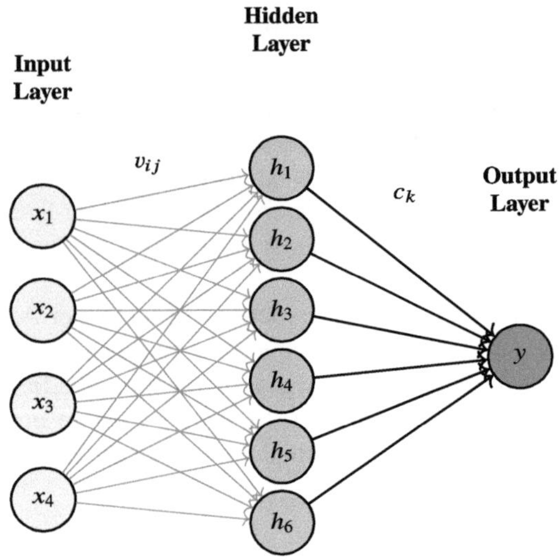

Fig. 6.7 Structure of a two-layer neural network

We can check that the assumptions of this chapter are satisfied by the super-resolution random field. The spacing test is given by

$$S_\sigma = \frac{\sigma(\alpha_1\lambda_1 + \alpha_2)\,\phi(\lambda_1/\sigma) + (\alpha_1\sigma^2 - \alpha_3^2)\,(1-\Phi)(\lambda_1/\sigma)}{\sigma(\alpha_1\lambda_2 + \alpha_2)\,\phi(\lambda_2/\sigma) + (\alpha_1\sigma^2 - \alpha_3^2)\,(1-\Phi)(\lambda_2/\sigma)},$$

where $\alpha_1, \alpha_2, \alpha_3$ have explicit forms given in Azaïs et al. (2020, Proposition 10), λ_1 is the maximum of $X(\cdot)$ and λ_2 the maximum of $X^{|t_1}(\cdot)$. The t-spacing is explicitly described in Azaïs et al. (2020, Proposition 11).

6.7.4 *Two-Layer Neural Networks with Smooth Rectifier*

Consider a design of N points on $\mathbb{R}^n : x_1, \ldots, x_N$, which is the input of our neural network. The output Y is real. The neural network consists of two layers:

- **(hidden layer)** a layer of r neurons with activation a C^2 function $\rho(\cdot)$;
- **(output layer)** a layer which is simply a mean; and

See Fig. 6.7 for an illustration. As a consequence the output is

$$\frac{1}{r}\sum_{k=1}^{r} c_k \rho(\langle v_k, x_i\rangle).$$

The unknown parameters are the weights of the neurons c_k, v_k, for $k \in [r]$. We assume that v_k belongs to a compact C^2 manifold without boundaries, denoted T.

Using the trick that the mean of r reals is the least squares estimator, we have to minimize

$$\frac{r}{N}\sum_{i=1}^{N}\Big(Y_i-\frac{1}{r}\sum_{k=1}^{r}c_k\rho(\langle v_k,x_i\rangle)\Big)^2=\frac{1}{N}\sum_{i=1}^{N}\sum_{k=1}^{r}\Big(Y_i-c_k\rho(\langle v_k,x_i\rangle)\Big)^2.$$

To enter into the framework of Sect. 6.7.1 we need to perform the following change of parameterization:

$$c_k\rho(\langle v_k,x_i\rangle)=\lambda_0\frac{c_kV}{\lambda_0}\frac{\rho(\langle v_k,x_i\rangle)}{V}=:\lambda_0a_k\frac{\rho(\langle v_k,x_i\rangle)}{V},$$

where $V^2:=\sum_{k=1}^{r}\rho^2(\langle v_k,x_i\rangle)$ and $\lambda_0^2:=V^2(c_1^2+\cdots+c_r^2)$ are normalizing constants. We can check that $t:=(a_1,\ldots,a_r,v_1,\ldots,v_r)\in S^{r-1}\times T^r$ with S^{r-1} the Euclidean sphere of $\mathbb{R}^r$ and that $\mathrm{M}=S^{r-1}\times T^r$. We hence define

$$\psi_t:=\left(\frac{a_k\rho(\langle v_k,x_i\rangle)}{\sqrt{\sum_{i'}\rho^2(\langle v_k,x_{i'}\rangle)}}\right)_{\substack{k\in[r]\\ i\in[N]}}\quad\text{and } c(s,t):=\langle\psi_s,\psi_t\rangle,$$

to uncover the framework described in Sect. 6.7.1.

Chapter 7
Shot Noise Processes

Shifting focus to non-Gaussian models, this chapter introduces and applies the previous results to shot noise processes, which are traditionally used to model the superimposition of random physical disturbances. First, we present regularity properties of these processes, including the existence and uniform boundedness of their marginal densities. Furthermore, we prove the uniform boundedness of moments using our previous results. As a consequence, we obtain central limit theorems for both the number of one-dimensional crossings and the volume of multidimensional level sets.

7.1 Introduction

The shot noise process was introduced by S.O. Rice in the monograph "Mathematical analysis of random noise" (Rice 1944). Below we present some references on this subject, without aiming to provide an exhaustive review. In relation to moment calculations, representation as an infinite divisible process or a stochastic integral w.r.t. a Poisson measure, J. Rice obtained an expression for the characteristic function, moments, cumulants, and a central limit theorem when the intensity of the jumps tends to infinity (Rice 1977). Breton (2010) studied regularity properties (such as the existence of the density) of shot noise processes. Torrisi and Leonardi (2022) have made a recent contribution regarding limit approximations of the process, when some of the parameters involved tend to infinity.

The first paper devoted to crossings of the shot noise process is Orsingher and Battaglia (1982). There, amongst other results, the exponential kernel is considered to analyze the crossings. In this framework, a formula for the upcrossing and downcrossing rates is given, and used to compute these rates when the impulses have exponential distribution. Although not mentioned explicitly, the formula obtained can be considered a KRF for the crossings. More recent contributions referring to

D. Armentano et al., *Geometry of Level Sets of Random Fields, Kac–Rice Formulas, Hermite Expansions and Applications*, Probability Theory and Stochastic Modelling 111, https://doi.org/10.1007/978-3-032-29144-8_7

crossings and level sets can be found in the papers Biermé and Desolneux (2012, 2016, 2011).

As for applications, it should be mentioned that the introduction of the process was motivated by a desire to model the "superposition of a great number of disturbances which occur at random", with the main example being the arrival of electrons to the anode of a vacuum tube (Rice 1944). More recent applications include the paper Klüppelberg and Mikosch (1995) devoted to applications in risk reserves, and the paper Klüppelberg and Kühn (2004) containing limit results involving fractional Brownian motion and applications to finance.

The rest of the chapter is organized as follows: after defining the shot noise process in Sect. 7.2, we consider the one-dimensional case in Sect. 7.3, where we give conditions for obtaining a central limit theorem for the crossings. In Sect. 7.4 we develop a similar exposition for the multidimensional shot noise process, obtaining a central limit theorem for the measure of the volume of the level sets.

7.2 Basic Definitions

We now return to definition (2.41), under minimal conditions for the existence of the process.

Definition 7.1 The stationary shot noise process $X(\cdot) = \{X(t) : t \in \mathbb{R}^d\}$ is defined by

$$X(t) = \sum_i \beta_i g(t - \tau_i), \tag{7.1}$$

where

- The sequence $\{\beta_i\}$ is conformed by independent real random variables with a common distribution F. A variable β with the same distribution is called the "impulse" and it is assumed to have finite first moment, i.e. $\mathbb{E}|\beta| < \infty$.
- The sequence $\{\tau_i\}$ comprises the marks of a Poisson process on $\mathbb{R}^d$ with constant intensity λ, independent of the sequence $\{\beta_i\}$.
- The function $g : \mathbb{R}^d \to \mathbb{R}$ is measurable and assumed to be integrable, i.e. $\int_{\mathbb{R}^d} |g(s)| \mathrm{d}s < \infty$. It is called the "kernel" of the shot noise process.

The above conditions ensure the a.s. convergence of the series in (2.41). More precisely, considering the sum for a convenient partition of the space $\mathbb{R}^d$ and applying the Cauchy criteria for the absolute convergence of series, the result is obtained, see Daley (1971, Theorem 1). Regarding the differentiability of the sample paths, it follows from the differentiability of the kernel under mild assumptions (see (7.3) below). A key issue to apply Theorem 1.22 is the verification of condition **(A.1.3)** with $h = 0$, that requires the boundedness of a density. This is a delicate issue, as was previously noted in Biermé and Desolneux (2012, 2016).

7.3 Shot Noise Process in $\mathbb{R}$

We consider below the real-valued case, e.g. $d = 1$. The case $d > 1$ will be considered in Sect. 7.4. By application of Theorem 1.22 we study the finiteness of moments of the process. Let us recall the main hypothesis.

(A.1.1) The sample paths of $X(\cdot)$ are $C^k(I)$ for some $k \geq 2$.
(A.1.2) For some $m = 1, 2, \ldots$, there exists a constant D_m such that

$$\mathbb{E}\left(|X^{(k)}|_\infty^m\right) \leq D_m.$$

(A.1.3) There exist $0 \leq h \leq k$ and a constant $C > 0$ such that the joint density of the random vector $X(t), X'(t), \ldots, X^{(h)}(t)$ is bounded by C uniformly in $t \in I$, and on a neighborhood of $(u, 0, \ldots, 0)$.

Corollary 7.1 *Consider a stationary shot noise process as in Definition 7.1 and satisfying condition* ***(A.1.1)*** *for some* $k \geq 2$, ***(A.1.2)*** *for some* $m \geq 1$ *and* k *given in* ***(A.1.1)****, and* ***(A.1.3)*** *for* $h = 0$*. Then*

$$\mathbb{E}(N_u^p) < \infty$$

for

$$p < \frac{m}{m+1}(k-1). \tag{7.2}$$

The rest of this section is devoted to obtaining sufficient conditions to allow us to verify this corollary. The differentiability of the paths follows directly from the differentiability of the kernel g, under the condition

$$\int_{\mathbb{R}} \sup_{|t|<\varepsilon} |g'(t-s)| \mathrm{d}s < \infty, \qquad \text{for some } \varepsilon > 0. \tag{7.3}$$

Furthermore, we have

$$X'(t) = \sum_i \beta_i g'(t - \tau_i),$$

see Biermé and Desolneux (2011, Proposition 2). Analogous results are obtained for the k-th derivative of the process provided that the k-th derivative of g satisfies (7.3). Note that if the corresponding derivative is monotonous, condition (7.3) is equivalent to the integrability of the derivative.

7.3.1 Uniform Boundedness of Moments

We now consider condition **(A.1.2)**. A formula for the moments of one-dimensional coordinates of a shot noise random field has been obtained by Bassan and Bona (1990), from where direct conditions for the finiteness of the moments can be obtained, in terms of the kernel and the impulse of the process, see Biermé and Desolneux (2011).

Note that condition **(A.1.2)** requires boundedness of the infinite norm of the trajectories, i.e. a more stringent condition. Below we give conditions on the impulse and the kernel in order to verify this hypothesis **(A.1.2)** for given k and m. To simplify our exposition, we assume that the compact interval in the hypothesis is $I = [-1, 1]$, the general case can be treated in the same way. We consider the partition of the real line $I_1 = I$ and $I_n = [-n, -n+1) \cup (n-1, n]$ $(n \geq 2)$. In this way we can write

$$X^{(k)}(t) = \sum_{n=1}^{\infty} \sum_{i:\tau_i \in I_n} \beta_i g^{(k)}(t - \tau_i).$$

Proposition 7.2 *Consider the shot noise in* (2.41) *with impulse s.t.* $\mathbb{E}(|\beta|^m) < \infty$ *for given* $m \geq 1$. *Define*

$$d_{k,n} = \sup_{t \in I_1, s \in I_n} |g^{(k)}(t - s)|,$$

for $k \geq 1$, *and assume that*

$$D_k = \sum_{n=1}^{\infty} d_{k,n} < \infty. \tag{7.4}$$

Then, the condition ***(A.1.2)*** *holds true for m and k as above, i.e.*

$$\mathbb{E}\left(|X^{(k)}|_{\infty}^{m}\right) < \infty.$$

Proof We have

$$\begin{aligned}
|X^{(k)}|_{\infty} &= \max_{-1 \leq t \leq 1} |X^{(k)}(t)| \\
&\leq \max_{-1 \leq t \leq 1} \sum_{n=1}^{\infty} \sum_{i:\tau_i \in I_n} |\beta_i| |g^{(k)}(t - \tau_i)| \\
&\leq \sum_{n=1}^{\infty} \sum_{i:\tau_i \in I_n} |\beta_i| \max_{-1 \leq t \leq 1} |g^{(k)}(t - \tau_i)|
\end{aligned}$$

$$\le \sum_{n=1}^{\infty} \sum_{i:\tau_i \in I_n} |\beta_i| \sup_{-1\le t\le 1, s\in I_n} |g^{(k)}(t-s)|$$

$$= \sum_{n=1}^{\infty} d_{k,n} Z_n,$$

where $Z_n := \sum_{\tau_i \in I_n} |\beta_i|$. As the sets I_n are disjoint and have the same length, the random variables Z_n are i.i.d. Each one has a compound Poisson distribution. As $\mathbb{E}(|\beta|^m) < \infty$ we obtain $\mathbb{E}(Z_1^m) < \infty$. We now use Jensen's inequality, based on the convergence in (7.4), to obtain the following bound:

$$\left(\sum_{n=1}^{\infty} d_{k,n} Z_n\right)^m = D_k^m \left(\sum_{n=1}^{\infty} \frac{d_{k,n}}{D_k} Z_n\right)^m \le D_k^{m-1} \sum_{n=1}^{\infty} d_{k,n} Z_n^m.$$

Now, as the random variables Z_n^m are i.i.d. and have finite moments,

$$\mathbb{E}\left(|X^{(k)}|_\infty^m\right) \le D_k^{m-1} \sum_{n=1}^{\infty} d_{k,n} \mathbb{E}\left(Z_n^m\right) = D_k^m \mathbb{E}\left(Z_1^m\right) < \infty,$$

implying that condition **(A.1.2)** holds true for k and m.

Remark 7.1 Given a kernel $g^{(k)}\colon \mathbb{R} \to \mathbb{R}$ we define by $G^{(k)}\colon [0,\infty) \to [0,\infty)$ the monotone hull of

$$\max(|g^{(k)}(t)|, |g^{(k)}(-t)|),\ t \ge 0,$$

as the smallest non-increasing function that dominates $|g^{(k)}(t)|$ and $|g^{(k)}(-t)|$ for $t \ge 0$. More precisely, $G^{(k)}(t) = \sup_{s>t}(|g^{(k)}(s)|, |g^{(k)}(-s)|)$. Then condition (7.4) is implied by the integrability of $G^{(k)}$. Then, when the kernel $g^{(k)}$ decreases monotonously for large positive and large negative values, condition (7.4) follows automatically from the integrability of the kernel $g^{(k)}$.

7.3.2 Boundedness of Marginal Densities

We now consider the third condition **(A.1.3)**. In Biermé and Desolneux (2012, Section 3.2), it is shown that when $\beta = 1$ a.s., in the following two particular situations the stationary shot noise process has a bounded density:

(a) The kernel is $g(t) = \mathrm{e}^{-t} \mathbb{1}_{\{t\ge 0\}}$, and the intensity $\lambda > 1$.
(b) The kernel satisfies $g(t) = t^{-\alpha}$ for $t \ge A$ for some $A > 0$ and $\alpha > 1/2$.

We present below a generalization of the results of Biermé and Desolneux (2012). Alternative results for the existence of a density can be found in Breton (2010, Corollary 2.1): interestingly, these do not depend on the law of β.

Proposition 7.3 *Consider a shot noise process* (2.41) *with a differentiable kernel. Denote by T an exponential random variable with parameter λ. Assume that one of the following conditions is met*

(A) *The impulse β has a density bounded by B and at least one of the following expectations $\mathbb{E}(1/|g(-T)|)$ or $\mathbb{E}(1/|g(T)|)$ is finite.*
(B1) *The expectation $\mathbb{E}(1/|\beta|)$ is finite, $g(x)$ has a strictly negative derivative in $(0, +\infty)$, and the function*

$$g_*^+(t) := \inf_{0 \le s \le t} |g'(s)|, \ (t \ge 0),$$

satisfies $\mathbb{E}(1/g_*^+(T)) < \infty$.
(B2) *The same as (B1) but replace $g(x)$ by $g(-x)$ and $g_*^+(t)$ by*

$$g_*^-(t) := \inf_{0 \le s \le t} |g'(-s)|, \ (t \ge 0).$$

Then $X(0)$ has a bounded density.

Note that Proposition 7.3 implies **(A.1.3)** with $h = 0$. The proof requires the following simple result.

Lemma 7.4 *Let X, Y be two independent random variables, where X has a density bounded by B. Then:*

(a) *The sum $X + Y$ has a density bounded by B.*
(b) *If $\mathbb{E}(1/|Y|) < \infty$, the product XY has a density bounded by $B\mathbb{E}(1/|Y|)$.*

Proof (a) Denote by F_X, F_Y, F_{X+Y} the respective cumulative distribution functions of $X, Y, X + Y$, and by f_X the density of X. The convolution formula states for the distribution of $X + Y$ that

$$F_{X+Y}(b) = \int_{\mathbb{R}} F_X(b - y)\mathrm{d}F_Y(y).$$

So

$$\begin{aligned}
\mathbb{P}(X + Y \in (a, b]) &= \int_{\mathbb{R}} (F_X(b - y) - F_X(a - y))\,\mathrm{d}F_Y(y) \\
&= \int_{\mathbb{R}} \int_a^b f_X(z - y)\mathrm{d}z\mathrm{d}F_Y(y) \\
&\le B(b - a),
\end{aligned}$$

completing the proof of (a). To see (b), we have

$$\begin{aligned}
\mathbb{P}(a < XY \le b) &= \int_{\mathbb{R}} \mathbb{P}(a < Xy \le b) \mathrm{d}F_Y(y) \\
&= \int_{-\infty}^{0} \mathbb{P}(a < Xy \le b) \mathrm{d}F_Y(y) + \int_{0}^{\infty} \mathbb{P}(a < Xy \le b) \mathrm{d}F_Y(y) \\
&= \int_{-\infty}^{0} \mathbb{P}\left(\frac{b}{y} \le X \le \frac{a}{y}\right) \mathrm{d}F_Y(y) \\
&\quad + \int_{0}^{\infty} \mathbb{P}\left(\frac{a}{y} < X \le \frac{b}{y}\right) \mathrm{d}F_Y(y) \\
&\le B(b-a) \int_{-\infty}^{0} \frac{1}{|y|} \mathrm{d}F_Y(y) + B(b-a) \int_{0}^{\infty} \frac{1}{y} \mathrm{d}F_Y(y) \\
&= B(b-a) \mathbb{E}\left(\frac{1}{|Y|}\right),
\end{aligned}$$

□

Proof of Proposition 7.3 Consider (A), assuming that $\mathbb{E}(1/|g(-T)|) < \infty$ (the other case is analogous). Based on Lemma 7.4 and the decomposition

$$X(0) = \sum_{i:\, \tau_i > 0} \beta_i g(-\tau_i) + \sum_{i:\, \tau_i \le 0} \beta_i g(-\tau_i) =: X^+(0) + X^-(0),$$

as $X^+(0)$ and $X^-(0)$ are independent, it is enough to see that $X^+(0)$ has a bounded density. Define by T_1 and T_2 the first two positive occurrences of the Poisson process. Condition on T_2 and apply (b) in Lemma 7.4 to obtain that $\beta_1 g(-T_1)$ has a conditional density bounded by $B\mathbb{E}(1/|g(-T_1)| \mid T_2)$. Applying now (a) in Lemma 7.4, we obtain that the sum $X^+(0)$ has a conditional density, denote it by $f_{X^+(0)|T_2}(x)$ with the same bound. Finally, integrating,

$$f_{X^+(0)}(x) = \mathbb{E}(f_{X^+(0)|T_2}(x)) \le \mathbb{E}(B\mathbb{E}(1/|g(-T_1)| \mid T_2)) = B\mathbb{E}(1/|g(-T_1)|),$$

concluding the proof in this case.

Let us consider now case (B). We see first that, conditional on T_2, the random variable $g(T_1)$ has a bounded density. In fact, by the change-of-variable formula

$$f_{g(-T_1)|T_2}(u) = \frac{1}{|g'(g^{-1}(u))|} \frac{1}{T_2} \mathbb{1}_{\{g^{-1}(u) \le T_2\}} \le \frac{1}{T_2 g_*^+(T_2)}.$$

By (b) in Lemma 7.4 the product $\beta_1 g(T_1)$ has a conditional density bounded by

$$\frac{1}{T_2 g_*^+(T_2)} \mathbb{E}(1/|\beta|). \tag{7.5}$$

Then, by conditional independence and (a) in Lemma 7.4, the sum $X^+(0)$ has a conditional density with the same bound. Finally, as the density of T_2 is $\lambda^2 t_2 \mathrm{e}^{-\lambda t_2}$, integrating the bound (7.5), we obtain that

$$\begin{aligned} f_{X^+(0)}(u) &= \mathbb{E}(f_{X^+(0)|T_2}(u)) \\ &\leq \int_0^\infty \frac{1}{t_2 g_*^+(t_2)} \mathbb{E}(1/|\beta|)\lambda^2 t_2 \mathrm{e}^{-\lambda t_2} \mathrm{d}t_2 \\ &= \lambda \mathbb{E}(1/|\beta|)\mathbb{E}(1/g_*^+(T)). \end{aligned}$$

This concludes the proof of the proposition, giving the respective bounds.

Below, we present two different situations in which the above results can be applied.

Corollary 7.5 *Consider a shot noise process with a C^4 kernel g such that for every $k \leq 4$ conditions (7.4) and (7.3) hold. Assume that the impulse has a finite third absolute moment, i.e. $\mathbb{E}(|\beta|^3) < \infty$. Assume, finally, one of the conditions referring to the density of the impulse and the integrability of the kernel of Proposition 7.3. Then*

$$\mathbb{E}(N_u^2) < \infty, \quad \textit{for any } p \geq 1.$$

Proof By Proposition 7.3 assumption **(A.1.3)** holds with $h = 0$. By the properties of g and β, Proposition 7.2 gives assumption **(A.1.2)**. Condition (7.3) gives the differentiability up to order 4. Theorem 1.22 then states the finiteness of the second moment.

Corollary 7.6 *Consider a shot noise process with a $C^\infty(\mathbb{R})$ kernel g such that for every $k \geq 1$*

$$g^{(k)}(t) \simeq c_k(t)\mathrm{e}^{\alpha t} \ \textit{as } t \to -\infty \quad \textit{and} \quad g^{(k)}(t) \simeq C_k(t)\mathrm{e}^{-\alpha t} \ \textit{as } t \to +\infty, \tag{7.6}$$

where $\simeq$ means equivalence and c_k, C_k are polynomials of an arbitrary degree. Assume furthermore that $\lambda > \alpha$ and $\mathbb{E}(|\beta|) < \infty$. If in addition either

(i) *β has bounded density, or*
(ii) *$\mathbb{E}(1/|\beta|) < \infty$ and $g(x)$, or $g(-x)$, have a strictly negative derivative on $(0, +\infty)$,*

then

$$\mathbb{E}(N_u^p) < \infty, \quad \textit{for any } p \geq 1.$$

Proof We apply Corollary 7.1 with $m = 1$ and arbitrarily large k. In view of (7.2) this gives the result for arbitrary p. The kernel is differentiable for any k. The relation (7.6) ensures condition (7.4), as it gives the integrability of the derivatives of any order of the kernel. It remains to see that the density of $X(0)$ is bounded, and this follows in case (i) from the fact that $\lambda > \alpha$, giving $\mathbb{E}(1/|g(-T_1)|) < \infty$, and the boundedness of the density follows by (A) in Proposition 7.3. In case (ii) we apply (B) in Proposition 7.3. We thus conclude our proof.

7.3.3 Central Limit Theorem for Crossings

We will now consider the study of the asymptotic behavior of the crossings of the shot noise process for large times, presenting a central limit theorem. Consider then the one-dimensional shot noise process defined in (2.41). Furthermore, in order to obtain the necessary asymptotic independence, assume that g is compact supported. For simplicity of exposition, we assume

$$\text{support}(g) \subset [-1, 1]. \tag{7.7}$$

As a consequence of this fact, the shot noise process is 2-dependent. This property is inherited by the crossings, considered as a process:

$$N(t) := N_u(X, [0, t]), \qquad t \geq 0, \quad u \in \mathbb{R} \text{ fixed}.$$

We also denote $N_u(X, [a, b])$ by $N(a, b)$. Based on the stationarity and on the 2-dependence, we compute the expectation and the variance of this process.

Proposition 7.7 *For the expectation and the variance of the process we have*

$$\mathbb{E}(N(t)) = t\mathbb{E}(N(1))$$

and

$$\begin{aligned}\text{Var}(N(n)) &= n\text{Var}(N(1)) + 2(n-1)\text{Cov}(N(1), N(1,2)) \\ &\quad + 2(n-2)\text{Cov}(N(1)N(2,3)) \\ &= cn - d \end{aligned} \tag{7.8}$$

where the constants are

$$\begin{aligned} c &= \text{Var}(N(1)) + 2\text{Cov}(N(1), N(1,2)) + \text{Cov}(N(1)N(2,3)), \\ d &= 2\text{Cov}(N(1), N(1,2)) + 4\text{Cov}(N(1)N(2,3)). \end{aligned}$$

Theorem 7.8 *Assume that the shot noise process as in Definition 7.1 with $d = 1$ has a kernel with bounded support as in (7.7), and that, for a fixed level, the crossings have second-order finite moment, i.e. $\mathbb{E}(N(t))^2 < \infty$. Then, the number of crossings with a fixed level satisfies a central limit theorem when $t \to \infty$. That is,*

$$\frac{N(t) - \mathbb{E}(N(t))}{\sqrt{\mathrm{Var}(N(t))}} \Rightarrow \mathcal{N}(0,1), \qquad \text{as } t \to \infty, \tag{7.9}$$

where $\Rightarrow$ denotes the convergence in distribution.

Remark 7.2 Corollary 7.5 gives a set of minimal hypothesis in order to obtain the finiteness of the second moment of the crossings.

Proof of Theorem 7.8 We first prove that we can compute the limit through a sequence of random variables. Let

$$X_k = N(k-1,k) - \mathbb{E}(N(k-1,k)) = N(k-1,k) - \mathbb{E}N(1),$$

and, for $n = \lfloor t \rfloor$:

$$\sigma_n = \sqrt{\mathrm{Var}(N(n))}, \qquad \sigma_t = \sqrt{\mathrm{Var}(N(t))}.$$

We see that

$$\frac{N(t) - \mathbb{E}(N(t))}{\sigma_t} - \frac{N(n) - \mathbb{E}(N(n))}{\sigma_n} \xrightarrow{L^2} 0. \tag{7.10}$$

To verify this fact we follow the arguments in Ibragimov and Linnik (1971, 18§7):

$$\begin{aligned}\left|\frac{1}{\sigma_n}\sum_{k=1}^n X_k - \frac{1}{\sigma_t}N(t)\right|^2 &= \left|\frac{1}{\sigma_n}\sum_{k=1}^n X_k - \frac{1}{\sigma_t}\sum_{k=1}^n X_k + \frac{1}{\sigma_t}\sum_{k=1}^n X_k - \frac{1}{\sigma_t}N(t)\right|^2 \\ &\le 2\left(\frac{1}{\sigma_n} - \frac{1}{\sigma_t}\right)^2\left(\sum_{k=1}^n X_k\right)^2 + \frac{2}{\sigma_t^2}N(n,t)^2.\end{aligned}$$

Taking expectations

$$\begin{aligned}\mathbb{E}\left|\frac{1}{\sigma_n}\sum_{k=1}^n X_k - \frac{1}{\sigma_t}N(t)\right|^2 &\le 2\left(1 - \frac{\sigma_n}{\sigma_t}\right)^2 + \frac{2}{\sigma_t^2}\mathbb{E}\left(N(n,t)^2\right) \\ &\le \left(1 - \frac{\sigma_n}{\sigma_t}\right)^2 + \frac{2}{\sigma_t^2}\mathbb{E}\left(N(1)^2\right).\end{aligned}$$

Here we used, as the process $N(n, t)$ is increasing in $t \in [n, n+1]$, that

$$\mathbb{E}\left(N(n,t)^2\right) \leq \mathbb{E}\left(N(n,n+1)^2\right) = \mathbb{E}\left(N(1)^2\right). \tag{7.11}$$

Finally

$$\begin{aligned}\sigma_t^2 &= \mathbb{E}\left(N(t)^2\right) = \mathbb{E}\left((N(n) + N(n,t))^2\right)\\ &= \mathbb{E}\left(N(n)^2 + N(n,t)^2 + 2N(n)N(n,t)\right)\\ &= \mathbb{E}\left(N(n)^2 + N(n,t)^2 + 2N(n-2,n)N(n,t)\right)\\ &= \sigma_n^2 + O(1),\end{aligned}$$

by similar monotonicity arguments as in (7.11). This concludes the proof of (7.10). The above verified moment conditions, taking into account the 2-dependence of the process, allow us to obtain a central limit theorem based on a Bernstein scheme for the sequence $\{N(n)\}$, applying for example Ibragimov and Linnik (1971, Theorem 19.2.1) that we state below. In view of (7.10), the proof of the theorem is concluded.

Theorem 7.9 (Ibragimov and Linnick) *Let X_j be a stationary m-dependent sequence with $\mathbb{E}X_0^2 < \infty$. Then*

$$\sigma^2 = \mathbb{E}X_0^2 + 2\sum_{j=1}^{\infty} \mathbb{E}X_0X_j$$

converges, and if $\sigma^2 \neq 0$,

$$\lim_{n\to\infty} \mathbb{P}\Big(\frac{1}{\sigma n^{\frac{1}{2}}}\sum_{j=1}^{n} X_j < z\Big) = \Phi(z).$$

7.4 Shot Noise Random Field in $\mathbb{R}^d$

Consider a shot noise random field $X\colon \mathbb{R}^d \to \mathbb{R}$ as in Definition 7.1, where we now consider the case $d > 1$. We are interested in the finiteness of high-order moments of the measure of level sets of this differentiable random field, restricted to the closed centered ball $B_d(0, a)$. For the computation of expectations of excursion sets of shot noise random fields with realizations with bounded variation (possibly discontinuous), see Biermé and Desolneux (2016). In order to obtain our results we apply Theorem 2.37. Let us recall its hypotheses.

(A.2.11) The sample paths of $X(\cdot)$ are $C^k(B_d(0,a))$, $k \geq 2$.

(A.2.12) For some $m = 1, 2, \ldots$ there exists a constant D_m such that

$$\max_{|v|=1} \mathbb{E}\left(\left|\frac{\partial^k X}{\partial v^k}\right|_\infty^m\right) \leq D_m,$$

where $\frac{\partial^k}{\partial v^k}$ denotes the k-th directional derivative w.r.t. the vector v.

(A.2.13) For some $0 \leq h \leq k$ there exists a constant $C > 0$ such that the joint density of

$$X(t), \frac{\partial X(t)}{\partial v}, \ldots, \frac{\partial^h X(t)}{\partial v^h}$$

is bounded by C uniformly in $t \in B_d(0,a)$, $v \in S^{d-1}$, and $(u_1, \ldots, u_{h+1}) \in \mathbb{R}^{h+1}$. When $h > 0$, we assume in addition that density of $X(t)$ is bounded uniformly in t.

The following result is a direct application of Theorem 2.37. Recall that $\mathcal{L}_u(B_d(0,a))$ is the u-level set restricted to the ball $B_d(0,a)$.

Corollary 7.10 *Consider a shot noise random field* (2.41) *satisfying condition* ***(A.2.11)*** *for some* $k \geq 1$, ***(A.2.12)*** *for some* $m \geq 1$ *and* k *above, and* ***(A.2.13)*** *for* $h = 0$. *Then*

$$\mathbb{E}(\sigma_{d-1}(\mathcal{L}_u(B_d(0,a)))^p) < \infty \tag{7.12}$$

for

$$p < \frac{m}{m+1}(k-1).$$

In the following subsections we obtain sufficient conditions to verify this corollary. The differentiability of the trajectories follows directly from the differentiability of the kernel g. Consider a multi-index $\alpha = (\alpha_1, \ldots, \alpha_d)$ of non-negative integers such that $|\alpha| = \sum_i \alpha_i = k$ and the usual notation for the partial derivatives. Then

$$\frac{\partial^k X(t)}{\partial t^\alpha} = \sum_i \beta_i \frac{\partial^k g(t)}{\partial t^\alpha}(t - \tau_i)$$

is a shot noise provided that the α-th derivative of g is integrable.

7.4.1 Boundedness of Moments

For simplicity of exposition we take $a = 1$, i.e. the unit ball $B(0, 1)$. Define the partition of $\mathbb{R}^d$ given by $A_1 = B_d(0, 1)$ and

$$\{A_n = B_d(0, r_n) \setminus B_d(0, r_{n-1})\colon n = 2, 3, \dots\},$$

where the sequence (r_n) has $r_1 = 1$ and is such that the volume of each set A_n equals the volume of A_1. In this way we can write

$$\frac{\partial^k X}{\partial v^k}(t) = \sum_{n=1}^{\infty} \sum_{i:\tau_i \in A_n} \beta_i \frac{\partial^k g}{\partial v^k}(t - \tau_i).$$

We now present a result useful for verifying the moment condition **(A.2.12)**. The proof follows the same lines as that of Proposition 7.2 in dimension one, and is omitted.

Proposition 7.11 *Consider a shot noise with impulse s.t.* $\mathbb{E}(|\beta|^m) < \infty$ *for given* m*, and kernel s.t.*

$$D'_k = \sum_{n=1}^{\infty} d'_{k,n} < \infty, \text{ for given } k,$$

with

$$d'_{k,n} = \sup_{|v|=1} \sup_{t \in A_1, s \in A_n} \left| \frac{\partial^k g}{\partial v^k}(t - s) \right|.$$

Then, the condition **(A.2.12)** *holds true for* m *and* k *as above, i.e.*

$$\sup_{|v|=1} \mathbb{E}\left(\left| \frac{\partial^k X}{\partial v^k} \right|_\infty^m \right) < \infty.$$

7.4.2 Boundedness of Marginal Densities

Here we present the conditions that allow us to verify condition **(A.2.13)**. Introduce the volume of the d-dimensional ball of radius r by

$$V(r) = k_d r^d, \quad k_d = \pi^{d/2} / \Gamma(1 + d/2).$$

Proposition 7.12 *Consider a shot noise random field* (2.41) *with a differentiable kernel. Denote by* T_1 *the occurrence of the Poisson field closest to the origin with the Euclidean distance (a.s. defined). Assume that either*

(A) *The impulse* β *has a density bounded by* B *and* $\mathbb{E}(1/|g(-T_1)|) < \infty$.
(B) $\mathbb{E}(1/|\beta|) < \infty$ *and the function*

$$G_*(r) := \sup_{u\in\mathbb{R}} \int_{t\in g^{-1}(u)\cap B_r} \frac{1}{|\nabla g(t)|} \mathrm{d}\mathcal{H}_{d-1}(t)$$

satisfies $\mathbb{E}(G_*(|T_1|)) < \infty$.

Then $X(0)$ *has a bounded density.*

The following auxiliary result is needed in the proof under (B). Its proof is elementary and has thus been omitted here.

Lemma 7.13 *Denote by* (T_n) *the occurrence of the Poisson field ordered by the Euclidean distance to the origin. Then, the random variables* $|T_1|$ *and* $|T_2|$ *have the following densities:*

$$f_{|T_1|}(x) = d\lambda k_d \mathrm{e}^{-\lambda k_d x^d} x^{d-1}, \quad f_{|T_2|}(x) = d\lambda^2 k_d^2 \mathrm{e}^{-\lambda k_d x^d} x^{2d-1}.$$

Proof of Proposition 7.12 This proof follows the same lines as the proof of Proposition 7.3. To prove case (A), we write

$$X(0) = \beta_1 g(-T_1) + \sum_{i=2}^{\infty} \beta_i g(-T_i).$$

Applying (b) of Lemma 7.4 we obtain that $\beta_1 g(-T_1)$ has a conditional density bounded by

$$B\,\mathbb{E}\Big(1/|g(-T_1)| \,\big|\, T_2\Big).$$

Applying now (a) of Lemma 7.4, by conditional independence we obtain that the sum $X(0)$ has a conditional density: $f_{X(0)|T_2}(x)$ with the same bound. Finally, integrating

$$f_{X(0)}(x) = \mathbb{E}(f_{X(0)|T_2}(x)) \le \mathbb{E}(B\mathbb{E}(1/|g(-T_1)| \mid T_2)) = B\mathbb{E}(1/|g(-T_1)|),$$

concluding the proof in this case.

Let us now consider case (B). Applying the co-area formula we see that, conditional on T_2, the random variable $g(-T_1)$ has a bounded density:

$$f_{g(-T_1)|T_2}(u) = \int_{t\in g^{-1}(u)\cap B_d(0,|T_2|)} \frac{1}{|\nabla g(t)|} \frac{1}{k_d|T_2|^d} \mathcal{H}_{d-1}(\mathrm{d}t) \le \frac{G_*(|T_2|)}{k_d|T_2|^d}.$$

By (b) of Lemma 7.4 the product $\beta_1 g(-T_1)$ has a conditional density bounded by

$$\frac{G_*(|T_2|)}{k_d|T_2|^d}\mathbb{E}(1/|\beta|). \tag{7.13}$$

Then, by conditional independence and (a) in Lemma 7.4, the sum $X(0)$ has a conditional density with the same bound (7.13). Integrating the bound (7.13), with the density of $|T_2|$ in Lemma 7.13, we obtain that

$$\begin{aligned} f_{X(0)}(x) = \mathbb{E}(f_{X(0)|T_2}(x)) &\le \frac{\mathbb{E}(1/|\beta|)}{k_d}\int_0^\infty \frac{G_*(r)}{r^d}\mathrm{e}^{-\lambda k_d r^d}\lambda^2 d k_d^2 r^{2d-1}\mathrm{d}r \\ &= dk_d\lambda^2\mathbb{E}(1/|\beta|)\int_0^\infty G_*(r)\mathrm{e}^{-\lambda k_d r^d} r^{d-1}\mathrm{d}r \\ &= \lambda\mathbb{E}(1/|\beta|)\mathbb{E}(G_*(|T_1|)), \end{aligned}$$

which is finite because of our hypotheses, thus concluding the proof.

Example 7.1 (Radial Kernel) Consider a shot noise with a radial kernel of the form $g(t) = \mathrm{e}^{-|t|^{2q}}$ $(q = 1, 2, \dots)$, and with impulse s.t. $\mathbb{E}(|\beta|) < \infty$. Assume further that

(i) $2q < d$, or $2q = d$ and $\lambda k_d > 1$,
(ii) β has a bounded density, or $\mathbb{E}(1/|\beta|) < \infty$.

Under these conditions, the conclusions of Theorem 2.37 are fulfilled, and consequently the moments of the level sets in (7.12) are finite for all p. It is easy to verify **(A.2.11)** and **(A.2.12)**. To see **(A.2.13)** assume first that β has a bounded density. Then, as g is radial, we have

$$\mathbb{E}\left(\frac{1}{g(-T_1)}\right) = \mathbb{E}\left(\frac{1}{g(|T_1|)}\right) = \int_0^\infty \frac{1}{g(r)} f_{|T_1|}(r)\mathrm{d}r < \infty, \tag{7.14}$$

due to the form of the kernel, Lemma 7.13, and condition (i).

Assume now that $\mathbb{E}(1/|\beta|) < \infty$. The sup in the function $G_*(r)$ in (B) in Proposition 7.12 is attained when $u = \mathrm{e}^{-r^{2q}}$, giving $G_*(r) = \mathrm{e}^{r^{2q}} r^{d-2q}$. A computation similar to (7.14) gives $\mathbb{E}G_*(|T_1|) < \infty$. In conclusion, the hypotheses of Corollary 7.10 are valid with arbitrary k, obtaining that the shot noise random field (2.41) satisfies (7.12) with arbitrary p.

7.4.3 Application: Central Limit Theorem in $\mathbb{R}^d$

As in the one-dimensional case, under some assumptions on the asymptotic independence of the shot noise field, the boundedness of moments allow us to obtain a central limit theorem on accumulated measure of the level sets when the domain where this level set is considered tends to infinity. We begin with a corollary similar to Corollary 7.10.

Corollary 7.14 *Consider the shot noise process defined in Sect. 7.4. Assume that the kernel $g(\cdot)$ is of class C^5, the impulse β has a finite moment of order 4, and the random variable β and the kernel g are such that hypothesis (A) or (B) of Proposition 7.12 hold. Then*

$$\mathbb{E}(\sigma_{d-1}(\mathcal{L}_u(B_d(0,a)))^3) < \infty.$$

Proof As the paths are C^5, using Proposition 7.11 we obtain

$$\sup_{|v|=1} \mathbb{E}\left(\left|\frac{\partial^k X}{\partial v^k}\right|_\infty^4\right) < \infty,$$

and using the results of Proposition 7.12, the distribution of $X(t)$ has a bounded density. This implies by Theorem 2.37 that, for $u \neq 0$, $\sigma_{d-1}(\mathcal{L}_u(B_d(0,a)))$ admits finite moments of order 3.

Based on these result, we obtain a central limit theorem. Since we use Theorem 7.16 we state it for a family of increasing rectangles:

$$\mathcal{T} = \prod_{i=1}^{d}[0, T_i].$$

Theorem 7.15 *Assume that the kernel $g(\cdot)$ has compact support included in, say, $(-\eta/2, \eta/2)^d$, and that the conditions of Corollary 7.14 hold. Then, for $u \neq 0$*

$$\frac{\sigma_{d-1}(\mathcal{L}_u(\mathcal{T}) - \mathbb{E}(\sigma_{d-1}(\mathcal{L}_u(\mathcal{T})))}{\sqrt{\mathrm{Var}(\sigma_{d-1}(\mathcal{L}_u(\mathcal{T})))}} \Rightarrow \mathcal{N}(0,1) \text{ as } \inf_i T_i \to \infty. \tag{7.15}$$

Proof Clearly the random field $X(\cdot)$ is η dependent in the sense that if $A_1, \ldots, A_m$ are m Borel sets with pairwise distance greater than η then $\{X(t); t \in A_1\} \ldots \{X(t); t \in A_m\}$ are independent. Corollary 7.14 gives the necessary finiteness of the moments. The proof is concluded using the following central limit theorem for η dependent random fields, see Leonenko (1975).

Theorem 7.16 *Let $\xi(\mathbf{x})$ be the random field described above. Assume that there exists a $\delta > 0$ such that*

$$\mathbb{E}[|\xi(\mathbf{x})|^{2+\delta}] < \infty.$$

Define

$$S(\mathcal{T}) := \frac{1}{(\prod_{i=1}^{d} T_i)^{\frac{1}{2}}} \int_{\prod_{i=1}^{d}[0,T_i]} \xi(\mathbf{x})\mathrm{d}\mathbf{x}.$$

Then we have

$$\mathrm{Var}(S(\mathcal{T})) \to \sigma^2 \neq 0, \ \textit{when} \ \inf_i T_i \to \infty.$$

Moreover,

$$\frac{1}{\sigma}S(\mathcal{T}) \Rightarrow N(0,1) \ \textit{when} \ \inf_i T_i \to \infty.$$

Chapter 8
Related Works

This final chapter serves as an expansive bibliographic review, synthesizing findings on related topics that use the Kac–Rice formula and Hermite expansions. It explores the winding turns of planar Gaussian processes, introducing a representation based on marked up-crossings. It also reviews the extensive literature surrounding the roots of random algebraic and trigonometric polynomials, and the zero sets of Kostlan–Shub–Smale random polynomial systems, discussing the asymptotic variances and distributions of their measure. Finally, it briefly outlines the geometric properties of random waves (such as Berry's Random Wave model), and non-local topological events like the number of connected components.

8.1 Introduction

Writing a monograph means making some choices. In this chapter, we discuss four themes that could have been the subjects of entire chapters in their own right, but which we consider to be less central to our exploration of Kac–Rice formulas. In place of an extended and detailed treatment, a short bibliographic review of each is presented below. Details and proofs can be found in the papers referenced. Finally, we briefly present some related works dealing with topological events that are addressed using different techniques.

- Section 8.2 is devoted to the study of winding turns: the number of times a complex valued process turns around its origin. We introduce a representation of this number using marked up-crossings, before using the Kac–Rice formula and Hermite expansion to establish a central limit theorem.
- Section 8.3 presents estimation and testing of anisotropy for Gaussian fields using integrals along the level set and the Kac–Rice formulas.

D. Armentano et al., *Geometry of Level Sets of Random Fields, Kac–Rice Formulas, Hermite Expansions and Applications*, Probability Theory and Stochastic Modelling 111, https://doi.org/10.1007/978-3-032-29144-8_8

- Section 8.4 presents the various models of random polynomials that have been studied in the existing literature: algebraic or trigonometric; Gaussian or non-Gaussian; with independent or dependent coefficients. As in the seminal works of Kac (1943) and Maslova (1974, 1975), we consider the computation of expectation and variance, before turning our attention to central limit theorems.
- Section 8.5 discusses random waves, a classical physics model. Different results on the nodal sets are listed here.
- Section 8.6 is devoted to the study of random polynomial systems, in particular the Kostlan–Shub–Smale system.
- Section 8.7 presents some topological characteristics which are non -local and, in particular, cannot be represented in an integral form, unlike the previous ones.

8.2 Winding Turns

The notion of the number of winding turns for a planar Gaussian process was first studied in Spitzer (1991), who considered the planar Brownian motion.

When dealing with processes with irregular paths, the argument in the complex plane is defined and studied with the help of stochastic calculus. The asymptotic behavior of the argument, as the time horizon tends to infinity, has been determined in some important cases. Among a vast number of results, it is worth mentioning again the theorem of Spitzer (1991), which states that, with a normalization by $\log t$, the argument of the Brownian curve tends in law towards a Cauchy distribution. Later on, a similar result was obtained for the Ornstein–Uhlenbeck process, see Vakeroudis (2012). Messulam and Yor (1982) proved that the occurrence of the Cauchy distribution as a limit (and consequently, the non-existence of the mean of the limit random variable) is due to the fact that Brownian motion makes many turns when it is close to the origin. If the process is restricted so that it remains outside a neighborhood of the origin, then the limiting distribution has moments of all orders.

In the 1980s, intense research was carried out on these questions. Readers may consult the recent paper Comtet and Tourigny (2015), written in memory of Marc Yor, for a modern view of the existing results and their relations with a deep problem from mathematical physics. Another key reference is Le Doussal et al. (2009), which provides a very interesting discussion of recent advances and applications to physics whilst also providing a significant number of references on such themes. Applications of winding turns in the field of physics include the physics of polymers, magnetic fields in superconductors, and the quantum Hall effect.

Following these results for non-differentiable random processes, the study of the winding number was extended to stationary Gaussian processes whose paths are sufficiently smooth. We can hence use the classical formula of complex analysis for the argument function to define the winding number N_W. If $A(t)$ denotes the argument at time t, we define N_W by the following formula:

$$N_W = \left\lfloor \frac{A(T)}{2\pi} \right\rfloor - \left\lfloor \frac{A(0)}{2\pi} \right\rfloor, \tag{8.1}$$

where $\lfloor\cdot\rfloor$ is the floor function. Using the complex analysis representation for $A(t)$ and techniques for complex-valued processes, Buckley and Feldheim (2018) studied the asymptotic behavior of the variation of the argument $\Delta(T)$, see Buckley and Feldheim (2018, Eq. (1)), of a centered complex Gaussian random process $\mathbb{X}(t)$ following the circularly symmetric model (i.e.: $\mathbb{E}(\mathbb{X}(0)\mathbb{X}(t)) = 0$) over the interval $[0, T]$ as $T \to \infty$. They computed the moments and obtained a central limit theorem.

In the paper Azaïs et al. (2023), a completely different approach was used. There, the number of winding turns around the origin was counted by subtracting the down-crossings from the up-crossings of the half line $\{x_1 > 0, x_2 = 0\}$. This approach may seem less intuitive and more involved, since a continuous functional is replaced by a discrete one that is a priori more difficult to handle. This approach does however offer an advantage in that it allows us to deploy the extensive theoretical machinery we have developed for the study of crossings, as described in the preceding chapters

Let us introduce the model. Consider a stationary mean-zero vectorial Gaussian process defined on $\mathbb{R}$,

$$\mathbf{X}(\cdot) = (X_1(\cdot), X_2(\cdot)) \in \mathbb{R}^2.$$

Let $r_i(\cdot)$ denote the covariance function of X_i, $i = 1, 2$ and $r_{12}(\cdot)$ their cross-covariance function:

$$r_i(t) = \mathbb{E}(X_i(t)X_i(0)), i = 1, 2; \quad r_{12}(t) = \mathbb{E}(X_1(t)X_2(0)), \quad t \in \mathbb{R}.$$

Without loss of generality, after a spatial scaling, we can assume that for each t the variance-covariance matrix of $\mathbf{X}(t)$ is the identity matrix I_2. That is, $r_i(0) = 1,\ i = 1, 2$ and $r_{12}(0) = 0$. In addition, note that

$$\mathbb{E}(X_1(0)X_2(t)) = r_{12}(-t); \quad \mathbb{E}(X_1(0)X_2'(t)) = -r_{12}'(-t).$$

Clearly, r_1, r_2 and r_{12} determine the distribution of $\mathbf{X}(\cdot)$.

To be able to compute winding turns, we assume that one of the coordinates, say X_2, is differentiable in quadratic mean. The fact that X_2 has a derivative is equivalent to

$$-r_2''(0) =: \lambda_{2,2} < \infty.$$

After a scaling in time, we can assume w.l.o.g. that $\lambda_{2,2} = 1$.

Define the number of winding turns around the origin by

$$\begin{aligned} N_W([0,T]) = \#\{t \leq T : X_1(t) > 0,\ X_2(t) = 0,\ X_2'(t) > 0\} \\ - \#\{t \leq T : X_1(t) > 0,\ X_2(t) = 0,\ X_2'(t) < 0\}. \end{aligned} \tag{8.2}$$

Thus, $N_W([0,T])$ is just the number of up-crossings minus the number of down-crossings of X_2 conditioned on the event $X_1(t) > 0$.

Consider the following conditions.

Assumption (A.8.1)

(i) Assume that $X_2(\cdot)$ satisfies the Geman condition for finiteness of the second moment of the number of zeros, see Proposition 1.18 and Lemma 1.19.

(ii) Set

$$\psi(t) = \max\left\{|r_2(t)|, |r_2''(t)|, |r_1(t)|, |r_{12}(t)|, |r_{12}'(t)|\right\},$$

and assume that $\psi \in \mathbb{L}^2([0,\infty))$ and $\psi(t) \to 0$ as $t \to \infty$.

(ii)' Assume that $r_1(t), r_2(t) \to 0$ as $t \to +\infty$ and

$$\int_{\mathbb{R}} r_2^2 + (r_{12}')^2 + (r_2')^2 + |r_1 r_2''| < +\infty. \tag{8.3}$$

These mixing conditions differ slightly from the one introduced by Assumption **(A.4.3)**. In fact, this is usually the case when we attempt to specialize the analysis to the particular model at hand. Note also that (ii') is weaker than (ii).

Set

$$N_W^*([0,T]) := \frac{N_W([0,T]) - \mathbb{E}N_W([0,T])}{\sqrt{T}}. \tag{8.4}$$

Theorem 8.1 *Consider* $\mathbf{X}(\cdot)$ *as described above. Let* $N_W([0,T])$ *and* $N_W^*([0,T])$ *be defined as in* (8.2) *and* (8.4) *respectively. Then,*

1. *For each* $T > 0$, $\mathbb{E}\big(N_W([0,T])\big) = -\frac{T}{2\pi} r_{12}'(0)$.
2. *Assume Assumption **(A.8.1)*** (i) *and* (ii'). *Then, there exists* $V_\infty < \infty$ *s.t.*

$$\lim_{T\to\infty} \frac{\operatorname{Var}\big(N_W([0,T])\big)}{T} = V_\infty. \tag{8.5}$$

3. *Assume Assumption **(A.8.1)*** (i) *and* (ii). *Then, as* $T \to \infty$, *the distribution of* $N_W^*([0,T])$ *converges towards the centered normal distribution with variance* V_∞.

In the particular case of independent coordinates, we have:

Theorem 8.2 *Assume that X_1 and X_2 are independent and that the covariance $r_1(\cdot)$ is differentiable outside the origin. Then, a sufficient condition so that $N_W^*([0, T])$ has a finite asymptotic variance is*

$$I := \int_0^\infty \frac{r_1'(t)}{\sqrt{1-r_1^2(t)}} \frac{r_2'(t)}{\sqrt{1-r_2^2(t)}} dt \text{ is convergent in the Riemann sense.}$$

The asymptotic variance takes the value

$$\lim_{T\to\infty} \frac{\operatorname{Var}(N_W([0, T]))}{T} = \frac{1}{\pi}\Big(\pi/2 + I\Big).$$

Note that in both theorems $X_1(\cdot)$ is not assumed to be differentiable and this is a notable difference with the cited papers.

8.3 Testing Anisotropy of Gaussian Random Fields

The paper Azaïs et al. (2025) presents an application of Kac–Rice formula s to detection of anisotropy. The problem is the following: *we observe partially a single trajectory of a Gaussian stationary random field $X(\cdot)$, in fact we observe the excursion set:*

$$\mathcal{E}(\mathcal{T}) := \big\{t \in \mathcal{T} : X(t) > u\big\}, \tag{8.6}$$

where $\mathcal{T}$ is the observation window and u is some (unknown) fixed level. This corresponds to a thresholding of the original image. Following the pioneering work of Cabaña (1987), our approach is based on the distribution of the normal vector to the level set. The computations are performed using the so-called *Palm distributions*. The Palm distribution of the normal to the level set is simply the distribution of the normal at a point "taken uniformly at random" on the level set. It is computed by means of the ratio of the results of two Kac–Rice formulas. To go further we must clarify our framework.

8.3.1 The Anisotropy Framework

We assume that $X(\cdot)$ admits C^2 paths and denote by Λ the variance-covariance matrix of the gradient $X'(0)$. By diagonalization we get:

$$\Lambda := \operatorname{Var}(X'(0)) = \mathbf{P}^\top \operatorname{diag}(\kappa_1^2, \ldots, \kappa_d^2)\, \mathbf{P}, \tag{8.7}$$

where $\mathbf{P}$ is a $d \times d$ unitary matrix, diag is a diagonal matrix and $\kappa_1^2 \leq \cdots \leq \kappa_d^2$ are the eigenvalues. The isotropy, which will be the null hypothesis of our test, corresponds to

$$\kappa_1^2 = \cdots = \kappa_d^2.$$

Remark 8.1 In the special case of dimension $d = 2$, this takes the form:

$$\Lambda = P_{-\theta_0} \begin{pmatrix} \kappa_1^2 & 0 \\ 0 & \kappa_2^2 \end{pmatrix} P_{\theta_0}, \tag{8.8a}$$

where P_{θ_0} is the rotation matrix with angle θ_0. Additionally, the anisotropy can be measured by a single parameter κ defined as:

$$\kappa = \sqrt{1 - \frac{\kappa_2^2}{\kappa_1^2}} \in [0, 1), \tag{8.8b}$$

where $\kappa = 0$ corresponds to the null hypothesis (isotropy).

8.3.2 *The Palm Distribution and Its Inversion*

The first main result is that the Palm distribution of the normal to the level curve, the normalized gradient, has a density w.r.t. the uniform probability measure η on $\mathbb{S}^{d-1}$ proportional to

$$\frac{\mathrm{d}\mathbb{P}(\zeta)}{\mathrm{d}\eta(\zeta)} = C_\Lambda \left(\zeta^\top \Lambda^{-1} \zeta\right)^{-\frac{d+1}{2}} = C_\Lambda \left\|\mathbf{D}^{-\frac{1}{2}}\mathbf{z}\right\|^{-(d+1)} = C_\Lambda \left(\frac{z_1^2}{\kappa_1^2} + \cdots + \frac{z_d^2}{\kappa_d^2}\right)^{-\frac{d+1}{2}}, \tag{8.9}$$

where $C_\Lambda > 0$ is a normalizing constant and $\mathbf{z} := (z_1, \ldots, z_d) \in \mathbb{S}^{d-1}$ are the coordinates of the vector ζ in the orthonormal basis $\mathbf{P}$.

Remark 8.2 In dimension $d = 2$, we uncover that the Palm density of the normal angle Θ on $(-\pi, \pi]$ is:

$$f_\Theta(\theta) = C_\kappa (1 - \kappa^2 \cos^2(\theta - \theta_0))^{-\frac{3}{2}}, \tag{8.10a}$$

where $C_\kappa = \left(\int_{-\pi}^{\pi} (1 - \kappa^2 \cos^2 \varphi)^{-3/2} \mathrm{d}\varphi\right)^{-1}$. This explicitly shows that the distribution depends only on the anisotropy parameter κ and the direction θ_0.

From the empirical variance matrix of the normal observed on $\mathcal{E}(\mathcal{T})$ it is possible, by diagonalization, to estimate eigenvalues κ_i^2 and the anisotropy directions (eigenvectors) in $\mathbf{P}$. From this information we can estimate anisotropy and construct some asymptotic test (as the size of the domain tends to infinity). It is proved in Azaïs et al. (2025, Theorem 4) that the inversion from observation of the empirical variance matrix eigenvalues (8.9) to the anisotropy parameter of (8.7) is one-to-one (up to a scaling factor), which is a consequence of the strict convexity of the mapping (8.9). Hence the anisotropy parameter can be obtained by a standard gradient descent with linear convergence rates as prescribed by Nesterov's theory. Explicit formulas for the gradient descent inversion and a practical algorithm are given in Azaïs et al. (2025).

Remark 8.3 In dimension $d = 2$ some simplification occurs; the computation of the empirical variance matrix is equivalent to the computation of:

$$C(\mathcal{T}) := \int_{\mathcal{L}(\mathcal{T})} \cos\big(2\Theta(t)\big)\mathrm{d}\mathcal{H}^1(t) \qquad \text{the cosine integral,}$$

$$\mathcal{S}(\mathcal{T}) := \int_{\mathcal{L}(\mathcal{T})} \sin\big(2\Theta(t)\big)\mathrm{d}\mathcal{H}^1(t) \qquad \text{the sine integral,}$$

where $\Theta(t)$ is the angle of the normal. Computing the expectations of these integrals normalized by the expected length of the level curve cancels out nuisance parameters (like the unknown level, mean, and variance), yielding:

$$\frac{\mathbb{E}(C(\mathcal{T}))}{\mathbb{E}(|\mathcal{L}(\mathcal{T})|)} = \cos(2\theta_0)g(\kappa) \quad \text{and} \quad \frac{\mathbb{E}(\mathcal{S}(\mathcal{T}))}{\mathbb{E}(|\mathcal{L}(\mathcal{T})|)} = \sin(2\theta_0)g(\kappa), \tag{8.10b}$$

where $g(\kappa)$ is an explicit elliptic function. The function $g(\cdot)$ is proven to be strictly invertible. Because these normalized integrals represent the coordinates of a vector with angle $2\theta_0$ and norm $g(\kappa)$, the anisotropy angle is recovered via $\hat{\theta}_0 = \frac{1}{2}\arctan\left(\frac{\mathcal{S}(\mathcal{T})}{C(\mathcal{T})}\right)$, and the anisotropy parameter κ is estimated by inverting the function g:

$$\hat{\kappa} = g^{-1}\left(\frac{\sqrt{C(\mathcal{T})^2 + \mathcal{S}(\mathcal{T})^2}}{|\mathcal{L}(\mathcal{T})|}\right), \tag{8.10c}$$

see an illustration in Fig. 8.1.

8.3.3 *Contour and LKC Methods*

The paper performs a comparison of the *contour method* above with the method of Biermé and Desolneux (2025) based on *Lipchitz–Killing curvatures* (LKC). A Monte Carlo experiment suggest that the contour method is more accurate, especially when the observation level u is close to the expectation of $X(0)$. A

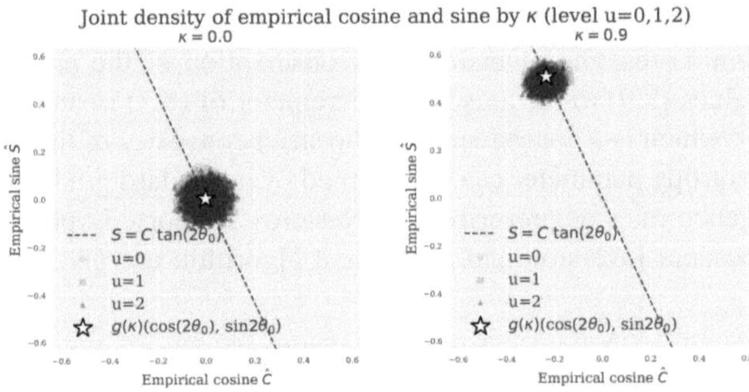

Fig. 8.1 Scatter plots of $(\hat{C}, \hat{S})$ over many realizations for $\kappa \in \{0, 0.9\}$. Under isotropy ($\kappa = 0$), the cloud is "distributed" around $(0, 0)$. Under anisotropy, the cloud is "distributed" around $g(\kappa)(\cos 2\theta_0, \sin 2\theta_0)$, enabling estimation of θ_0 via $\frac{1}{2}\arctan(\hat{S}, \hat{C})$ and of κ via g^{-1}

comparison is also performed with full observation, showing that the loss of information is minimal.

Using the Wiener Chaos method of Chap. 4 and especially results of Berzin (2021), in the case of a large window of observation, we obtain asymptotic normality of estimators. From that a test of anisotropy is proposed in Azaïs et al. (2025, Theorem 10).

8.4 Random Polynomials

The study of the roots of random polynomials is among the most important and the most popular topics in mathematics and in some areas of physics. It was also the object of the seminal work on Kac–Rice formulas (Kac 1943). For almost a century, a considerable amount of literature about this problem has emerged from fields such as probability, geometry, algebraic geometry, algorithmic complexity, quantum physics, etc. Even after this extensive study, it remains an extremely active field. Here, we limit our attention to the question of the number of roots of random polynomials. For a picture concerning the location of the roots see Bharucha-Reid and Sambandham (1986), Ibragimov and Zaporozhets (2013), Butez (2018) and references therein.

There are several reasons why researchers study random polynomials, and several ways to randomize them. The literature shows an extremely rich variety of possible behaviors of the roots of random polynomials. Indeed, there are models of degree d polynomials with as few as $(\frac{2}{\pi} + o(1)) \log d$ real roots as $d \to \infty$; while there are also models with exactly $\sqrt{d}$ real roots or with $\frac{2d}{\sqrt{3}} + o(d)$ real roots as $d \to \infty$. Thus, the asymptotic expected number of roots (both absolutely and relative to the maximal possible number of roots) is extremely sensitive to the precise model we are concerned with.

The last decade or so has seen a rapid increase in new results and methods on this topic. First Granville and Wigman (2011) and Azaïs and León (2013) established the Central Limit Theorem in the case of Gaussian Qualls' trigonometric polynomials; Azaïs et al. (2016) extended this result to classical trigonometric polynomials. At the same time, Dalmao (2015) established a Central Limit Theorem for Kostlan polynomials; and finally Do and Vu (2020) did the same for the number of real roots of Weyl polynomials. For a general early-stage picture of the field see e.g. Bharucha-Reid and Sambandham (1986), Bogomolny et al. (1996), Edelman and Kostlan (1995) while for a recent quick overview see Aguirre et al. (2024) and references therein.

In the rest of the section we briefly review results on algebraic and trigonometric random polynomials.

8.4.1 Random (Algebraic) Polynomials

The case of algebraic polynomials $P_d(x) = \sum_{k=1}^{d} a_k x^k$ with i.i.d. standardized coefficients, known as Kac's polynomials, was the first to be extensively studied and completely understood by the end of the 1970s. We owe the first asymptotically sharp result on the expected number of real roots to Kac (1943). He proved that as $d \to \infty$:

$$\mathbb{E}(N_0(P_d, \mathbb{R})) \sim \frac{2}{\pi} \log d.$$

Thus, Kac's polynomials have, on average, surprisingly few real zeros. In Maslova (1974), the asymptotic variance of $N_0(P_d, \mathbb{R})$ was established:

$$\mathrm{Var}(N_0(P_d, \mathbb{R})) \sim \frac{4}{\pi}\left(1 - \frac{2}{\pi}\right) \log d,$$

as $d \to \infty$, under mild conditions on the random coefficients. Moreover, Maslova (1975) extends this result proving the astonishing Central Limit Theorem for $N_0(P_d, \mathbb{R})$ as the degree $d \to \infty$.

For Kostlan's polynomials, that is, for independent centered normal coefficients such that $\mathrm{Var}(a_k) = \binom{d}{k} : k = 0, \ldots, d$, we have,

$$\mathbb{E}(N_0(P_d, \mathbb{R})) = \sqrt{d},$$

see Kostlan (2002). Moreover, the asymptotic variance is also of order $\sqrt{d}$ and a CLT holds as $d \to \infty$, see Dalmao (2015). In the close case of Weyl's polynomials, when the coefficients' variances are $\mathrm{Var}(a_k) = \frac{1}{\sqrt{k!}} : k = 0, \ldots, d$, the situation is similar but the expectation and the variance of the number of real roots are of order d, see Do and Vu (2020).

For cases where the coefficients are not Gaussian or not i.i.d., and for other ensembles of random polynomials, much more is known today than ten or twenty years ago, see for instance Aguirre et al. (2024).

8.4.2 Random Trigonometric Polynomials

Random trigonometric polynomials (RTPs for short) are an important class of random polynomials: roughly speaking, they are the polynomials obtained by replacing the argument t by its complex exponentiation $e^{\iota t}$. Part of the motivation for studying RTPs stems from the fact that they can be seen as partial sums of random Fourier series, and that they appear naturally in applications. For instance, the distribution of their number of roots occurs in nuclear physics and in statistical and quantum mechanics.

There are two main variants of RTPs:

$$T_d(x) := \frac{1}{\sqrt{d}} \sum_{k=1}^{d} a_k \cos(kx) + b_k \sin(kx),$$

$$Q_d(x) := \frac{1}{\sqrt{d}} \sum_{k=1}^{d} a_k \cos(kx), \quad x \in [0, 2\pi]. \tag{8.11}$$

Here, $d \in \mathbb{N}$ is the degree of the RTP and $(a_k)_k, (b_k)_k$ are two independent sequences of random variables defined on a common probability space. The classical setting is that of Q_d but T_d was introduced afterwards in Qualls (1970) since it has the advantage of being wide sense stationary when $a_k, b_k : k \geq 1$ are i.i.d. random variables.

Nowadays, we have a fairly complete picture of the behavior of the number of roots of RTPs and of how the joint distribution of the coefficients influence it. We present here a quick picture of recent results concerning different aspects of $N_0(T_d, [0, 2\pi])$. Nevertheless, to give some context, let us mention that

$$\mathbb{E}(N_0(T_d, [0, 2\pi])) = \sqrt{\frac{(d+1)(2d+1)}{6}} \sim_{d\to\infty} \frac{2d}{\sqrt{3}}.$$

The same asymptotic holds for $\mathbb{E}(N_0(Q_d, [0, 2\pi]))$, see Wilkins (1991). Besides, Bogomolny et al. (1996) argued informally that

$$\mathrm{Var}(N_0(T_d, [0, 2\pi])) \sim cd,$$

as $d \to \infty$ and gave an approximation of the value of the constant c.

I.i.d. Gaussian Coefficients

In the case where the coefficients a_k, b_k in (8.11) are i.i.d. standard Gaussian random variables, Granville and Wigman (2011) proved that there exists a constant $0 < V < \infty$ such that

$$\lim_{d\to\infty} \frac{\mathrm{Var}(N_0(T_d, [0, 2\pi]))}{d} = V.$$

This result is obtained through a careful manipulation of the Kac–Rice formula, which gives as a byproduct an integral expression for V. Furthermore, the authors state a CLT for $N_0(T_d, [0, 2\pi])$ as $d \to \infty$ using a result from Berk (1973) for triangular arrays which requires bounds for moments of order $2 + \delta$, $\delta > 0$, and for the variance of partial sums. Furthermore, Granville and Wigman (2011) observes that the covariance function r_d of T_d has a scaling limit r_∞ and that this fact should clear the way for the use of the number of roots of a limit process.

Following this lead, Azaïs and León (2013) generalized the results in Granville and Wigman (2011) to the number of crossings through a general level u by T_d. Their (much shorter) proof implemented a coupling between a scaled version of the RTP $T_d : d \geq 1$ and the Paley process[1] Y which is its scaling limit in some sense. The CLT is obtained via Wiener Chaos and the L^2 contiguity of the chaotic components of the respective numbers of roots.

Let us give a few details of this method of proof. The authors begin by performing the scale change $x \mapsto \frac{x}{d}$, thus considering the sequence of centered stationary Gaussian processes

$$Y_d(x) := T_d\Big(\frac{x}{d}\Big), \quad x \in [0, 2d\pi].$$

It is clear that $N_u(Y_d, [0, 2d\pi]) = N_u(T_d, [0, 2\pi])$. The advantage of this scaling is evident when considering the covariance function of Y_d, which is given by

$$r_d(x) = \frac{1}{d}\sum_{k=1}^{d} \cos\left(\frac{kx}{d}\right), \quad x \in [0, 2d\pi].$$

It is easy to show that

$$r_d(x) = \cos\left(\frac{d+1}{2d}x\right)\frac{\sin\left(\frac{x}{2}\right)}{d\sin\left(\frac{x}{2d}\right)} \to_{d\to\infty} \mathrm{sinc}(x) := \frac{\sin(x)}{x}, \quad x \in [0, \infty),$$

[1] That is, the centered stationary Gaussian process whose covariance function $r(t)$ equals the cardinal sine function $\mathrm{sinc}(t) := \frac{\sin(t)}{t}$.

uniformly in compact sets not containing 0. The first and second derivatives of r_d converge in the same sense to the first and second derivatives of sinc. The cardinal sine function sinc is the covariance function of the centered stationary Gaussian Paley process Y defined on $[0, \infty)$. The CLT for $N_u(Y, [0, 2d\pi])$ as $d \to \infty$ is obtained by an adaptation of Malevich's classical method (Malevich 1969) and the idea is, as advanced in Granville and Wigman (2011), to derive the CLT for $N_u(Y_d, [0, 2d\pi])$ from that for $N_u(Y, [0, 2d\pi])$.

The next step is to define these processes on a common probability space (coupling). Consider the Hilbert space

$$\mathcal{H} := \left\{ \boldsymbol{h} = (h_1, h_2) : \int_{\mathbb{R}} h_1^2(\lambda)\mathrm{d}\lambda + \int_{\mathbb{R}} h_2^2(\lambda)\mathrm{d}\lambda < \infty \right\}$$

with the inner product

$$\langle \boldsymbol{h}, \boldsymbol{g} \rangle = \int_{\mathbb{R}} h_1(\lambda) g_1(\lambda)\mathrm{d}\lambda + \int_{\mathbb{R}} h_2(\lambda) g_2(\lambda)\mathrm{d}\lambda.$$

The map

$$\boldsymbol{h} \mapsto W(\boldsymbol{h}) := \int_{\mathbb{R}} h_1(\lambda)\mathrm{d}W_1(\lambda) + \int_{\mathbb{R}} h_2(\lambda)\mathrm{d}W_2(\lambda),$$

where $W_!$ and W_2 are two independent standard B.m., defines an isonormal process associated with $\mathcal{H}$ allowing us to consider all the processes involved in the same probability space. Indeed, we can write

$$Y_d(x) = W\Big((h_{1,d}(x, \cdot), h_{2,d}(x, \cdot)) \Big), \quad d \geq 1,$$

$$Y(x) = W\Big(\mathbb{1}_{[0,1]}(\cdot)(\cos(x \cdot), \sin(x \cdot)) \Big),$$

with

$$h_{1,d}(x, \lambda) = \sum_{k=1}^{d} \cos\Big(\frac{kx}{d}\Big) \mathbb{1}_{[\frac{k-1}{d}, \frac{k}{d}]}(\lambda) \text{ and}$$

$$h_{2,d}(x, \lambda) = \sum_{k=1}^{d} \sin\Big(\frac{kx}{d}\Big) \mathbb{1}_{[\frac{k-1}{d}, \frac{k}{d}]}(\lambda).$$

The following step is to expand $N_u(Y_d, [0, 2d\pi]) = \sum_{k=0}^{\infty} I_{q,d}([0.2d\pi])$ and $N_u(Y, [0, 2d\pi]) = \sum_{k=0}^{\infty} I_q([0.\infty))$ using the method described in Chap. 4. The variance of the tail of the expansion is controlled using Arcones' inequality, see Lemma 4.10, combined with the explicit verification of Geman's condition 1.18

with uniform bounds w.r.t. d and the uniform bounds for $x \in [0, d\pi]$

$$|r_d(x)| \leq \frac{\pi}{x}, \quad |r_d'(x)| \leq \frac{\pi}{2x} + \frac{\pi^2}{4x^2}, \quad |r_d''(x)| \leq (\texttt{const})\Big(\frac{1}{x} + \frac{1}{x^2} + \frac{1}{x^3}\Big).$$

Finally, the asymptotic normality of each remaining chaotic component $I_{q,d}([0, d\pi])$ is deduced from that of $I_q([0, d\pi])$ using the fact, deduced from the coupling, that $\|I_{q,d}([0, d\pi]) - I_q([0, d\pi])\|_{L^2(\Omega)}$ tends to 0 as $d \to \infty$. A little more work is needed to get the CLT on the whole domain $[0, 2d\pi]$ as $d \to \infty$.

Later on, Azaïs et al. (2016) and Do et al. (2021) adapted this method to extend these results to $N_0(Q_d, [0, 2\pi])$ and to the number of roots of a family of orthogonal polynomials respectively. Here the situation is a little more complicated since the processes are no longer stationary. It is further worth mentioning that Maslova's method is replaced by the use of the Fourth Moment Theorem via the so-called contractions, see e.g. Peccati and Taqqu (2011, Th.11.4.1.), which in practice amount to bounding integrals such as those in (4.14).

8.4.2.1 I.i.d. Non-Gaussian Coefficients

Once the i.i.d. Gaussian case is well understood, the question naturally arises as to whether or not the same results hold true for different choices of the joint law of the coefficients $a_k, b_k : k \geq 1$ in (8.11). The expressions *universality* or *non-universality* are used in this context. Here we omit results about local universality, i.e. concerning the behavior on microscopic intervals of size $\frac{(\texttt{const})}{d}$, see e.g. Azaïs et al. (2015), Iksanov et al. (2016), Coutin and Peralta (2023).

The paper Flasche (2017), reducing the study of the number of roots to that of sign changes, proved that for i.i.d. standardized $a_k, b_k : k \geq 1$ one has for $0 \leq a < b \leq 2\pi$ and $u \in \mathbb{R}$ that

$$\lim_{d\to\infty} \frac{\mathbb{E}(N_u(T_d, [a, b]))}{d} = \frac{b-a}{\pi\sqrt{3}} \mathrm{e}^{-u^2/2}.$$

In other words, the asymptotic mean number of crossings is universal in the i.i.d. case.

Naturally the next step is to consider the asymptotic variance of $N_u(T_d, [0, 2\pi])$. Assuming in addition the existence of a high enough order moment of a_1, Bally et al. (2019) and Do et al. (2022) proved that the limit variance of $N_0(T_d, [0, 2\pi])$ differs from that in the Gaussian case by the term $\frac{2}{15}(\mathbb{E}a_1^4 - 3)$. Hence, in contrast to the behavior of the asymptotic expectation, the asymptotic variance is not universal in the i.i.d. case. Though these papers in fact proved a slightly more general result, the elegance and simplicity of the above comparison to the Gaussian case in terms of the kurtosis of the coefficients is appealing.

8.4.2.2 Dependent Gaussian Coefficients

A different approach to the question of universality entails keeping the Gaussianity of the coefficients but introducing some dependence. Over the last five years, a fairly complete picture of the asymptotic expectation in this framework has emerged.

Assume that the coefficients $(a_k)_k$ and $(b_k)_k$ in (8.11) form two independent stationary sequences of standard Gaussian random variables with correlation

$$\rho(k-\ell) := \mathbb{E}(a_k a_\ell) = \mathbb{E}(b_k b_\ell), \quad k, \ell \geq 1.$$

A key rôle in the sequel is played by the spectral function ψ_ρ of the correlation sequence ρ, which is defined as

$$\psi_\rho(x) := \sum_{k\in\mathbb{Z}} \rho(|k|)\mathrm{e}^{\imath kx}, \quad x \in (0, 2\pi).$$

Indeed, the variances and covariances of $T_d(x)$ and its derivatives $T_d'(x)$, $T_d''(x)$ can be written as the convolution of ψ_ρ with some well known kernels. For instance,

$$\mathrm{Var}(T_d(x)) = F_d * \psi_\rho(x), \quad x \in [0, 2\pi],$$

where F_d stands for the Féjer kernel. This expression is well suited to the study of the asymptotic expectation of $N_0(T_d, [0, 2\pi])$.

Under the hypothesis that $\psi_\rho \in L^2((0, 2\pi), \mathrm{d}x)$ is continuous and bounded away from zero, Angst et al. (2019) proved that

$$\lim_{d\to\infty} \frac{\mathbb{E}(N_0(T_d, [0, 2\pi]))}{d} = \frac{2}{\sqrt{3}},$$

that is, the same limit value as in the i.i.d. Gaussian case. It bears mentioning that this framework includes the case where the coefficients are given by the unit increments of a fractional Brownian motion $B^H = \{B_t^H : t \in [0, \infty)\}$ of any Hurst index $H \in (0, 1)$, namely, $a_k = B_k^H - B_{k-1}^H$ (taking $(b_k)_k$ as an independent copy of $(a_k)_k$). In this case,

$$\rho(k) = \frac{1}{2}\left(|1+k|^{2H} + |1-k|^{2H} - 2|k|^{2H}\right) \sim (\mathtt{const})|k|^{2H-2}, \quad \text{as } |k| \to \infty.$$

This shows that the decreasing rate of the correlation has no real influence in the asymptotic mean number of roots of T_d.

However, the papers Pautrel (2020) and Angst et al. (2022) showed a plethora of possible limit behavior of $N_0(T_d, [0, 2\pi])$ depending on the type of spectral measure μ_ρ. Naturally, the case of Angst et al. (2019) corresponds to the absolutely continuous case $\mu_\rho(\mathrm{d}x) = \psi_\rho(x)\mathrm{d}x$.

For some discrete measures μ_ρ we have

$$\liminf_{d\to\infty} \frac{\mathbb{E}(N_0(T_d,[0,2\pi]))}{d} \neq \limsup_{d\to\infty} \frac{\mathbb{E}(N_0(T_d,[0,2\pi]))}{d},$$

and a whole interval of possible limit values for this expectation according to the chosen subsequence.

On the other hand, if μ_ρ admits an absolutely continuous component with (possibly defective) density function ψ_ρ, we have

$$\lim_{d\to\infty} \frac{\mathbb{E}(N_0(T_d,[0,2\pi]))}{d} = \frac{\mathrm{Leb}(\{\psi_\rho^{-1}(0)\})}{\pi\sqrt{2}} + \frac{2\pi - \mathrm{Leb}(\{\psi_\rho^{-1}(0)\})}{\pi\sqrt{3}}.$$

Thus, if $\mathrm{Leb}(\psi_\rho^{-1}(0)) = 0$, we recover the same limit as in the i.i.d. case.

Finally, if ψ_ρ admits a logarithmic moment, we also recover the i.i.d. case limit

$$\lim_{d\to\infty} \frac{\mathbb{E}(N_0(T_d,[0,2\pi]))}{d} = \frac{2}{\sqrt{3}}$$

in the (stronger) almost sure sense.

8.5 Nodal Sets of Random Waves

Random waves are another popular class of random fields. A huge amount of literature, first in physics and then in mathematics, has been dedicated to the study of their nodal (i.e. level zero) sets. As we briefly describe below, Kac–Rice formulas can help to understand the asymptotic behavior of the nodal set, though it is Hermite expansions that are particularly helpful for this task, see the monograph Wigman (2024) for an up-to-date review of the field. In this section Δ stands for the Laplace operator in the space variables.

Rice (1945) proposed the following model to describe the noise in an electrical current

$$\xi(t) = \sum_n a_n \cos(k_n t + \varepsilon_n),$$

where $(\frac{k_n}{2\pi})$ denotes the different frequencies, (a_n) are independent and identically distributed Gaussian random variables, and (ε_n) are uniformly distributed on $[0, 2\pi]$. We have

$$E[\xi(t)\xi(s)] = 2\pi \sum_n E[a_n^2] \cos(k_n(t-s)).$$

Later, for modeling the movement of the surface of the sea, Longuet-Higgins (1957) used

$$\xi(t, x, y) = \sum_n a_n \cos(u_n x + v_n y + k_n t + \varepsilon_n),$$

where (u_n, v_n) are the spatial frequencies in each coordinate. Note that if the *dispersion relation* $k_n^2 = u_n^2 + v_n^2$ holds true, this process satisfies the wave equation $\partial_{tt}\xi = \Delta\xi$, justifying qualifying ξ as a random wave. This definition can be extended for considering waves in $\mathbb{R}^d$. Let $k_n = (k_1^n, k_2^n, \ldots, k_d^n)$ be a sequence of vectors of spatial frequencies, and denote by $|k_n|$ their norms. Let (a_n, b_n) be a sequence of independent standard Gaussians, and let (ε_n) be a sequence of independent random variables uniformly distributed on $[0, 2\pi]$. Then, if $x \in \mathbb{R}^d$, the random waves are now

$$\xi(t, x) = \Re\left(\sum_n (a_n + i b_n) \mathrm{e}^{i(\langle x, k_n\rangle + |k_n| t + \varepsilon_n)}\right).$$

Indeed, each expression above represents a stationary Gaussian process with a discrete spectral measure. More generally, the Bochner representation allows for an extension of these models to other types of spectral measures, verifying the dispersion relation.

The seminal paper Berry (1977) considers another random model defined on R^2 by

$$u_N^k(x_1, x_2) = \sqrt{\frac{2}{N}} \sum_{n=1}^{N} \cos(k x_1 \cos\theta_n + k x_2 \cos\theta_n + \phi_n),$$

where (θ_n, ϕ_n) are pairs of independent uniform random variables defined on $[0, 2\pi]$ and $(x_1, x_2) \in R^2$. The central limit theorem implies that, when $N \to \infty$, these processes converge in the sense of finite-dimensional distributions towards an isotropic Gaussian process b_k whose covariance satisfies

$$\begin{aligned} E[b_k(x) b_k(y)] &= \lim_{N\to\infty} \operatorname{Cov}(u_N^k(x_1, x_2), u_N^k(y_1, y_2)) \\ &= J_0(k\|x - y\|), \end{aligned}$$

where J_0 denotes the zero-order Bessel function of the first kind. Alternatively, b_k can be described by its spectral measure, which happens to be uniform on the unit circle. Using a series representation for b_k it is shown that b_k is the eigenfunction of the Helmholtz equation with eigenvalue $-k^2$, that is,

$$\Delta b_k + k^2 b_k = 0.$$

In the literature this model is called *Berry's random wave model*. Since the work of Berry in Berry (1977), this model has been believed to represent the behavior of deterministic Laplace eigenfunctions on generic chaotic surfaces in the high energy limit. There has been recent interest in the study of the nodal sets of such processes motivated by some cancellations obtained in Berry (2002), which yield a lower order of growth than anticipated for the variance.

Define the nodal set as

$$b_k^{-1}(0) = \{x \in R^2 : b_k(x) = 0\}.$$

The behavior of $b_k^{-1}(0)$ is well understood, see Wigman (2024), Nourdin et al. (2019) and references therein. First, the regularity of the field b_k implies that the nodal set is a smooth manifold (a union of smooth curves) and that the different Kac–Rice formulas apply. In particular, if $\ell_k(\mathcal{D}) = \mathcal{H}_1(b_k^{-1}(0) \cap \mathcal{D})$ is the length of the set restricted to a set $\mathcal{D}$, then $E(\ell_k(\mathcal{D})) = \frac{k\sigma_2(\mathcal{D})}{2\sqrt{2}}$. Moreover, the second moment of this functional is finite and a Hermite expansion can be obtained:

$$\ell_k(\mathcal{D}) = \sum_{q=0}^{\infty} \pi_q(\ell_k(\mathcal{D})),$$

and the following asymptotic for the variance, when $k \to \infty$, holds

$$\mathrm{Var}\,(\ell_k(\mathcal{D})) \sim O(\log k).$$

As argued in Berry (2002) this growth order was unexpected and is due to some cancellations which are apparent from the Hermite expansion. Indeed, the proof can be outlined by saying, by symmetry, the odd order chaoses do not contribute to the asymptotic variance. The variance of the second chaos is bounded, and in fact vanishes asymptotically. The variance of the fourth chaos, which needs careful examination, is $O(\log k)$ and in addition the sum of the contributions of the components of superior order grows more slowly. Finally, the central limit theorem is obtained by the asymptotic behavior of the fourth chaos component.

In Dalmao et al. (2021) and Dalmao (2023) Berry's model is studied in dimension three and similar results for its nodal sets are established. Nevertheless, some differences depending on the dimension are evidenced there. For instance, regarding the relative order of growth of expectation and variance we have

$$\mathrm{Var}(\ell_k(\mathcal{D})) \sim (\mathtt{const}) \log(\mathbb{E}(\ell_k(\mathcal{D})))$$

in the 2-dimensional case while

$$\mathrm{Var}(\ell_k(\mathcal{D})) \sim (\mathtt{const})\sqrt{\mathbb{E}(\ell_k(\mathcal{D}))}$$

in the 3-dimensional case. Moreover, while in the 2-dimensional case there is one fixed chaos dominating the asymptotic behavior of the nodal length, in the 3-dimensional case every chaos contributes to the limit.

There exist some related models of random waves defined on manifolds. Their nodal sets have a similar behavior to that of Berry's Random Wave model. We present two important examples here, see Wigman (2024) and references therein for a deeper view of random waves.

1. **Arithmetic Random Waves**
   Set $\mathcal{S} := \{n \in \mathbb{N} : n = a^2 + b^2, \text{ for some } a, b \in \mathbb{Z}\}$ and for $n \in \mathcal{S}$,

$$\Lambda_n := \{\lambda = (a, b) \in \mathbb{Z}^2 : a^2 + b^2 = n\}.$$

   The Laplace eigenvalues on the torus $\mathbb{T}^2$ are the numbers $4\pi^2 n : n \in \mathcal{S}$ and the corresponding (complex) eigenspace is generated by the exponentials $e_\lambda(\cdot) := \mathrm{e}^{2\pi\imath\langle\lambda,\cdot\rangle} : \lambda \in \Lambda_n$ where $\langle\cdot,\cdot\rangle$ stands for the Euclidean inner product in $\mathbb{R}^2$. The Arithmetic Random Wave is defined by

$$f_n(x) := \frac{1}{\sqrt{2|\Lambda_n|}} \sum_{\lambda\in\Lambda_n} a_\lambda e_\lambda(x), \qquad x \in \mathbb{T}^2,$$

   where the coefficients $(a_\lambda)_{\lambda\in\Lambda_n}$ are standard complex Gaussian i.i.d. except for the symmetries $a_{-\lambda} = \bar{a_\lambda}$, which make f_n real-valued.
2. **Random Spherical Harmonics**
   The Laplace eigenvalues on the sphere S^2 are the numbers $\lambda_\ell := \ell(\ell+1) : \ell \in \mathbb{N}$. The corresponding Laplace eigenspaces $\mathcal{H}_\ell$ have dimension $2\ell + 1$. Denote by $(Y_k)_{k=1}^{2\ell+1}$ a real-valued arbitrary orthogonal basis of $\mathcal{H}_\ell$. A Random Spherical Harmonic is defined by

$$T_\ell(x) := \sum_{k=1}^{2\ell+1} a_k Y_k(x), \qquad x \in \mathbb{S}^2,$$

   where the coefficients $(a_k)_{k=1}^{2\ell+1}$ are i.i.d. standard complex Gaussian.

8.6 Systems of Polynomial Equations

The case of systems of polynomial equations has received less attention. The main ensemble, which we describe below, is that of Kostlan–Shub–Smale random polynomial systems, introduced in the early 1990s by Shub and Smale (1993) and Kostlan (2002). Kostlan argues that this is the most natural distribution for a polynomial system. The exact value of the expectation was obtained in the early 1990s by geometric means, see Edelman and Kostlan (1995) for the one-dimensional case and Shub and Smale (1993) for the multi-dimensional one. A decade later Azaïs and Wschebor (2005b) and Wschebor (2005) obtained by probabilistic methods the asymptotic variance as the number of equations and variables tends to infinity. A decade ago, Dalmao (2015) obtained the asymptotic variance and a CLT for the number of zeros as the degree d goes to infinity in the case of one equation in one variable. Letendre (2019) studied the asymptotic behavior of the volume of random real algebraic sub-manifolds. His results include the finiteness of the limit variance, when the degree tends to infinity, of the volume of the zero sets of Kostlan–Shub–Smale systems with strictly fewer equations than variables. This result was generalized to the case of co-dimension 0 by Letendre and Puchol (2019). In parallel Armentano et al. (2018, 2021, 2022) obtained the same results and the central limit theorem, restricted to the Kostlan–Shub–Smale systems. Some results for the expectation and variance of related models are included in Azaïs and Wschebor (2005b).

In the present section we state the main results of the papers Armentano et al. (2018, 2021, 2022) which, as the degree goes to infinity, study the behavior of the normalized number of real roots of a Kostlan–Shub–Smale square random system with m equations and m variables or the measure of the nodal set of a system with r equations and m variables with $r < m$.

These papers use Kac–Rice formulas to show the finiteness of the limit variance and Hermite expansions as in Chap. 4 to show that it is strictly positive. Furthermore, they strongly exploit the invariance under isometries of the distribution of the polynomials.

8.6.1 Square Case

Let us start with some notation. Consider a square system $\mathbf{P}$ of m polynomial equations in m variables with common degree $d > 1$. More precisely, let $\mathbf{P} = (P_1, \dots, P_m)$ with

$$P_\ell(t) = \sum_{|\mathbf{j}| \le d} a_{\mathbf{j}}^{(\ell)} t^{\mathbf{j}},$$

where we use multi-index notation as in (4.2):

1. $\mathbf{j} = (j_1, \dots, j_m) \in \mathbb{N}^m$ and $|\mathbf{j}| = \sum_{k=1}^m j_k$;
2. $a_{\mathbf{j}}^{(\ell)} = a_{j_1 \cdots j_m}^{(\ell)} \in \mathbb{R}$, $\ell = 1, \dots, m$, $|\mathbf{j}| \le d$;
3. $t = (t_1, \dots, t_m)$ and $t^{\mathbf{j}} = \prod_{k=1}^m t_k^{j_k}$.

Let us say that $\mathbf{P}$ has the Kostlan–Shub–Smale (KSS for short) distribution if the coefficients $a_{\mathbf{j}}^{(\ell)}$ are independent centered normally distributed random variables with variances

$$\mathrm{Var}\left(a_{\mathbf{j}}^{(\ell)}\right) = \binom{d}{\mathbf{j}} = \frac{d!}{j_1! \dots j_m!(d - |\mathbf{j}|)!}.$$

We are interested in the number of real roots of $\mathbf{P}$, which we denote by $N_d^{\mathbf{P}}$. Shub and Smale (1993)[2] and Kostlan (2002) proved that $\mathbb{E}(N_d^{\mathbf{P}}) = d^{m/2}$. The result is the following.

Theorem 8.3 *Let* $\mathbf{P}$ *be a KSS random polynomial system with m equations, m variables and degree d. Then, as* $d \to \infty$*, we have*

- $\lim_{d \to \infty} \frac{\mathrm{Var}(N_d^{\mathbf{P}})}{d^{m/2}} = V_\infty^2$*, where* $0 < V_\infty^2 < \infty$.
- *The distribution of*

$$\tilde{N}_d^{\mathbf{P}} := \frac{N_d^{\mathbf{P}} - d^{m/2}}{d^{m/4}}$$

converges towards a centered normal random variable with variance V_∞.

Let us give some details of the proof of these results. The first step is to homogenize the system, i.e., to consider the homogenization $Y_\ell : \mathbb{R}^{n+1} \to \mathbb{R}$ defined by

$$Y_\ell(t_0, t_1, \dots, t_n) = t_0^d P_\ell(t_1/t_0, \dots, t_m/t_0), \qquad (\ell = 1, \dots, m).$$

In particular, $Y_\ell(1, \cdot) = P_\ell(\cdot)$. Now $\mathbf{Y} = (Y_1, \dots, Y_m)$ is a homogeneous system of polynomial equations of equal degree d. Its covariance structure is given by

$$r_d(s, t) := \mathbb{E}(Y_\ell(s)Y_\ell(t)) = \langle s, t\rangle^d; \quad s, t \in \mathbb{R}^{m+1},$$

where $\langle \cdot, \cdot \rangle$ is the Euclidean inner product of $\mathbb{R}^{m+1}$. This is another way of defining the KSS system. As a consequence, the distribution of the system $\mathbf{Y}$ is invariant under the action of the orthogonal group in $\mathbb{R}^{m+1}$.

Note that $\mathbf{Y}(\lambda t) = \lambda^d \mathbf{Y}(t)$ for every $\lambda \in \mathbb{R}$, where $t = (t_0, \dots, t_m) \in \mathbb{R}^{m+1}$. In particular, the roots of $\mathbf{Y}$ consist of lines through 0 in $\mathbb{R}^{m+1}$. Hence, a natural place to study N_d^P is to consider the solution set of $\mathbf{Y}(t) = 0$ where t lies in the sphere $S^m \subset \mathbb{R}^{m+1}$, since each root of $\mathbf{P}$ in $\mathbb{R}^m$ corresponds exactly to two (opposite) roots of $\mathbf{Y}$ on the unit sphere S^m of $\mathbb{R}^{m+1}$. (A different natural space to study this problem is the one considered by Kostlan, Shub and Smale, namely, the associated projective space of $\mathbb{R}^{m+1}$). Furthermore, one can prove that the subset

[2] In fact, the Shub–Smale results include the case where the degrees of each P_ℓ may be different and equal to d_ℓ. Their result is that $\mathbb{E}(N_d^{\mathbf{P}}) = (d_1 \cdots d_m)^{1/2}$.

of homogeneous polynomials $\mathbf{Y}$ with roots lying in the hyperplane $t_0 = 0$ has Lebesgue measure zero. Then, denoting by $N_{\mathbf{Y}}$ the number of roots of $\mathbf{Y}$ on S^m, we have $N_d^{\mathbf{P}} = \frac{N_{\mathbf{Y}}}{2}$ almost surely.

Now, we proceed to compute the Hermite expansion of the normalized version of the number of roots, i.e., we write

$$\widetilde{N_{\mathbf{Y}}} = \frac{N_{\mathbf{Y}} - 2d^m/2}{2d^{m/4}} = \sum_{q=1}^{\infty} I_{q,d},$$

where $I_{q,d}$ is written as a stochastic integral with respect to the Brownian motion as described in Chap. 4. Taking advantage of this expansion, the CLT is easily obtained for each term in the expansion as well as for any finite sum of them.

The challenging aspect of the proof involves assessing the negligibility of the variance tail resulting from the degeneracy of the covariance of $\mathbf{Y}$ at the diagonal $\{(s,t) \in S^m \times S^m : s = t\}$. To address this challenge, we employ a technique commonly used under stationarity in the Euclidean scenario, which entails covering a neighborhood of the diagonal with small, isometric regions. The variance of the number of roots within each such region is managed using the Kac–Rice formula. Achieving a balance between the number of these regions and the bound becomes essential.

While isometric regions cannot be straightforwardly chosen on the sphere, a meticulous construction allows us to cover a significant portion of the sphere with regions whose orthogonal projections onto the tangent space at a suitable point become isometric in the limit. Assuming the existence of a common local limit process on the tangent spaces, we can uniformly bound the tail of the variance of the roots in each region by approximating it with the corresponding tail of the number of roots of the local limit process. For more details, see Armentano et al. (2018, 2021).

8.6.2 *Rectangular Case*

Similar results can be extended to the rectangular case. Let us consider the system $\mathbf{Y}_r = (Y_1, \ldots, Y_r)$ of r homogeneous polynomial equations in $m+1$ variables of common degree $d > 1$ (where $r < m$). As in the previous section, the sphere $S^m \subset \mathbb{R}^{m+1}$ is a natural space to consider the level set of zero, which we denote by $\mathcal{L}_{\mathbf{Y}_d}$. However, in this case, $\mathcal{L}_{\mathbf{Y}_d}$ is a smooth submanifold of dimension $(m-r)$ (almost surely). This result follows from Proposition 2.10, but in this particular case, exploiting the algebraic structure of the problem yields the same conclusion.[3] We denote by $\sigma_{m-r}(\mathcal{L}_{\mathbf{Y}_d})$ the $(m-r)$-volume of the zero level set (on the sphere).

Shub and Smale (1993) and Kostlan (2002) proved that $\mathbb{E}(\sigma_{m-r}(\mathcal{L}_{\mathbf{Y}_d})) = 2d^{r/2}c_{m,r}$, $r \leq m$, where $c_{m,r}$ is the geometric measure of the sphere S^{m-r} as a submanifold of S^m. Letendre (2019) proved that there exists $0 < V_\infty^r < \infty$ such that

$$\lim_{d\to\infty} \frac{\mathrm{Var}(\sigma_{m-r}(\mathcal{L}_{\mathbf{Y}_d}))}{d^{r-m/2}} = V_\infty^r. \tag{8.12}$$

A different proof can be found in the arXiv version (Armentano et al. 2017).

Theorem 8.4 *Let $\mathbf{Y}_d$ be an $r \times (m+1)$ KSS homogeneous system. Then, if $r < m$, the standardized $(m-r)$-volume of the zero level set*

$$\frac{\sigma_{m-r}(\mathcal{L}_{\mathbf{Y}_d}) - \mathbb{E}[\sigma_{m-r}(\mathcal{L}_{\mathbf{Y}_d})]}{d^{\frac{r}{2}-\frac{m}{4}}}$$

converges in distribution as $d \to \infty$ towards a centered normal random variable with positive variance.

The proof of Theorem 8.4 follows the same lines as that of the square case ($r = m$), but some technical points must be adapted for this case. For more details, see Armentano et al. (2017).

8.7 Component Count and Topological Events

So far, we have considered "additive" or "local" functionals of level or excursion sets of stochastic processes or random fields such as the number of crossings, the number of critical points, the measure of level sets and the Euler characteristic. It is worth mentioning that another class of functionals involving "topological events" of level or excursion sets such as the number of connected components, box crossings or Betti numbers have also been studied recently, see for instance Nazarov and Sodin (2016), Wigman (2021), Gayet (2022), Beliaev et al. (2024) and references therein.

Functionals concerning topological events aim at intrinsic, or invariant, properties of level or excursion sets in contrast to metric properties such as their volume. Usually, these functionals are "non-additive", in the sense that their value observed at the union of two domains does not coincide with the sum of their values observed

[3] More precisely, a classical result of *algebraic geometry* says that the set of systems of polynomial equations with degenerate solutions (i.e. roots where the derivative is rank deficient), is the zero set of a non-trivial system of polynomial equations over the coefficients $\{a^{(\ell)}_{j0\ldots j_m}\}$, namely the *discriminant variety*. Hence, it has measure zero on the space of coefficients, and therefore almost surely $\mathbf{Y}_r$ is not in the discriminant variety.

at each domain, or "non-local", in the sense that their value on a domain cannot be inferred from subdomains.

Here, we briefly describe an important example: the number of connected components of the excursion set of a smooth Gaussian field, that we refer to as the component count. In dimension one it essentially coincides with (half) the number of level crossings and in higher dimension it can be bounded above by the number of critical points.

Since the seminal paper Bogomolny and Schmit (2002), the methods used to attack these problems have usually been inspired by percolation theory. More precisely, after sampling in a mesh, connected components of the excursion set can be identified with clusters in the graph. Naturally, technical difficulties arise regarding this discretization. When the covariance function decays fast enough the inherent independence in Percolation models can be essentially recovered.

Let $X(\cdot)$ be a smooth stationary centered Gaussian field on $\mathbb{R}^d$. For $u \in \mathbb{R}$ and $T > 0$, consider the excursion set $E = E(T, u) := \{X \geq u\}$ and define its component count $N_u(T)$ as the number of connected components of E contained in $[-T, T]^d$ which do not intersect the border of the box. As usual, the interest relies on the asymptotic behavior of the component count as the box grows to $\mathbb{R}^d$.

Following a non-rigorous prediction by Bogomolny and Schmit (2002), Nazarov and Sodin (2009, 2016) proved under mild hypotheses that there exists a positive constant c_{NS}, the so-called Nazarov–Sodin (or Bogomolny–Schmit) constant, such that, as $T \to \infty$ we have

$$\frac{N_u(T)}{T^d} \to c_{NS},$$

almost surely and in L^1. In other words, the component count satisfies a Law of Large Numbers. It is worth mentioning that the actual value of the Nazarov–Sodin constant is unknown except in trivial cases. Special attention has been paid to Gaussian and non-Gaussian random fields defined on the sphere, see Nazarov and Sodin (2009), Ingremeau and Rivera (2019), Sartori (2024), Kabluchko and Wigman (2020) and references therein.

This result was later complemented with bounds on the variance of the component count, see the discussion in Beliaev et al. (2024), concentration results, see Priya (2020) in the case of Gaussian Laplace eigenfunctions for instance, and with a Central Limit Theorem in Beliaev et al. (2024).

In order to state the CLT under the assumptions that the covariance function (and its first derivatives) of the underlying Gaussian field defined on $\mathbb{R}^d$ decays polynomially with exponent greater than $9d$, Beliaev et al. (2024) uses a martingale-theoretic approach inspired by Penrose (2001) in the context of percolation theory, see also Beffara and Gayet (2017). The key point is to represent the component count $N_u(T)$ as a martingale difference. After dealing with the perturbation arising from the discretization scheme, this representation allows the authors to exploit an ad hoc version of the CLT for martingale difference arrays.

The authors claim that, in comparison with the classical method of Hermite expansions and Fourth Moment Theorem, their method can be extended naturally to other topological functionals, additive or otherwise, and provide an alternative way to set the strict positivity of the limit variance.

References

Absil, P.-A., Mahony, R., and Sepulchre, R. (2008). *Optimization Algorithms on Matrix Manifolds*. Princeton University Press.

Absil, P.-A., Mahony, R., and Trumpf, J. (2013). An extrinsic look at the Riemannian Hessian. In *International Conference on Geometric Science of Information*, pages 361–368. Springer.

Adler, R. J. (1981). *The Geometry of Random Fields*. Wiley Series in Probability and Mathematical Statistics. John Wiley & Sons, Ltd., Chichester.

Adler, R. J. (2000). On excursion sets, tube formulas and maxima of random fields. *Annals of Applied Probability*, pages 1–74.

Adler, R. J. and Hasofer, A. (1976). Level crossings for random fields. *The Annals of Probability*, 4(1):1–12.

Adler, R. J. and Taylor, J. E. (2009). *Random Fields and Geometry*. Springer Science & Business Media.

Aguirre, A., Nguyen, H., and Wang, J. (2024). Concentration of the number of real roots of random polynomials. *Electron. J. Probab.*, 30:1–43.

Ancona, M. and Letendre, T. (2025). Multijet bundles and application to the finiteness of moments for zeros of Gaussian fields. *Analysis & PDE*, 18(6):1433–1476.

Angst, J., Dalmao, F., and Poly, G. (2019). On the real zeros of random trigonometric polynomials with dependent coefficients. *Proc. Amer. Math. Soc.*, 147(1):205–214.

Angst, J., Pautrel, T., and Poly, G. (2022). Real zeros of random trigonometric polynomials with dependent coefficients. *Trans. Amer. Math. Soc.*, 375(10):7209–7260.

Angst, J. and Poly, G. (2020). On the absolute continuity of random nodal volumes. *The Annals of Probability*, 48(5):2145–2175.

Angst, J. and Poly, G. (2022). On the zeros of non-analytic random periodic signals. *Int. Math. Res. Not. IMRN*, 2022(7):4931–4968.

Arcones, M. A. (1994). Limit theorems for nonlinear functionals of a stationary Gaussian sequence of vectors. *Ann. Probab.*, 22(4):2242–2274.

Armentano, D., Azaïs, J., Ginsbourger, D., and León, J. R. (2019). Conditions for the finiteness of the moments of the volume of level sets. *Electronic Communications in Probability*, 24(17):1–8.

Armentano, D., Azaïs, J.-M., Dalmao, F., and León, J. R. (2018). Asymptotic variance of the number of real roots of random polynomial systems. *Proc. Amer. Math. Soc.*, 146(12):5437–5449.

D. Armentano et al., *Geometry of Level Sets of Random Fields, Kac–Rice Formulas, Hermite Expansions and Applications*, Probability Theory and Stochastic Modelling 111, https://doi.org/10.1007/978-3-032-29144-8

Armentano, D., Azaïs, J.-M., Dalmao, F., and León, J. R. (2021). Central limit theorem for the number of real roots of Kostlan–Shub–Smale random polynomial systems. *Amer. J. Math.*, 143(4):1011–1042.

Armentano, D., Azaïs, J.-M., Dalmao, F., and León, J. R. (2022). Central limit theorem for the volume of the zero set of Kostlan–Shub–Smale random polynomial systems. *J. Complexity*, 72:Paper No. 101668, 22.

Armentano, D., Azaïs, J. M., Dalmao, F., León, J. R., and Mordecki, E. (2023). On the finiteness of the moments of the measure of level sets of random fields. *Braz. J. Probab. Stat.*, 37(1):219–245.

Armentano, D., Azaïs, J.-M., and León, J. R. (2025). On a general Kac–Rice formula for the measure of a level set. *The Annals of Applied Probability*, 35(3):1828–1851.

Armentano, D., Azaïs, J.-M., Dalmao, F., and León, J. R. (2017). On the asymptotic variance of the number of real roots of random polynomial systems. *arXiv preprint arXiv:1703.08163*.

Auffinger, A. and Ben Arous, G. (2013). Complexity of random smooth functions on the high-dimensional sphere. *The Annals of Probability*, 41(6):4214–4247.

Auffinger, A., Ben Arous, G., and Černý, J. (2013). Random matrices and complexity of spin glasses. *Communications on Pure and Applied Mathematics*, 66(2):165–201.

Azaïs, J.-M. (1989). Approximation des trajectoires et temps local des diffusions. *Annales de l'IHP Probabilités et statistiques*, 25(2):175–194.

Azaïs, J.-M., Cierco-Ayrolles, C., and Croquette, A. (1999). Bounds and asymptotic expansions for the distribution of the maximum of a smooth stationary Gaussian process. *ESAIM: Probability and Statistics*, 3:107–129.

Azaïs, J.-M., Dalmao, F., and De Castro, Y. (2024). Second maximum of a Gaussian random field and exact (t-)spacing test. arXiv preprint.

Azaïs, J.-M., Dalmao, F., and Delmas, C. (2024). Multivariate CLT for critical points. *arXiv preprint arXiv 2401.09117*.

Azaïs, J.-M., Dalmao, F., and León, J. R. (2016). CLT for the zeros of classical random trigonometric polynomials. *Ann. Inst. Henri Poincaré Probab. Stat.*, 52(2):804–820.

Azaïs, J.-M., Dalmao, F., and León, J. R. (2023). Winding number for stationary Gaussian processes using real variables. *Annales de l'Institut Henri Poincare (B) Probabilites et statistiques*, 59(3):1183–1202.

Azaïs, J.-M., Dalmao, F., León, J. R., Nourdin, I., and Poly, G. (2015). Local universality of the number of zeros of random trigonometric polynomials with continuous coefficients. *ArXiv preprint*, abs/1512.05583.

Azaïs, J.-M., De Castro, Y., and Gamboa, F. (2015). Spike detection from inaccurate samplings. *Applied and Computational Harmonic Analysis*, 38(2):177–195.

Azaïs, J.-M., De Castro, Y., and Mourareau, S. (2020). Testing Gaussian process with applications to super-resolution. *Applied and Computational Harmonic Analysis*, 48(1):445–481.

Azaïs, J.-M. and Delmas, C. (2002). Asymptotic expansions for the distribution of the maximum of Gaussian random fields. *Extremes*, 5(2):181–212.

Azaïs, J.-M. and Delmas, C. (2022). Mean number and correlation function of critical points of isotropic Gaussian fields and some results on GOE random matrices. *Stochastic Processes and their Applications*, 150:411–445.

Azaïs, J.-M., Delmas, C., and Rabier, C.-E. (2014). Likelihood ratio test process for quantitative trait locus detection. *Statistics*, 48(4):787–801.

Azaïs, J.-M. and León, J. R. (2013). CLT for crossings of random trigonometric polynomials. *Electron. J. Probab.*, 18:no. 68, 17.

Azaïs, J.-M. and León, J. R. (2020a). Necessary and sufficient conditions for the finiteness of the second moment of the measure of level sets. *Electronic Journal of Probability*, 25:1–15.

Azaïs, J.-M. and León, J. R. (2020b). Necessary and sufficient conditions for the finiteness of the second moment of the measure of level sets. *Electronic Journal of Probability*, 25(107):1–15.

Azaïs, J.-M. and Wschebor, M. (2004). Upper and lower bounds for the tails of the distribution of the condition number of a Gaussian matrix. *SIAM Journal on Matrix Analysis and Applications*, 26(2):426–440.

Azaïs, J.-M. and Wschebor, M. (2005a). On the distribution of the maximum of a Gaussian field with d parameters. *The Annals of Applied Probability*, 15(1A):254–278.

Azaïs, J.-M. and Wschebor, M. (2005b). On the roots of a random system of equations. The theorem of Shub and Smale and some extensions. *Foundations of Computational Mathematics*, 5(2):125–144.

Azaïs, J.-M. and Wschebor, M. (2008). A general expression for the distribution of the maximum of a Gaussian field and the approximation of the tail. *Stochastic Processes and Their Applications*, 118(7):1190–1218.

Azaïs, J.-M. and Wschebor, M. (2009). *Level Sets and Extrema of Random Processes and Fields*. John Wiley & Sons, Inc., Hoboken, NJ.

Azaïs, J.-M. and Wschebor, M. (2010). Erratum to: A general expression for the distribution of the maximum of a Gaussian field and the approximation of the tail [*Stochastic Process. Appl.* 118 (7)(2008) 1190–1218]. *Stoch. Process. Appl*, 120:2100–2101.

Azaïs, J.-M. and Cierco-Ayrolles, C. (2002). An asymptotic test for quantitative gene detection. *Annales de l'Institut Henri Poincare (B) Probability and Statistics*, 38(6):1087–1092.

Azaïs, J.-M., Dalmao, F., and Castro, Y. D. (2025). Gaussian random field's anisotropy using excursion sets. *arXiv preprint arXiv 2512.11085*.

Azaïs, J.-M. and Delmas, C. (2026). Landscape k-complexity of isotropic centered Gaussian fields. *arXiv preprint arXiv 2503.20397v2*.

Azmoodeh, E., Malicet, D., Mijoule, G., and Poly, G. (2016). Generalization of the Nualart–Peccati criterion. *Ann. Probab.*, 44(2):924–954.

Bally, V., Caramellino, L., and Poly, G. (2019). Non universality for the variance of the number of real roots of random trigonometric polynomials. *Probab. Theory Related Fields*, 174(3–4):887–927.

Banach, S. (1925). Sur les lignes rectifiables et les surfaces dont l'aire est finie. *Fundamenta Mathematicae*, 7:225–236.

Bassan, B. and Bona, E. (1990). Moments of stochastic processes governed by Poisson random measures. *Commentationes Mathematicae Universitatis Carolinae*, 31(2):337–343.

Beffara, V. and Gayet, D. (2017). Percolation of random nodal lines. *Publ. Math. Inst. Hautes Études Sci.*, 126:131–176.

Beliaev, D., Cammarota, V., and Wigman, I. (2019). Two point function for critical points of a random plane wave. *International Mathematics Research Notices*, 2019(9):2661–2689.

Beliaev, D., Cammarota, V., and Wigman, I. (2020). No repulsion between critical points for planar Gaussian random fields. *Electronic Communications in Probability*, 25:1–13.

Beliaev, D., McAuley, M., and Muirhead, S. (2024). A central limit theorem for the number of excursion set components of Gaussian fields. *Ann. Probab.*, 52(3):882–922.

Belyaev, Y. K. (1966). On the number of intersections of a level by a Gaussian stochastic process. I. *Theory of Probability & Its Applications*, 11(1):106–113.

Belyaev, Y. K. and Piterbarg, V. (1972). The asymptotic formula for the mean number of a-points of excursions of Gaussian fields above high levels. *Bursts of Random Fields*, pages 62–89.

Benzaquen, S. and Cabaña, E. M. (1982). The expected measure of the level sets of a regular stationary Gaussian process. *Pacific J. Math.*, 103(1):9–16.

Berk, K. N. (1973). A central limit theorem for m-dependent random variables with unbounded m. *Ann. Probability*, 1:352–354.

Berry, M. V. (1977). Regular and irregular semiclassical wavefunctions. *J. Phys. A*, 10(12):2083–2091.

Berry, M. V. (2002). Statistics of nodal lines and points in chaotic quantum billiards: perimeter corrections, fluctuations, curvature. *J. Phys. A*, 35(13):3025–3038.

Berzin, C. (2021). Estimation of local anisotropy based on level sets. *Electronic Journal of Probability*, 26:1–72.

Berzin, C., Latour, A., and León, J. R. (2022). Kac–Rice formula: A contemporary overview of the main results and applications. *ArXiv preprint*, abs/2205.08742.

Bharucha-Reid, A. T. and Sambandham, M. (1986). *Random Polynomials*. Probability and Mathematical Statistics. Academic Press, Inc., Orlando, FL.

Biermé, H. and Desolneux, A. (2011). Regularity and crossings of shot noise processes. *MAP5 2010-20. 2011 (hal-00484118v2).*

Biermé, H. and Desolneux, A. (2012). Crossings of smooth shot noise processes. *The Annals of Applied Probability*, 22(6):2240 – 2281.

Biermé, H. and Desolneux, A. (2016). On the perimeter of excursion sets of shot noise random fields. *The Annals of Probability*, 44(1):521 – 543.

Biermé, H. and Desolneux, A. (2025). The anisotropy of 2D or 3D Gaussian random fields through their Lipschitz–Killing curvature densities. *Annals of Applied Probability*, 35(3):–. Preprint available at hal-04112024.

Bogomolny, E., Bohigas, O., and Leboeuf, P. (1996). Quantum chaotic dynamics and random polynomials. *J. Statist. Phys.*, 85(5–6):639–679.

Bogomolny, E. and Schmit, C. (2002). Percolation model for nodal domains of chaotic wave functions. *Physical Review Letters*, 88(11):114102.

Borodin, A. N. (2017). *Stochastic Processes*. Springer.

Breton, J.-C. (2010). Regularity of the laws of shot noise series and of related processes. *Journal of Theoretical Probability*, 23(1):21–38.

Brillinger, D. R. (1972). On the number of solutions of systems of random equations. *The Annals of Mathematical Statistics*, 43(2):534–540.

Buckley, J. and Feldheim, N. (2018). The winding of stationary Gaussian processes. *Probability Theory and Related Fields*, 172:583–614.

Bulinskaya, E. V. (1961). On the mean number of crossings of a level by a stationary Gaussian process. *Theory of Probability & Its Applications*, 6(4):435–438.

Butez, R. (2018). The largest root of random Kac polynomials is heavy tailed. *Electron. Commun. Probab.*, 23:Paper No. 20, 9.

Cabaña, E. (1986). Integrals on random sets. https://drive.google.com/file/d/1DyUHlV2FrUanUua6wBTUmIrJOhmvfT8x/view.

Cabaña, E. M. (1987). Affine processes: a test of isotropy based on level sets. *SIAM Journal on Applied Mathematics*, 47(4):886–891.

Candès, E. J. and Fernandez-Granda, C. (2014). Towards a mathematical theory of super-resolution. *Communications on pure and applied Mathematics*, 67(6):906–956.

Cesàro, E. (1896). *Lezioni di geometria intrinseca*. Presso l'autore-editore.

Cheng, D. and Schwartzman, A. (2018). Expected number and height distribution of critical points of smooth isotropic Gaussian random fields. *Bernoulli: official journal of the Bernoulli Society for Mathematical Statistics and Probability*, 24(4B):3422.

Chizat, L. (2022). Sparse optimization on measures with over-parameterized gradient descent. *Mathematical Programming*, 194(1):487–532.

Comtet, A. and Tourigny, Y. (2015). Explicit formulae in probability and in statistical physics. *In Memoriam Marc Yor-Séminaire de Probabilités XLVII*, pages 505–519.

Coutin, L. and Peralta, L. (2023). Rates of convergence for the number of zeros of random trigonometric polynomials. *Bernoulli*, 29(3):1983–2007.

Cramér, H. and Leadbetter, M. (1967). *Stationary and Related Stochastic Processes – Sample Function Properties and their Applications.* John Wiley and Sons, Inc.

Cuzick, J. (1976). A central limit theorem for the number of zeros of a stationary Gaussian process. *The Annals of Probability*, 4(4):547–556.

Daley, D. J. (1971). The definition of a multi-dimensional generalization of shot noise. *J. Appl. Probability*, 8:128–135.

Dalmao, F. (2015). Asymptotic variance and CLT for the number of zeros of Kostlan–Shub–Smale random polynomials. *C. R. Math. Acad. Sci. Paris*, 353(12):1141–1145.

Dalmao, F. (2023). A note on 3d-monochromatic random waves and cancellation. *ALEA Lat. Am. J. Probab. Math. Stat.*, 20(2):1091–1109.

Dalmao, F., Estrade, A., and León, J. R. (2021). On 3-dimensional Berry's model. *ALEA Lat. Am. J. Probab. Math. Stat.*, 18(1):379–399.

De Castro, Y., Gadat, S., Marteau, C., and Maugis-Rabusseau, C. (2021). Supermix: sparse regularization for mixtures. *The Annals of Statistics*, 49(3):1779–1809.

De Castro, Y. and Gamboa, F. (2012). Exact reconstruction using Beurling minimal extrapolation. *Journal of Mathematical Analysis and applications*, 395(1):336–354.

Delmas, C. (1998). An asymptotic expansion for the distribution of the maximum of a class of Gaussian fields. *Comptes Rendus de l'Académie des Sciences-Series I-Mathematics*, 327(4):393–397.

Delmas, C. (2001). *Distribution du maximum d'un champ aléatoire et applications statistiques*. PhD thesis, Université Paul Sabatier.

Dembo, A. and Zeitouni, O. (1998). *Large Deviations Techniques and Applications*, volume 38. Springer.

Do, Y., Nguyen, H. H., and Nguyen, O. (2022). Random trigonometric polynomials: universality and non-universality of the variance for the number of real roots. *Ann. Inst. Henri Poincaré Probab. Stat.*, 58(3):1460–1504.

Do, Y., Nguyen, H. H., Nguyen, O., and Pritsker, I. E. (2021). Central limit theorem for the number of real roots of random orthogonal polynomials. *ArXiv preprint*, abs/2111.09015.

Do, Y. and Vu, V. (2020). Central limit theorems for the real zeros of Weyl polynomials. *Amer. J. Math.*, 142(5):1327–1369.

do Carmo, M. P. (2016). *Differential Geometry of Curves & Surfaces*. Dover Publications, Inc., Mineola, NY, second edition.

Döbler, C. and Peccati, G. (2018). Fourth moment theorems on the Poisson space: analytic statements via product formulae. *Electron. Commun. Probab.*, 23:Paper No. 91, 12.

Duval, V. and Peyré, G. (2015). Exact support recovery for sparse spikes deconvolution. *Foundations of Computational Mathematics*, 15(5):1315–1355.

Edelman, A., Arias, T. A., and Smith, S. T. (1998). The geometry of algorithms with orthogonality constraints. *SIAM journal on Matrix Analysis and Applications*, 20(2):303–353.

Edelman, A. and Kostlan, E. (1995). How many zeros of a random polynomial are real? *Bull. Amer. Math. Soc. (N.S.)*, 32(1):1–37.

Estrade, A. and Fournier, J. (2016). Number of critical points of a gaussian random field: condition for a finite variance. *Statistics & Probability Letters*, 118:94–99.

Estrade, A. and León, J. R. (2016). A central limit theorem for the Euler characteristic of a Gaussian excursion set. *The Annals of Probability*, 44(6):3849–3878.

Federer, H. (1969). *Geometric Measure Theory*. Springer.

Feller, W. (1991). *An Introduction to Probability Theory and its Applications, Volume 2*, volume 81. John Wiley & Sons.

Flasche, H. (2017). Expected number of real roots of random trigonometric polynomials. *Stochastic Process. Appl.*, 127(12):3928–3942.

Gass, L. (2025). Spectral criteria for the asymptotics of local functionals of Gaussian fields and their application to nodal volumes and critical points. https://arxiv.org/abs/2501.07356, pages 1–41.

Gass, L. and Stecconi, M. (2024). The number of critical points of a Gaussian field: finiteness of moments. *Probability Theory and Related Fields*, 190(3):1167–1197.

Gayet, D. (2022). Asymptotic topology of excursion and nodal sets of Gaussian random fields. *J. Reine Angew. Math.*, 790:149–195.

Geman, D. (1972). On the variance of the number of zeros of a stationary Gaussian process. *The Annals of Mathematical Statistics*, pages 977–982.

Granville, A. and Wigman, I. (2011). The distribution of the zeros of random trigonometric polynomials. *Amer. J. Math.*, 133(2):295–357.

Guillemin, V. and Pollack, A. (2010). *Differential Topology*. AMS Chelsea Publishing, Providence, RI. Reprint of the 1974 original.

Hasofer, A. (1976a). The mean number of maxima above high levels in Gaussian random fields. *Journal of Applied Probability*, 13(2):377–379.

Hasofer, A. (1976b). The mean number of maxima above high levels in Gaussian random fields. *Journal of Applied Probability*, 13(2):377–379.

Ibragimov, I. and Zaporozhets, D. (2013). On distribution of zeros of random polynomials in complex plane. In *Prokhorov and Contemporary Probability Theory*, volume 33 of *Springer Proc. Math. Stat.*, pages 303–323. Springer, Heidelberg.

Ibragimov, I. A. and Linnik, Y. V. (1971). *Independent and Stationary Sequences of Random Variables*. Wolters-Noordhoff Publishing.

Iksanov, A., Kabluchko, Z., and Marynych, A. (2016). Local universality for real roots of random trigonometric polynomials. *Electron. J. Probab.*, 21:Paper No. 63, 19.

Ingremeau, M. and Rivera, A. (2019). A lower bound for the Bogomolny–Schmit constant for random monochromatic plane waves. *Math. Res. Lett.*, 26(4):1179–1186.

Itô, K. (1963). The expected number of zeros of continuous stationary Gaussian processes. *Journal of Mathematics of Kyoto University*, 3(2):207–216.

Janson, S. (1997). *Gaussian Hilbert Spaces*, volume 129 of *Cambridge Tracts in Mathematics*. Cambridge University Press, Cambridge.

Kabluchko, Z. and Wigman, I. (2020). Asymptotics for the expected number of nodal components for random lemniscates. *International Mathematics Research Notices*, 2022(3):2337–2375.

Kac, M. (1943). On the average number of real roots of a random algebraic equation. *Bulletin of the American Mathematical Society*, 49(4):314–320.

Kergin, P. (1980). A natural interpolation of C^k functions. *Journal of Approximation Theory*, 29(4):278–293.

Klüppelberg, C. and Kühn, C. (2004). Fractional Brownian motion as a weak limit of Poisson shot noise processes—with applications to finance. *Stochastic Processes and their Applications*, 113(2):333–351.

Klüppelberg, C. and Mikosch, T. (1995). Explosive Poisson shot noise processes with applications to risk reserves. *Bernoulli*, 1(1/2):125–147.

Kostlan, E. (2002). On the expected number of real roots of a system of random polynomial equations. In *Foundations of Computational Mathematics*, pages 149–188. World Scientific.

Kratz, M. F. and León, J. R. (1997). Hermite polynomial expansion for non-smooth functionals of stationary Gaussian processes: crossings and extremes. *Stochastic Process. Appl.*, 66(2):237–252.

Kratz, M. F. and León, J. R. (2000). Central limit theorems for the number of maxima and an estimator of the second spectral moment of a stationary Gaussian process, with application to hydroscience. *Extremes*, 3(1):57–86.

Kratz, M. F. and León, J. R. (2001). Central limit theorems for level functionals of stationary Gaussian processes and fields. *J. Theoret. Probab.*, 14(3):639–672.

Kratz, M. F. and León, J. R. (2006a). On the second moment of the number of crossings by a stationary Gaussian process. *The Annals of Probability*, 34(4):1601–1607.

Kratz, M. F. and León, J. R. (2006b). On the second moment of the number of crossings by a stationary Gaussian process. *The Annals of Probability*, 34(4):1601–1607.

Ladgham, S. and Lachièze-Rey, R. (2023). Local repulsion of planar Gaussian critical points. *Stochastic Processes and their Applications*, 166:104221.

Le Doussal, P., Etzioni, Y., and Horovitz, B. (2009). Winding of planar Gaussian processes. *Journal of Statistical Mechanics: Theory and Experiment*, 2009(07):P07012.

Leonenko, N. N. (1975). Central limit theorem for m-dependent random fields in schemes related to series schemes. *Ukrainian Mathematical Journal*, 27(5):556–559.

Lerario, A. and Stecconi, M. (2019). Differential topology of Gaussian random fields. *arXiv preprint arXiv:1902.03805*.

Letendre, T. (2019). Variance of the volume of random real algebraic submanifolds. *Transactions of the American Mathematical Society*, 371(6):4129–4192.

Letendre, T. and Puchol, M. (2019). Variance of the volume of random real algebraic submanifolds II. *Indiana University Mathematics Journal*, 68(6):1649–1720.

Lifshits, M. A. (1983). On the absolute continuity of distributions of functionals of random processes. *Theory of Probability & Its Applications*, 27(3):600–607.

Lindgren, G. (1972). Local maxima of Gaussian fields. *Arkiv för matematik*, 10(1–2):195–218.

Longuet-Higgins, M. S. (1957). The statistical analysis of a random, moving surface. *Philosophical Transactions of the Royal Society of London. Series A, Mathematical and Physical Sciences*, 249(966):321–387.

Maillard, A., Arous, G. B., and Biroli, G. (2020). Landscape complexity for the empirical risk of generalized linear models. In *Mathematical and Scientific Machine Learning*, pages 287–327. PMLR.

Major, P. (1981). *Multiple Wiener–Itô Integrals – With Applications to Limit Theorems*, volume 849 of *Lecture Notes in Mathematics*. Springer, Berlin.

Malevich, T. (1969). Asymptotic normality of the number of crossings of level zero by a Gaussian process. *Theory of Probability & Its Applications*, 14(2):287–295.

Marcus, M. B. (1977). Level crossings of a stochastic process with absolutely continuous sample paths. *The Annals of Probability*, pages 52–71.

Maslova, N. B. (1974). On the variance of the number of real roots of random polynomials. *Theory of Probability & Its Applications*, 19(1):35–52.

Maslova, N. B. (1975). On the distribution of the number of real roots of random polynomials. *Theory of Probability & Its Applications*, 19(3):461–473.

Mehta, M. L. (2004). *Random Matrices*. Elsevier.

Mehta, M. L. and Normand, J.-M. (2001). Moments of the characteristic polynomial in the three ensembles of random matrices. *Journal of Physics A: Mathematical and General*, 34(22):4627.

Messulam, P. and Yor, M. (1982). On D. Williams' "pinching method" and some applications. *Journal of the London Mathematical Society*, 2(2):348–364.

Monge, G. (1785). *Mémoire sur les développées, les rayons de courbure, et les différens genres d'inflexions des courbes a double courbure*, volume X in Mémoires de Mathematique et de Physique, Présentés à l'Academie des Sciences par divers Savans. De l'Imprimerie Royale.

Morgan, F. (2016). *Geometric Measure Theory – A Beginner's Guide*. Elsevier/Academic Press, Amsterdam, fifth edition. Illustrated by James F. Bredt.

Nash, J. (1956). The imbedding problem for Riemannian manifolds. *Annals of mathematics*, 63(1):20–63.

Nazarov, F. and Sodin, M. (2009). On the number of nodal domains of random spherical harmonics. *Amer. J. Math.*, 131(5):1337–1357.

Nazarov, F. and Sodin, M. (2016). Asymptotic laws for the spatial distribution and the number of connected components of zero sets of Gaussian random functions. *J. Math. Phys. Anal. Geom.*, 12(3):205–278.

Nicolaescu, L. I. (2017). A CLT concerning critical points of random functions on a Euclidean space. *Stochastic Processes and their Applications*, 127(10):3412–3446.

Nicolaescu, L. I. (2023a). Counting zeros of random functions. *The American Mathematical Monthly*, 130(7):625–646.

Nicolaescu, L. I. (2023b). The Kac–Rice formula and the number of critical points of random functions. *Preprint*.

Nosko, V. P. (1969). Local structure of Gaussian random fields in the neighborhood of high-level shines. *Doklady Akademii Nauk*, 189(4):714–717.

Nourdin, I. (2011). Yet another proof of the Nualart–Peccati criterion. *Electron. Commun. Probab.*, 16:467–481.

Nourdin, I. and Peccati, G. (2009). Stein's method on Wiener chaos. *Probab. Theory Related Fields*, 145(1–2):75–118.

Nourdin, I. and Peccati, G. (2012). *Normal Approximations with Malliavin Calculus: From Stein's Method to Universality*, volume 192 of *Cambridge Tracts in Mathematics*. Cambridge University Press, Cambridge.

Nourdin, I., Peccati, G., and Rossi, M. (2019). Nodal statistics of planar random waves. *Comm. Math. Phys.*, 369(1):99–151.

Nualart, D. and Ortiz-Latorre, S. (2008). Central limit theorems for multiple stochastic integrals and Malliavin calculus. *Stochastic Process. Appl.*, 118(4):614–628.

Nualart, D. and Peccati, G. (2005). Central limit theorems for sequences of multiple stochastic integrals. *Ann. Probab.*, 33(1):177–193.

Nualart, D. and Wschebor, M. (1991). Intégration par parties dans l'espace de Wiener et approximation du temps local. *Probab. Theory Related Fields*, 90(1):83–109.

Orsingher, E. and Battaglia, F. (1982). Probability distributions and level crossings of shot noise models. *Stochastics*, 8(1):45–61.

Pautrel, T. (2020). New asymptotics for the mean number of zeros of random trigonometric polynomials with strongly dependent Gaussian coefficients. *Electron. Commun. Probab.*, 25:Paper No. 36, 13.

Peccati, G. (2014). Quantitative CLTs on a Gaussian space: a survey of recent developments. In *Journées MAS 2012*, volume 44 of *ESAIM Proc.*, pages 61–78. EDP Sci., Les Ulis.

Peccati, G. and Taqqu, M. S. (2011). *Wiener Chaos: Moments, Cumulants and Diagrams*, volume 1 of *Bocconi & Springer Series*. Springer, Milan; Bocconi University Press, Milan. A survey with computer implementation, Supplementary material available online.

Peccati, G. and Tudor, C. A. (2005). Gaussian limits for vector-valued multiple stochastic integrals. In *Séminaire de Probabilités XXXVIII*, volume 1857 of *Lecture Notes in Math.*, pages 247–262. Springer, Berlin.

Penrose, M. D. (2001). A central limit theorem with applications to percolation, epidemics and Boolean models. *Ann. Probab.*, 29(4):1515–1546.

Perry, A., Wein, A. S., and Bandeira, A. S. (2020). Statistical limits of spiked tensor models. *Annales de l'Institut Henri Poincaré, Probabilités et Statistiques*, 56(1):230–264.

Petters, A. O., Levine, H., and Wambsganss, J. (2012). *Singularity Theory and Gravitational Lensing*, volume 21. Springer Science & Business Media.

Petters, A. O., Rider, B., and Teguia, A. M. (2009). A mathematical theory of stochastic microlensing. II. Random images, shear, and the Kac–Rice formula. *Journal of mathematical physics*, 50(12).

Piterbarg, V. I. (1996). *Asymptotic Methods in the Theory of Gaussian Processes and Fields*, volume 148. American Mathematical Soc.

Priya, L. (2020). Concentration for nodal component count of Gaussian Laplace eigenfunctions. *ArXiv preprint*, abs/2012.10302.

Protter, P. E. (2005). *Stochastic Integration and Differential Equations*, volume 21 of *Stochastic Modelling and Applied Probability*. Springer-Verlag, Berlin. Second edition. Version 2.1, Corrected third printing.

Qualls, C. (1970). On the number of zeros of a stationary Gaussian random trigonometric polynomial. *J. London Math. Soc. (2)*, 2:216–220.

Rabier, C.-E. and Delmas, C. (2021). The SgenoLasso and its cousins for selective genotyping and extreme sampling: application to association studies and genomic selection. *Statistics*, 55(1):18–44.

Rice, J. (1977). On generalized shot noise. *Advances in Applied Probability*, 9(3):553–565.

Rice, S. O. (1944). Mathematical analysis of random noise. *Bell System Tech. J.*, 23(3):282–332.

Rice, S. O. (1945). Mathematical analysis of random noise. *Bell System Tech. J.*, 24:46–156.

Sartori, A. (2024). On the universality of the Nazarov–Sodin constant. *Electron. J. Probab.*, 29:Paper No. 27, 21.

Schoenberg, I. J. (1938). Metric spaces and positive definite functions. *Transactions of the American Mathematical Society*, 44(3):522–536.

Schoenberg, I. J. (1942). Positive definite functions on spheres. *Duke Mathematical Journal*, 9(1):96–108.

Shub, M. and Smale, S. (1993). Complexity of Bezout's theorem II. Volumes and probabilities. In *Computational Algebraic Geometry*, pages 267–285. Springer.

Slud, E. (1991a). Multiple Wiener–Itô integral expansions for level-crossing-count functionals. *Probability theory and related fields*, 87(3):349–364.

Slud, E. (1991b). Multiple Wiener–Itô integral expansions for level-crossing-count functionals. *Probab. Theory Related Fields*, 87(3):349–364.

Spitzer, F. (1991). Some theorems concerning 2-dimensional Brownian motion. *Random Walks, Brownian Motion, and Interacting Particle Systems: A Festschrift in Honor of Frank Spitzer*, pages 21–31.

Spivak, M. (1965). *Calculus on Manifolds. A Modern Approach to Classical Theorems of Advanced Calculus*. W. A. Benjamin, Inc., New York-Amsterdam.

Stecconi, M. (2022). Kac-Rice formula for transverse intersections. *Anal. Math. Phys.*, 12(2):Paper No. 44, 64.

Steinwart, I. and Christmann, A. (2008). *Support Vector Machines*. Springer Science & Business Media.

Switzer, P. (1976). Geometrical measures of the smoothness of random functions. *J. Appl. Probability*, 13(1):86–95.

Taylor, J. E. and Worsley, K. J. (2007). Detecting sparse signals in random fields, with an application to brain mapping. *Journal of the American Statistical Association*, 102(479):913–928.

Torres, S. (1994). Topological analysis of COBE-DMR cosmic microwave background maps. *Astrophysical Journal, Part 2-Letters, vol. 423, no. 1, p. L9–L12*, 423:L9–L12.

Torrisi, G. L. and Leonardi, E. (2022). Asymptotic analysis of Poisson shot noise processes, and applications. *Stochastic Processes and their Applications*, 144:229–270.

Tsirelson, V. (1976). The density of the distribution of the maximum of a Gaussian process. *Theory of Probability & Its Applications*, 20(4):847–856.

Vakeroudis, S. (2012). On hitting times of the winding processes of planar Brownian motion and of Ornstein–Uhlenbeck processes via Bougerol's identity. *Theory of Probability & Its Applications*, 56(3):485–507.

Vogeley, M. S., Park, C., Geller, M. J., Huchra, J. P., and Gott III, J. R. (1994). Topological analysis of the CfA redshift survey. *The Astrophysical Journal, Part 1 (ISSN 0004-637X), vol. 420, no. 2, p. 525–544*, 420:525–544.

Wigman, I. (2021). On the expected Betti numbers of the nodal set of random fields. *Anal. PDE*, 14(6):1797–1816.

Wigman, I. (2024). On the nodal structures of random fields: a decade of results. *J. Appl. Comput. Topol.*, 8(6):1917–1959.

Wilkins, Jr., J. E. (1991). Mean number of real zeros of a random trigonometric polynomial. *Proc. Amer. Math. Soc.*, 111(3):851–863.

Worsley, K., Taylor, J., Tomaiuolo, F., and Lerch, J. (2004). Unified univariate and multivariate random field theory. neuroimage 23. *S189–S195*.

Worsley, K. J., Evans, A. C., Marrett, S., and Neelin, P. (1992). A three-dimensional statistical analysis for CBF activation studies in human brain. *Journal of Cerebral Blood Flow & Metabolism*, 12(6):900–918.

Wschebor, M. (1982). Formule de Rice en dimension d. *Zeitschrift für Wahrscheinlichkeitstheorie und Verwandte Gebiete*, 60(3):393–401.

Wschebor, M. (1985). *Surfaces aléatoires: mesure géométrique des ensembles de niveau*, volume 1147. Springer.

Wschebor, M. (2005). On the Kostlan–Shub–Smale model for random polynomial systems. Variance of the number of roots. *Journal of Complexity*, 21(6):773–789.

Ylvisaker, N. D. (1968). A note on the absence of tangencies in Gaussian sample paths. *Annals of Mathematical Statistics*, 39:261–262.

Zähle, U. (1984). A general Rice formula, palm measures, and horizontal-window conditioning for random fields. *Stochastic Processes and their Applications*, 17(2):265–283.

Index

D. Armentano et al., *Geometry of Level Sets of Random Fields, Kac–Rice Formulas, Hermite Expansions and Applications*, Probability Theory and Stochastic Modelling 111, https://doi.org/10.1007/978-3-032-29144-8

Index of Notation

D. Armentano et al., *Geometry of Level Sets of Random Fields, Kac–Rice Formulas, Hermite Expansions and Applications*, Probability Theory and Stochastic Modelling 111, https://doi.org/10.1007/978-3-032-29144-8

MIX
Papier aus verantwortungsvollen Quellen
Paper from responsible sources
FSC® C105338

If you have any concerns about our products,
you can contact us on
ProductSafety@springernature.com

In case Publisher is established outside the EU,
the EU authorized representative is:
Springer Nature Customer Service Center GmbH
Europaplatz 3, 69115 Heidelberg, Germany

Printed by Libri Plureos GmbH
in Hamburg, Germany